C0-ADX-361

TECTONICS AND SEDIMENTATION

Based on a Symposium
Sponsored by the Society of
Economic Paleontologists and Mineralogists

Edited by
William R. Dickinson
Stanford University
Stanford, California

SOCIETY OF ECONOMIC PALEONTOLOGISTS AND MINERALOGISTS
Special Publication No. 22
Tulsa, Oklahoma, U.S.A. November 1974

A Publication of

The Society of Economic Paleontologists and Mineralogists

a division of

The American Association of Petroleum Geologists

PREFACE

A symposium on *Tectonics and Sedimentation* was held in conjunction with the Annual Meeting of the American Association of Petroleum Geologists and the Society of Economic Paleontologists and Mineralogists in Anaheim, California, on May 15, 1973. Sponsored by the S.E.P.M. Research Committee, the symposium was organized by me as a group of fourteen invited papers for which abstracts were published in the A.A.P.G. *Bulletin*. Earle F. McBride of the University of Texas kindly consented to serve as co-chairman of the session. Of the twelve papers actually presented orally at the meeting, eight are included in this volume, together with an introductory article that I prepared subsequently.

The obvious impetus for a collection of papers on tectonics and sedimentation at this time is the sweeping impact of the recent development of plate-tectonic theory on geological concepts. Because of this current emphasis, the papers assembled here deal with aspects of the interplay between tectonic events and sedimentation for which or from which plate-tectonic implications can be drawn. The papers selected were chosen for their merits as solid documentation of actual relations between specific regional patterns of tectonic elements and specific sedimentary sequences related to them. This flavor of particularity stems from my view that the past few years have seen an abundant flowering of plate-tectonic generalizations, which cannot yield much more insight until better data can be brought to bear on outstanding problems. I judged, therefore, that a series of regional summaries, rich in detail and based upon an adequate foundation of patiently gathered data, could make a distinct contribution to currently developing trends in geologic thought while also adding to our fund of established geologic knowledge. My central guidance to each author was thus to communicate the facts of his case as fully as he understood them, regardless of what theoretical interpretations were placed upon them. As a natural result, some important connections between plate-tectonic theory and salient concepts of sedimentation are implicit rather than explicit in the papers included here. For this reason, I try to give an overview of significant relations in my introductory article, but the opinions expressed in it should not necessarily be attributed to the other authors!

Even as restricted to specific sedimentary situations with potential import for plate-tectonic theory and interpretations, no single volume could hope to address all pertinent facets of the complex relations between tectonics and sedimentation. In organizing the symposium, I chose to focus on the geologic history of characteristic sedimentary sequences exposed to view on the outcrop in North America, and mainly in the western United States. I thus deliberately excluded the direct fruits of modern oceanographic research, which is treated extensively elsewhere, and also work done overseas. I held the focus on largely unmetamorphosed sedimentary sequences where the full spectrum of stratigraphic and sedimentological methods can be applied in research. The papers included here thus constitute a representative catalogue of salient types of sedimentary sequences whose development is described and interpreted with reference to their tectonic settings in North America. This approach to tectonics and sedimentation is exemplary, rather than exhaustive. Accordingly, the reader will not find detailed commentary on all possible tectonic settings within these covers. I hope this evident failing in breadth is balanced favorably by the depth of treatment for the major kinds of tectonic environments discussed. I am sure the authors share with me the additional hope that future workers will find these summaries of pertinent geological relations to be a clarifying guide for future interpretations, even in cases where the specific interpretations offered prove to be faulty.

The North American sedimentary sequences discussed include examples of sedimentation in the following tectonic environments, as interpreted here: (1) the marginal blanket and flanking wedge of an inactive continental margin (Stewart and Poole for the early Paleozoic Cordilleran margin); (2) clastics shed toward the continent from highlands formed by the initial deformation of a previously inactive continental margin (Poole for the middle Paleozoic Cordilleran margin); (3) basins developed between a marginal orogenic belt and the craton (Bissell for the late Paleozoic Cordilleran margin); (4) cratonal successions (Sloss and Speed); (5) pre-orogenic and synorogenic flysch deposits (Morris for the Ouachita Mountains); (6) synorogenic and postorogenic molasse deposits (Eisbacher, Carrigy, and Campbell for the Canadian Cordillera); (7) fluvial deposits in a major intermontane basin coupled to an adjacent highland source (Andersen and Picard for the Uinta Basin); and (8) a belt of local basins along a complex transform cutting continental crust (Crowell for the San Andreas system).

I am indebted to Robert H. Shaver for his effective guidance on editorial practices and to Ruth Tener for additional advice on handling manuscripts. Michael Churkin, Peter J. Coney, Hubert Gabrielse, J. G. Johnson, Keith B. Ketner, Steven S. Oriel, Paul E. Potter, Norman J. Silberling, Robert J. Weimer, and Edward L. Winterer helped me review manuscripts.

WILLIAM R. DICKINSON, *Editor*
Geology Department
Stanford University
Stanford, California 94305

CONTENTS

PLATE TECTONICS AND SEDIMENTATION

WILLIAM R. DICKINSON
Stanford University, Stanford, California

ABSTRACT

The theory of plate tectonics offers a fresh opportunity to interpret the evolution of sedimentary basins in terms of changing plate interactions and shifting plate junctures. Although plate-tectonic theory lays primary emphasis on horizontal movements of the lithosphere, large vertical movements are also implied in response to changes in the thickness of crust, in the thermal condition of lithosphere, and in the isostatic balance of lithosphere over asthenosphere. As thick sedimentation requires either an initial depression or progressive subsidence to proceed, the auxiliary vertical movements largely control the evolution of sedimentary basins. Ancillary geographic changes related to the governing horizontal movements also affect patterns of sedimentation strongly.

The geosynclinal terminology used prior to the advent of plate tectonics is inadequate to describe fully the plate-tectonic settings of sedimentary basins. Basins can be described instead in terms of the type of substratum beneath the basin, the proximity of the basin to a plate margin, and the type of plate juncture nearest to the basin. Intraplate settings of oceanic or continental character contrast with zones of plate interaction, which include those of divergent, convergent, and transform motions and within each of which the underlying crustal structure is or may be complex. The evolution of a sedimentary basin thus can be viewed as the result of a succession of discrete plate-tectonic settings and plate interactions whose effects blend into a continuum of development.

Oceanic basins contain an assemblage of diachronous facies whose relations are controlled by thermal subsidence of the lithosphere as it moves away from midoceanic rises. Rifted continental margins undergo successive stages of structural evolution as the following features are formed: prerift arch, rift valley, proto-oceanic gulf, narrow ocean, and open ocean. Sedimentary phases related to each stage are components of the rifted-margin prism of strata that masks the continent-ocean interface beneath a continental terrace-slope-rise association or a progradational continental embankment. Marginal fracture ridges along marginal offsets and aulacogens along failed arms of triple junctions locally break the continuity of rifted-margin prisms. Sedimentary basins associated with arc-trench systems where oceanic lithosphere is consumed include trenches beyond the subduction complex beneath the trench slope break, forearc basins in the arc-trench gap, intra-arc basins within the magmatic arc, and interarc basins or retroarc basins in the backarc area. Interarc basins are oceanic basins between a migratory intraoceanic arc and a remnant arc, whereas retroarc basins rest on continental basement adjacent to a foreland fold-thrust belt behind a continental margin arc. Peripheral basins adjacent to suture belts formed by crustal collision occur in an analogous foreland setting between orogen and craton, but in front of a colliding magmatic arc. Retroarc basins and peripheral basins both imply partial subduction of continental lithosphere. Intracontinental basins include infracontinental types, beneath which incipient continental separation gave rise to crust of transitional thickness, as well as supracontinental types.

INTRODUCTION

The theory once termed the *new global tectonics* (Isacks and others, 1968), which postulates a segmented and mobile lithosphere, is no longer new. Most geologists apparently accept its main tenets as valid, together with the corallary concepts of continental drift and seafloor spreading, transform faults (Wilson, 1965) and subduction zones (White and others, 1970), although some geologists, notably Beloussov (1970) and Meyerhoff (Meyerhoff, 1972), have challenged these fundamental concepts. *Plate tectonics* (McKenzie, 1972a; Dewey, 1972) has become an alternate designation for the new global tectonics because the discrete spherical caps of essentially rigid lithosphere inferred to be in relative motion with respect to one another, and to the softer and weaker asthenosphere beneath, are commonly called *plates* (McKenzie and Parker, 1967). The characteristic patterns of lateral motion of these surficial slabs, curved to conform to the spherical outline of the earth, were described by Morgan (1968) as motions of crustal blocks. He indicated, however, that these fundamental tectonic entities are actually blocks of *tectosphere,* a layer thicker than crust in the ordinary sense of the layer above M and essentially synonymous with the lithosphere of others. Because of the large lateral dimensions of the main intact pieces of lithosphere, which are of the order of only 100 km thick, the passage of time has favored usage of the word plate, rather than block, as the basic descriptor.

As a comprehensive theory that purports to explain the global distribution of all belts of tectonic deformation within the crust as the loci of the boundaries or junctures between plates of lithosphere, plate tectonics has the flavor of a fresh paradigm that must be accepted or rejected almost in its entirety with only modest allowances made for deviant behavior. The evident tectonic complexity of the earth admittedly forces the recognition that unusually broad zones of deformation occur along some plate junctures (Atwater, 1970), that the motions of some small plates are controlled partly by the interaction of adjacent large plates (McKenzie, 1970, 1972b; Roman, 1973a, 1973b), and that intraplate deformation is possible on a limited scale or to a limited degree (Sykes, 1970; Doyle, 1971). These adjustments within the framework of plate-tectonic theory dilute its elegance somewhat but do not challenge its fundamental premises. Moreover, the history of mountain belts is better illuminated by plate tectonics than by any preceding theory (Coney, 1970; Dewey and Bird, 1970a).

Concepts derived from plate tectonics are used here as the basis for a discussion of general relations between tectonics and sedimentation. Plate tectonics offers fresh ways to explain the evolution of sedimentary basins, and many concepts of the past can be discarded or must be modified to conform to the new point of view. The development of plate-tectonic interpretations and models of sedimentary basins thus entails the mental exercise of changing outworn interpretations and unjustified conclusions without denying established facts. Application of plate-tectonic analysis to the evolution of a specific sedimentary basin also requires the uniformitarian assumption that present styles of plate-tectonic behavior are useful keys to plate-tectonic behavior during the time span represented by the evolution of the basin.

Unfortunately, there seem to be no clearcut means yet to judge when plate-tectonic behavior of the modern sort began, or whether somewhat different forms of plate interactions prevailed at different times in the past. Events that could conceivably mark times of tectonic transition when plate tectonics could have been initiated or could have undergone some change in kind include (a) the breakup of Pangaea starting roughly a quarter of a billion years ago (Dietz and Holden, 1970), (b) the formation of the oldest recognized blueschist belts (Ernst, 1972) and ophiolite sequences (Burke and Dewey, 1972) about a half billion years ago, (c) a null in the reported frequency of radiometric dates for orogenic granitic and metamorphic rocks at about three-quarters of a billion years ago, (d) the development of the oldest lasting cratons during the Precambrian, or (e) the formation of the first cratonic nuclei deep in the Precambrian (see also Burke and Dewey, 1973).

Perhaps the most revolutionary facet of plate-tectonic theory as applied to sedimentary basins is the startling light it sheds on the tempo of major geologic events. At ordinary rates of spreading along midoceanic rises and of plate consumption at trenches (Le Pichon, 1968; Chase, 1972), oceanic basins 5000 km wide can form or, once formed, can disappear within 50 to 100 my, a span of time representing only one or two periods of the standard geologic column. It follows that no sedimentary basin with a long history of deposition is likely to have remained in the same plate-tectonic setting throughout its evolution. Realization of this principle is a vital guard against oversimplified versions of local geologic history in terms of plate tectonics. From a plate-tectonic standpoint, the Phanerozoic alone is an immense span of time, nominally long enough to open and close an ocean as broad as the Atlantic five or ten times!

Among the many things that might be written about plate tectonics and sedimentation, this paper discusses the following topics: (a) the vertical movements of lithosphere that are inherent in plate tectonics and required to set the *conditions for sedimentation,* (b) the *ancillary effects* of horizontal movements of lithosphere described as continental drift and seafloor spreading, (c) the problem of translating the *basin terminology* employed by geosynclinal theory into terms compatible with plate-tectonic theory, (d) the main *plate-tectonic settings* important for sedimentation, and (e) the gross outlines of *basin evolution* implied by the concept of plate tectonics.

CONDITIONS FOR SEDIMENTATION

Thick sedimentation in a given place implies the prior existence of a deep hole into which sediment can be dumped or progressive subsidence of the substratum to accommodate successive increments of strata. The formation of either kind of sediment trap on a large scale requires pronounced vertical movements of the earth's crust. Plate-tectonic theory as a geometric paradigm to explain tectonic patterns lays special emphasis instead on grand horizontal translations of lithosphere with its capping of crust. However, major vertical motions of crust and lithosphere are required to accompany the horizontal motions by any feasible geologic interpretations of the mechanisms of plate motions and interactions. The vertical motions are related to changes in crustal thickness, in thermal regime, and in the conditions for isostatic balance. These three facets of plate-tectonic theory postulate inherent vertical motions of an order and on a scale that no previous tectonic theory can match in overall scope. Despite its quite proper formal emphasis on horizontal translations of lithosphere, plate-tectonic theory thus also affords the best theoretical basis yet devised to account for grand vertical movements of crust and lithosphere. From the standpoint of sedimentation, therefore, it is a mistake to view the emphasis on horizontal motions as a potential weakness of plate-tectonic theory. The ancillary vertical motions induced by plate interactions are fully equal to the demands of data on sedimentary basins and their provenances.

Crustal thickness.—New oceanic crust and lithosphere is formed continually at spreading centers along midoceanic rise crests and within marginal or interarc basins behind migrating arc structures. When continental blocks are rifted apart by extensions of spreading centers, thin oceanic crust of igneous origin is thus created adjacent to thick continental crust by submarine volcanism and associated intrusions. Frequency curves of crustal elevation show that the floor of such a new oceanic basin should stand typically about 4 km below the mean surface level of the two continental fragments formed by the rift. The process of continental separation can thus form a new crustal depression capable of serving as a receptacle for sediment, and in principle such a sediment trap can form adjacent to any part of a continental block as potential provenance. In detail, continental separation involves the development of a belt of transitional crust and lithosphere between each continental fragment and the adjacent oceanic basin. The width of the transitional region is not well known, but is probably 100 to 250 km wide in typical cases. Prior to sedimentary loading, the transitional crust will presumably stand at elevations intermediate between those of the continental block on one side and the oceanic basin on the other. Off Norway, Talwani and Eldholm (1972) described a broad transitional region in water depths of 1000 to 2500 m.

Studies of incipient continental separations suggest that two types of processes contribute to the development of belts of *transitional crust* of intermediate thickness (fig. 1):

(1) Attenuation of continental crust by stretching is accomplished by extensional faulting at upper crustal levels accompanied probably by pseudoplastic flowage at deep crustal levels (Lowell and Genik, 1972); sediment deposited on this type of transitional crust will rest upon basement rocks of continental character, but not upon a continental block of ordinary crustal thickness.

(2) Crust with oceanic affinities but unusual thickness forms where sedimentation contemporaneous with volcanism within an incipient rift depression helps construct a crustal profile of mingled sedimentary and igneous components in a complex of lavas, dikes, sills, and sediments (D. G. Moore, 1973); sediment deposited later on this type of transitional crust will rest upon a substratum of oceanic character having a crustal profile that may be of nearly continental thickness.

Plate interactions related to the consumption of lithosphere at arc-trench systems, rather than its creation at rise crests, can also produce crust of anomalous thickness that differs from both the normal oceanic value of 5 to 10 km and the normal continental value of 30 to 40 km. Such anomalous crust can be either of fundamentally oceanic or of fundamentally continental character in terms of its rock components. *Anomalous crust,* in the sense the term is used here, can form by either of two basically different mechanisms that are tectonically linked only as two contrasting facets of the geologic processes that operate within arc-trench systems (fig. 2):

(1) Igneous materials are added to the crustal structures of magmatic arcs, either as volcanic components of the surficial edifice or as intrusive components within the crustal roots of the arcs. Presumably by this process, the crustal thickness beneath intraoceanic arcs like Tonga-Kermadec (Shor and others, 1971, fig. 3) and the Marianas (Murauchi and

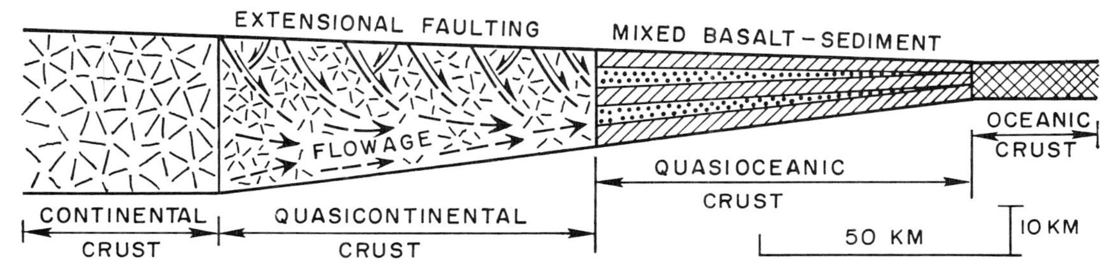

FIG. 1.—Schematic diagram at true scale to illustrate concepts of quasicontinental and quasioceanic types of transitional crust along a rifted continental margin (see text for discussion). Either type of transitional crust may be dominant to the near exclusion of the other type in specific cases (*e.g.,* see Lowell and Genik, 1972, figs. 3, 5 for quasicontinental and Moore, 1973a, fig. 10 for quasioceanic).

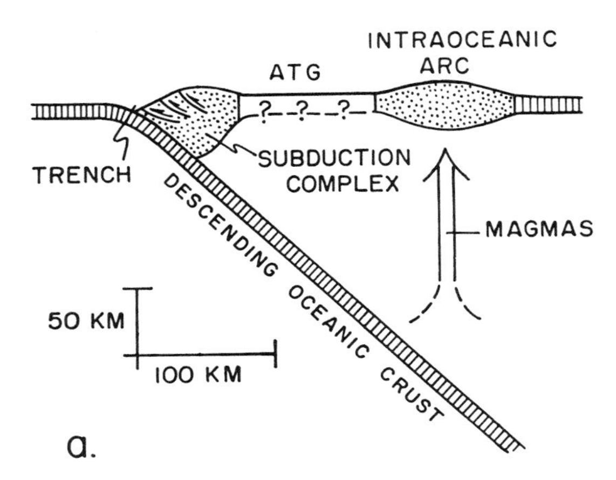

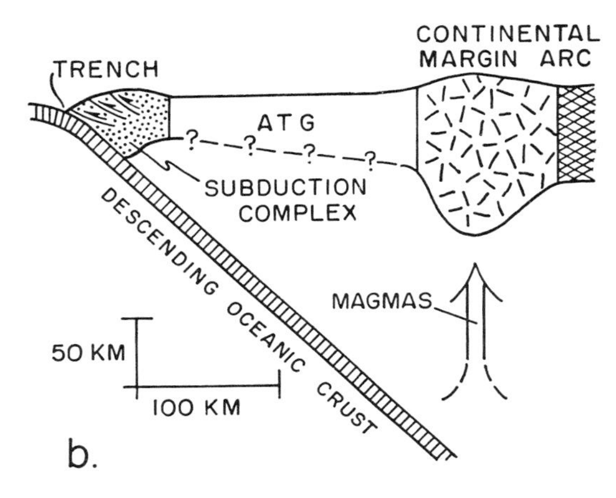

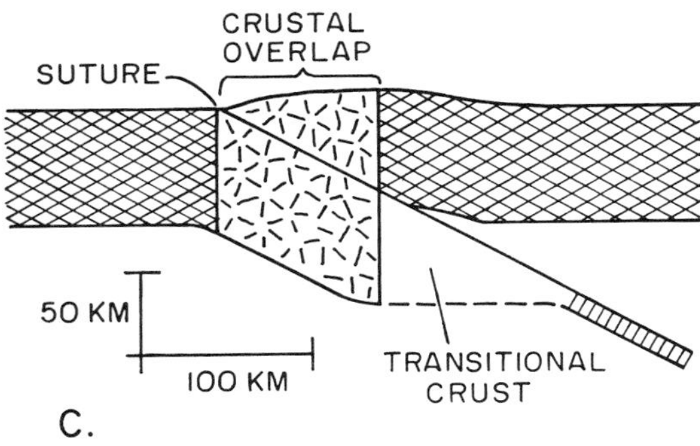

FIG. 2.—Schematic diagram at true scale to illustrate concepts of paraoceanic and paracontinental types of anomalous crust (see text for discussion): *a,* intraoceanic arc-trench system; *b,* continental margin arc-trench system; *c,* suture belt formed by crustal collision when continental block on descending plate encountered subduction zone. Oceanic crust is ruled and continental crust is cross-hatched; paraoceanic crust formed of overthickened oceanic elements is stippled and paracontinental crust formed of overthickened continental elements is jackstrawed. ATG denotes crust of variable and uncertain thickness within arc-trench gap.

others, 1968, fig. 3), whose deep underpinnings are probably oceanic crust, can be increased from the normal oceanic range to a thickness of 12 to 15 km, or perhaps even to 15 to 25 km, as argued by Markhinin (1968) for the Kuriles. Beneath the continental margin arc of the central Andes (James, 1971), the unusually thick crust of nearly 75 km likely also includes major contributions of magmatic rock injected from below into pre-existing continental crust during arc activity.

(2) Oceanic crustal slabs are stacked tectonically in subduction zones to produce thickened crust, or subduction can cause actual overlapping of continental blocks. In California, the subduction complex of the Franciscan assemblage (Ernst, 1970) is a structurally scrambled terrane of oceanic materials (Hamilton, 1969). Included are ophiolitic scraps, deformed seamounts, oceanic pelagites, and turbidite graywackes in mélanges (Hsu, 1968) and thrust slices with a total apparent tectonic thickness of 20 to 30 km (Hamilton and Myers, 1966). No basement rocks of continental character have been detected either within or beneath the complex. In Tibet, the unusual crustal thickness of as much as 75 km has been attributed by Dewey and Burke (1973) to crustal thickening that was effected by essentially doubling up the continental crust. A northern extension of the Indian subcontinent apparently was carried beneath the Tibetan plateau from a subduction zone marked now by the suture belt of ophiolitic mélanges and other deformed oceanic materials along the Indus line between the Himalayas and the Trans-Himalayan ranges.

Changes in crustal thickness within arc-trench systems where lithosphere is consumed thus tend uniformly, although in diverse fashion, to produce thicker crust. This trend fosters isostatic uplift and thus, potentially, the creation of elongate highlands as major sources of sediment. Intraplate crust can also thicken with time—certainly in oceanic regions and possibly in continental ones. In the oceans, the construction of volcanic chains like that of Hawaii on top of previously formed lithosphere can roughly double the thickness of the crust locally. By contrast, continental separations and arc migrations promote crustal attenuation and produce thinner crust associated with newly formed lithosphere. From considerations of the isostatic balance of crust, taken in isolation from other factors, spreading centers are thus generators of sites of potentially thick sedimentation.

Thermal regime.—Plate-tectonic behavior involves convective motions of asthenosphere and lithosphere (Elsasser, 1971), regardless of whether some sort of triggering perturbation of the system is induced primarily by tidal forces governed by astronomic relations (Bostrom, 1971; Knopoff and Leeds, 1972; G. W. Moore, 1973). Convectional overturn of mantle material causes relative uplift and subsidence of the surface of the lithosphere in places whose locations are partly independent of local crustal thickness.

The magnitude of the thermal effect is best understood for the elevation of oceanic crust, which stands at shallow depths beneath active rise crests and at progressively greater depths down the flanks of the rises (Sclater and others, 1971). As the age of the oceanic crust can be inferred from the correlation of magnetic anomalies, rates of subsidence can be estimated empirically. The crests of midoceanic rises have depths of 2.5 to 3 km, but all ocean floors that are roughly 75 my old and lack much sediment cover have a depth of about 5.5 km; oceanic crust of intermediate age stands at intermediate depths related in a regular fashion to age. Rates of subsidence are initially almost 100 m/my but decline with time towards a figure of 10 m/my. The observed subsidence can be explained well simply by the thermal contraction of a cooling lithosphere that is about 100 km thick. The calculations assume that isostatic compensation takes place at the base of the slab of lithosphere where it is in contact with the asthenosphere. Various assumptions for the conductivity and basal temperature of a slab of lithosphere 75 to 100 km thick allow subordinate contributions to crustal elevation from phase changes in the slab and from convective bulging of the asthenosphere beneath the slab.

Thermal tumescence along intracontinental rifts prior to continental breakup, and the succeeding thermal decay following continental separation, also cause major uplift and subsidence of continental basement rocks (Sleep, 1971). An initial thermal uplift of the order of 1 to 2.5 km can be inferred, and is matched well by the observed domal uplift of 1.5 km during the late Tertiary in central Kenya along the East African rift system (Baker and others, 1972). Crustal thinning by erosion of the arched region along an incipient rift may contribute significantly to the net crustal attenuation associated with continental separations (Hsu, 1972). The duration of purely thermal subsidence along a fresh continental margin newly formed by rifting is unlikely to persist for more than 100 my, and for typical continental ruptures the major effects probably occur within 50 my while the continental margin is within 1000 km of the rise crest (fig. 3).

Potential activators of crustal uplift and subsidence traceable to changing thermal regimes

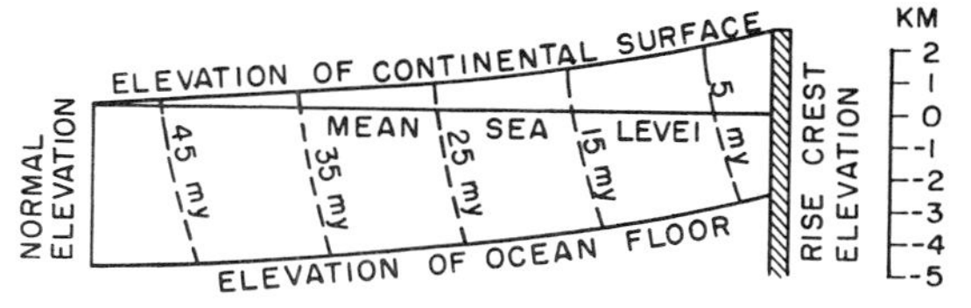

FIG. 3.—Schematic diagram to illustrate subsidence of a rifted continental margin as it moves away from a rise crest. Dashed lines show successive idealized positions of the continental slope at the intervals of time indicated. Width of diagram is about 1000 km for a half-spreading rate of 2 cm/yr; effects of sedimentation are ignored.

include the poorly understood hotspots, whose positions Morgan (1972) ascribed to fixed advective plumes or columns of hot material rising from the deep mantle. If hotspots form bumps on the upper boundary of the asthenosphere, the motions of mobile lithosphere may cause parts of the plates to bob up and down as they cross over the sites of the underlying hotspots (Menard, 1973a). Epeirogenic warping amounting to hundreds of meters vertically, and with wavelengths of the order of 1000 km or more, may be attributable to such a phenomenon. The eventual impact of recent analyses (Burke and others, 1973; Molnar and Atwater, 1973) showing that all supposed hotspots cannot be fixed in position relative to one another is not yet clear. In discussing midplate tectonics, Turcotte and Oxburgh (1973) have offered alternative explanations for the origin of the linear island and seamount chains whose relations the concept of hotspots purports to explain. With fixed hotspots, the unidirectional extension of these volcanic chains is interpreted as a result of the passage of plates of lithosphere over fixed hotspots below. However, the same general effects can be achieved in theory by postulating the development of propagating cracks in the lithosphere induced by thermal stresses from the cooling of slabs of lithosphere and by membrane stresses from changes in the radii of curvature of spherical caps of lithosphere as they change latitude on the globe. Regardless of how the hotspot controversy is resolved, the possibility of epeirogenic warping of lithosphere in irregular patterns as plates pass over a bumpy asthenosphere remains open (Menard, 1973b), unless the boundary between lithosphere and asthenosphere is assumed arbitrarily to be uniformly smooth.

Isostatic balance.—In the past, isostatic reasoning commonly has been applied by assuming the base of the crust at M to be the level of compensation. To the extent that slabs of litho-

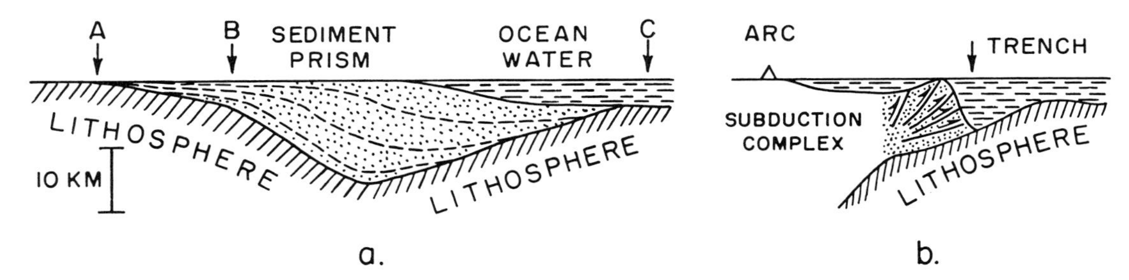

FIG. 4.—Schematic diagrams to illustrate subsidence of substratum by flexural bending of lithosphere under surficial loading: *a,* downbowing of continental margin owing to load of sediment prism deposited offshore (after Walcott, 1972, fig. 2) where A is inland line of flexure and B is initial edge of continental block before marginal subsidence (note that substratum at B is depressed to a depth at roughly the same level as that of normal ocean floor at C); *b,* hypothetical downbowing of ocean floor offshore from load of subduction complex stacked tectonically in subduction zone associated with trench beneath which oceanic substratum is depressed prior to final descent into the mantle (after Hamilton, 1973, fig. 1 and note added in proof); vertical exaggeration 10 ×.

sphere are literally rigid, the base of the lithosphere is a more appropriate level of compensation to choose. In reality, perhaps, given the limited strength of rock masses, assumption of partial compensation at both those levels, and likely at others as well, may prove the most useful stance to adopt. In any case, our past tendency to think of isostasy in terms of crustal balance alone is invalid.

If we may speak, nevertheless, of crustal isostasy in isolation from the broader context, then depressions associated with plate consumption and subduction zones are anisostatic, in the sense that the elevation of the top of the crust is not there a function of crustal thickness and density alone. Instead, the overall motion of a descending slab of lithosphere compels the crustal elements in the top tier of the slab to follow downward. Where oceanic lithosphere is consumed, the trenches are 2.5 to 5 km deeper than the floors of the adjacent oceanic basins; despite this marked difference in elevation, the crustal profile beneath the oceanward slope of the trench is demonstrably the same as in the open ocean. Beneath the axis of the trench, the ponding of turbidites may even increase the thickness of the crust there, although in cases where this effect is dominant a bathymetric trench may not be present as an expression of subduction.

McKenzie (1969) has argued convincingly that the presence of continental blocks prevents plate consumption by simple gravimetric resistance to plate descent. Even so, the attempted subduction of a continental margin, or of continental lithosphere generally, though arrested at a stage short of actual plate consumption, may be able to accomplish appreciable anisostatic subsidence, as that term was used above with reference to oceanic trenches. The local subsidence of a continental surface standing initially near sea level to nearly oceanic depths along a linear belt seems conceivable if the excess depth of 2.5 to 5 km noted for trenches can be extrapolated to this roughly analogous setting. Even if conditions of deep water were never attained, a linear belt of thick sediments deposited on subsiding continental basement might develop as a result of partial subduction. Such a region would presumably appear in the geologic record as a mobile pericratonic fringe bordering an otherwise stable craton.

Walcott (1972) has shown also that flexural bending of lithosphere under sedimentary loading of oceanic and transitional crust just offshore from a rifted continental margin can cause marked subsidence of continental basement along the adjacent edge of the continental block (fig. 4a). As sediments accumulate off the continental margin, the isostatically compensated lithosphere sags downward, and the upper surface of the continental block tilts seaward. The depressed belt along the continental margin may be 200 km wide inward from the initial continental edge to the line of no vertical displacement within the continental block, and the substratum at the initial edge of the continent can become buried beneath as much as 4 km of sediment deposited as an elongate wedge thickening seaward within the depressed belt. Landward of the depressed belt is a gentle linear upwarp (not shown on fig. 4a) parallel to the axis of the linear sag in the lithosphere offshore.

Elevation changes related to crudely analogous flexures of the lithosphere may occur around arc-trench systems in response to tectonic loading represented by the buildup of tectonically stacked subduction complexes. Hamilton (1973) argues that the weight of a seaward-thinning wedge of mélanges forming the subduction complex bows down the descending plate of lithosphere oceanward of the subduction zone (fig. 4b). This action would tend to reduce the angle of plate descent near the surface, as accretionary lateral growth of the welt of mélanges covered the initial site of plate consumption, and perhaps to increase the depth of the trench as a result of the sag developed in the lithosphere.

A complication of uncertain significance colors all detailed considerations of the behavior of lithosphere. Several kinds of data, and especially those on terrestrial heat flow, suggest that continental lithosphere is thicker than oceanic lithosphere and perhaps as much as twice as thick (Scalater and Francheteau, 1970). If so, important questions about the origin of continental lithosphere and about the motions of plates of lithosphere over the asthenosphere are raised. In any case, rigorous treatment of the isostatic balance of lithosphere over asthenosphere clearly cannot be attempted until possible variations in the thickness of the lithosphere are better known.

ANCILLARY EFFECTS

The sedimentary record is only partly a result of paleotectonic conditions suitable for sedimentation. Paleogeographic relations govern to a large extent the nature of the sediment accumulated in a given place at a given time. The factors that govern geographic variations with time are largely ancillary side effects of plate tectonics. The main influences of this kind are related to changes in latitude, changing patterns of geography, and eustatic changes.

Research on paleomagnetism (McElhinny, 1973) indicates that drifting continents have changed latitude drastically during the course of geologic history. Unless one supposes a wholly uniform climate from equator to pole at times in the past, this conclusion implies that each continental block has moved through fundamentally different climatic belts during its history. In general, reconstructed paleolatitudes also cross each continental block in different directions for different times. It follows that an adequate analysis of any sedimentary basin must include the recovery of the paleolatitudes of the basin for the times of interest. Where long periods of time are represented by the sedimentary sequence, a graph showing the trend of changing paleolatitudes for the center of the basin, or for the ends of an elongate basin, may well prove to be essential for a full interpretation of sedimentation.

Changing geographic patterns arising from continental drift may exert important influences on the distribution of potential sediment sources. Patterns of oceanic circulation should also be affected, as well as patterns of atmospheric circulation related to rain shadows and other important effects. Unfortunately, a full assessment of these types of influences on sedimentation in a particular basin at specific times in the past must await the development of a sequential atlas of paleogeography on a globally integrated basis. For much of the Phanerozoic our knowledge is still inadequate to shape this goal even with respect to the positions of all the parts of the present continental blocks. We may never be able to reconstruct well the configurations of the floors of vanished ocean basins, which may have harbored rises for which no clear evidence remains.

Eustatic changes in sea level stemming from ice storage in polar regions are probably modulated, in the long view, by the movement of continental blocks into and out of positions where they can support large glaciers; distributions of other continents so as to block latitudinal circulation in the oceans probably also favor extensive glaciation on the continents located at high latitudes (Crowell and Frakes, 1970). The remarkable display of cyclic sedimentation in the late Paleozoic sequences deposited at low paleolatitudes on coastal plains and in epeiric seas of North America and Europe can be ascribed tentatively to fluctuations of the glaciation at high paleolatitudes in Gondwana (e.g., Wanless and Shepard, 1936).

The glacial explanation of eustatic effects relies upon changing volumes of ocean water coupled with constant volume of the ocean basins. In recent years, various authors (*e.g.*, Valentine and Moores, 1972) have speculated that changes in the globally integrated spreading rate along midoceanic rises can cause eustatic effects by changing the volume of the ocean basins while the volume of ocean water remains constant. The root of the idea rests upon the principle that oceanic crust subsides with age; therefore, if the mean age of the oceanic crust changes with time, the mean depth will also vary. Evaluation of the effect is difficult (e.g., Johnson, 1971), both because we lack sufficient data to estimate globally integrated past spreading rates closely on an areal basis and because the available worldwide data on areal flooding of continental blocks at specific times in the past remains partly equivocal (but see the paper by Sloss and Speed in this volume for a fresh synthesis and a unique interpretation). The whole question of continental freeboard through time is discussed provocatively by Wise (1972).

BASIN TERMINOLOGY

Any field of human inquiry requires a sort of code of simple words or phrases to denote complex concepts. Without this aid to brevity all communication becomes too tedious to pursue. When the underlying framework of concepts changes, the established code begins to lose meaning. Something of this sort has occurred in the past few years as the geosynclinal theory for sedimentary basins and orogeny has given way to plate-tectonic theory. Although the rocks, which are the substance of the geologic record that we discuss, remain the same, the way that we view their evidence has changed. During the present period of transition in concepts, there are two fundamentally opposed ways to proceed with description. One course is the adoption of a wholly new terminology for sedimentary basins. The other course is the adaption of the old terminology to reflect the new concepts. In practice, the most likely path of thinking is a middle course that blends the two approaches by borrowing from the old where convenient and inventing the new where necessary. In practice, also, a quick consensus of views is unlikely, for many thoughtful workers will offer conflicting terminological schemes, each with its own flavor, strengths, and weaknesses (e.g., Mitchell and Reading, 1969; Dewey and Bird, 1970; Dickinson, 1971a).

The most challenging obstacle to a clean translation from geosynclinal to plate-tectonic terminology stems from the two meanings of the word *basin* in geological science. In one sense a basin is merely a bathymetric or topographic depression, but in a more significant sense a basin is the prism of rock forming a thick sedimentary succession. Various types of bathymetric basins can be related readily to the current global pattern of plate tectonics, but sedimentary basins can be related to plate tectonics only by deductive reasoning that postulates the historical dimensions of plate interactions. On the other hand, the existence of sedimentary basins is the starting point for geosynclinal theory, and their relationship to bathymetric basins must be inferred from inductive reasoning.

The dominant theme of geosynclinal theory is the recognition of parallel and adjacent miogeosynclinal and eugeosynclinal belts (Kay, 1951). Although several other kinds of geosynclines have been named, their designations arise as extensions of terminology to encompass sequences whose history does not conform to the ruling concepts for the recognition of these two basic stratigraphic elements of orogenic belts. In general, miogeosynclinal terranes are characterized by clearcut depositional contacts with continental basement and by a lack or paucity of turbidites and interstratified volcanic rocks. By contrast, eugeosynclinal terranes are characterized by equivocal contact relationships with continental basement and by an abundance of turbidites and volcanic rocks within the sedimentary sequence. As a first approximation, the former can thus be interpreted as thick accumulations of strata on the margins of continental blocks and the latter, as strata formed somewhere within an adjacent oceanic basin including its island arcs. This loose approach to the translation of geosynclinal terminology into terms compatible with plate tectonics is not entirely satisfactory. It does not allow for the considerable sophistication of geosynclinal theory in full flower, and results in the unnecessary lumping of things that the full geosynclinal terminology accords different status. Nor does it meet the need to relate various types of sedimentary basins to different kinds of plate interactions, rather than just to the two main kinds of substratum.

PLATE-TECTONIC SETTINGS

In terms of plate tectonics, the settings of basins can be described with reference to three fundamental factors: (1) the type of crust and lithosphere that serves as substratum for the basin; (2) the proximity of the basin to a plate margin, and (3) the type of plate juncture or junctures nearest to the basin.

Types of substratum.—In terms of immediate substratum, normal *continental* crust and standard *oceanic* crust are clearly two end members. For the *transitional* crust discussed in an earlier section, the term *quasicontinental* is applied here to the type characterized by attenuated continental basement, and the term *quasi-oceanic* is applied to the type characterized by an overthickened profile of plainly oceanic elements including both igneous and sedimentary materials (see fig. 1). The terms *paracontinental* and *paraoceanic* are applied, respectively, to *anomalous* crust of previously continental or quasicontinental, and previously oceanic or quasioceanic, character where crustal thickening has occurred by the addition of igneous materials in magmatic arcs and hotspot chains or through processes of tectonic stacking related to subduction zones (see fig. 2). There is inherent ambiguity between arc-related and subduction-related subtypes of paracontinental and paraoceanic anomalous crust because both magmatic arcs and subduction zones can migrate with respect to the intervening sliver of lithosphere (Dickinson, 1973). The different categories of crust are largely conceptual at present, for the operational means to distinguish between them with geophysical observations remain elusive for the most part. Moreover, the terms are unlikely to suffice as a full catalogue of significantly different types of crust and associated lithosphere. They would prove to be especially inadequate, and perhaps misleading to some extent, if it develops that significant exchange of substance can occur between crust and mantle, or between lithosphere and asthenosphere, in settings that lie within intact plates.

Proximity to junctures.—The degree of proximity of a basin to a plate margin must be understood in relative rather than absolute terms. The point is the extent to which tectonic effects related directly to plate interactions influence the setting of the basin. The thermal decay of the lithosphere as it moves away from a spreading center is one such effect, which will be confined within a certain distance from the rise crest depending upon the spreading rate. Similarly, the locus of arc magmatism along a certain trend parallel to the associated trench will occur at a distance that depends upon such parameters as the rate of plate consumption and the dip of the inclined seismic zone. In broad terms, basin settings can thus be divided into *intraplate* settings as opposed to *zones of plate interaction.*

Plate junctures.—There are three basic varieties of plate junctures: (1) *divergent,* where old lithosphere separates at spreading centers, and new lithosphere is built along midoceanic rise crests to fill the opening gap by accretion of material to the retreating edges of the separating plates; (2) *convergent,* where plate consumption carries old lithosphere downward into the asthenosphere along inclined seismic zones, and the processes that operate in the subduction zones and magmatic arcs modify the lithosphere of the overriding plate by adding both magmatic and tectonic increments to its profile; and (3) *transform,* where two plates slide past one another along a lateral fault zone without either forming new lithosphere or destroying old lithosphere.

The three kinds of junctures are geometric end members and are analogous, in terms of the geometry of strain indicated, to the three familiar classes of faults: normal (extensional), reverse-thrust (contractional), and strike-slip (lateral). There are variants of the three types of plate junctures where motions oblique to the trends of the junctures occur (Dickinson, 1972). Obliquity of convergence, as indicated independently by slip vectors determined for earthquakes and by calculations of relative plate motions from correlations of magnetic anomalies at sea, is common along modern subduction zones. Obliquity of divergence, however, is commonly resolved into a rectilinear lattice of rise crests and connecting transforms for apparently mechanical reasons (Lachenbruch and Thompson, 1972). Along transforms where the relative motion of the plates has a component of divergence, the same mechanical tendency evidently fosters a similar rectilinear lattice of transform segments connected by short rise segments. Along transforms where the relative motion of the plates has a component of convergence, the combined effect has been called transpression (Harland, 1971).

Some incipient plate junctures may become inactive before developing the full characteristics of their class. For example, an aborted divergent juncture within a continent might never undergo sufficient plate separation to develop a fully oceanic crust and lithosphere. A sedimentary basin with an unexposed floor of transitional crust might well form above the site of such a feature. Although clearly intracontinental in its setting, such a basin can be described as *infracontinental* as opposed to *supracontinental* basins deposited on a full complement of continental basement. Similarly, plate consumption might be arrested at some stage of partial subduction before a magmatic arc was developed. As the confident application of plate-tectonic logic depends upon the recognition of the full display of geologic features characteristic of each type of plate juncture, such partially developed junctures are apt to foster ambiguous interpretations.

Combined parameters.—Considering jointly the parameters of crustal substratum, proximity to plate interaction, and type of plate juncture, the gross settings of sedimentary basins can be grouped into a hierarchy consistent with plate tectonics. Initially, intraplate settings are contrasted with settings within zones of plate interaction, which include zones of divergent, convergent, and transform motion. For each of the four broad classes of plate settings so derived, the nature of the crustal substratum may vary, and subclasses of the settings can be recognized on the basis of the variations:

(a) For intraplate settings, the substratum need not be normal continental or standard oceanic in nature, for transitional or anomalous crust inherited from extinct plate junctures can be present.

(b) Zones of divergence include both intracontinental and intraoceanic types, although the two commonly are merely sequential stages in the evolution of a single plate juncture responsible also for the formation of transitional crust during intermediate stages of its evolution.

(c) Zones of convergence include types, or phases of development, in which either oceanic or continental (or transitional) crust is drawn into the subduction zone, and in parallel also include types in which the anomalous crust of the arc-trench system develops from crust of either oceanic or continental (or transitional) character initially; consequently, arc-trench systems embrace multiple settings with diverse overall relations.

(d) Transform zones include three basic types in which the two plates sliding past one another are both oceanic, both continental, or one continental and one oceanic, but transitional or anomalous crustal blocks may also be involved; moreover, each of the three basic types includes two variants where some auxiliary motion is either divergent or convergent.

The most important classes of geosynclines represent the net development of sedimentary successions over spans of time long enough for the plate-tectonic settings of the sites of deposition to change. Typical steps in the evolution of different classes of geosynclines thus apparently represent characteristic patterns in the evolution of plate interactions and the consequent changes in plate-tectonic setting. Examples of apparently anomalous geosynclinal evolution then represent the results of less common sequences of plate interactions with the different changes in plate-tectonic setting implied thereby.

BASIN EVOLUTION

Geosynclinal theory is forced by its inductive basis to address the analysis of basin evolution in terms of type examples. Where deviations from supposed norms of evolutionary trends occur, the theory is unable to offer clear insights. Plate tectonics, by contrast, approaches the problem of basin evolution in terms of alternate sequential patterns of plate interactions. Given the overall framework of varied plate motions, deductive reasoning from the theory has the potential to shed some insight on quite unfamiliar evolutionary trends, as well as to explain in coherent fashion a range of events that might issue in different circumstances from any particular stage of the evolutionary development of a given type of basin. By treating basin evolution as a function of plate motions and interactions, plate tectonics thus broadens the scope of theoretically controlled analysis and reduces the need for wholly intuitive suggestions.

Plate tectonics as a framework of thinking thus precludes the possibility of a neat catalogue of basin types. Each basin, seen in a developmental perspective of space and time, to some extent partakes of a unique flavor in principle. The constants in the equation of basin evolution are the types of plate interactions and settings, but the order in which they may be arranged in space and time is variable within wide limits. Geosynclinal theory, by contrast, assumes the overall trend of development as the standard, and views the incremental events that occur during evolution as variable within wide limits. In an analogical but very real sense, geosynclines as conceived by classic theory thus have an ontogeny, or life history, driven by tendencies akin in their effects to a vital force. Plate tectonics casts a more prosaic light on sedimentary basins, but allows naturally for more shadings of behavior without the need to modify any of its fundamental tenets.

From the standpoint of plate tectonics, the evolution of sedimentary basins is incidental to the formation and consumption of lithosphere. The major perturbations of a stable and level earth's surface are related to the opening of oceanic basins accompanied by the rifting and fragmentation of continental blocks, and to the closing of oceanic basins accompanied by the collision and assembly of continental blocks. The principal trends of basin evolution can thus be described as they pertain to the following realms of interplay between tectonics and sedimentation: (a) *oceanic basins* underlain by oceanic lithosphere, (b) *rifted continental margins* along the transitional interface between oceanic and continental lithosphere, (c) *arc-trench systems* where oceanic lithosphere is consumed beneath island arcs or continental margins, (d) *suture belts* where continental blocks are juxtaposed by crustal collision, and (e) *intracontinental* basins in the interior of continental blocks. For none of these realms of interplay should one assume invariant modes of evolution, but a discussion of each in order does afford the means to indicate the salient relationships of plate tectonics to basin evolution.

Oceanic Basins

Depending upon their size, the distribution of divergent plate junctures within them, and the distribution of convergent plate junctures within or around them, oceanic basins may lie at any given time either wholly within zones of plate interaction or in wholly intraplate settings; most typically, they lie partly within zones of plate interaction but otherwise in an intraplate setting. Each segment of oceanic crust and lithosphere typically experiences the following succession of plate-tectonic settings in order: (1) the zone of plate interaction along a divergent plate juncture where the oceanic substratum is formed; (2) the intraplate setting of a deep oceanic basin; and (3) the zone of plate interaction along a convergent plate juncture where the bulk of the oceanic lithosphere is consumed while variable and uncertain proportions of the oceanic crustal elements are caught up in the subduction zone. During either the initial or final phases of evolution in zones of plate interaction, selected segments of the oceanic crust may be subjected also to deformation along transform plate junctures associated with the divergent or convergent plate junctures.

Ignoring features related to rifted continental margins and to arc-trench systems, the principal settings of oceanic facies controlled by tectonic relations are the following (fig. 5): (a) rise crests where the layered igneous succession of ophiolite sequences (Vine and Moores, 1972) are formed along the trends of the spreading centers, (b) rise flanks where the oceanic substratum gradually subsides as it cools in moving away from spreading centers, and (c) deep basins beneath which the thermal contraction of lithosphere is essentially complete. This gross picture must be modified to allow for special conditions of shearing along active transforms near rise crests and for sharp topographic contrasts across and along fracture zones that mark the inactive extensions of transforms down the rise flanks. The outline of settings may also be inappropriate in detail for the oceanic basins generated by spreading centers within marginal seas or interarc basins behind migrating island arcs.

Nevertheless, the ideal triad of rise crest, rise flank, and deep basin serves to emphasize the main characteristic trends of evolution for an oceanic basin (see fig. 5). The igneous components of the ophiolite sequence formed at the spreading center are the first of a series of diachronous facies typical for oceanic sequences. The pelagic sediment that covers the igneous portion of the ophiolite sequence has a stratigraphy with facies relationships that reflect changing water depths (Berger, 1973). While the rise crest and flanks are above the carbonate compensation depth, calcareous sediment is deposited, but is succeeded by siliceous sediment deposited lower on the rise flanks and in the

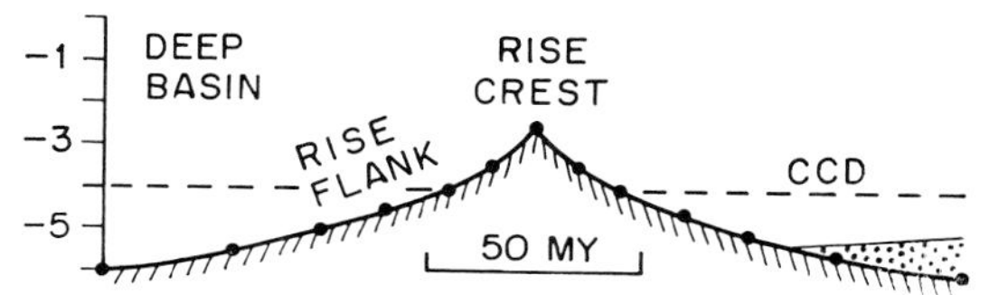

FIG. 5.—Sketch to illustrate principal settings of oceanic facies on transverse profile of typical mid-oceanic rise (depths after Sclater and others, 1971). Vertical scale is in km of water depth, but horizontal scale is in my because lateral dimensions of a mid-oceanic rise are dependent upon the spreading rate. CCD (dashed line) is typical level of carbonate compensation depth (Berger, 1973). Abyssal plain of turbidites indicated schematically, without showing isostatic compensation of substratum, by stippled area on right.

deep basins. In basins that tap turbidity currents from landmasses, the pelagic layers are covered eventually by turbidites of abyssal plains.

Where the oceanic basin changes latitude during its history or otherwise encounters different oceanic provinces, complexities are introduced into the diachronous sedimentary succession (Frakes and Kemp, 1972; Heezen and others, 1973). For example, owing to the high equatorial productivity of calcareous plankton, the carbonate compensation depth is lowered significantly below the lysocline in a narrow belt along the equator. This phenomenon has potential consequences for an oceanic sedimentary sequence formed on one side of the equator as a doublet of rise-crest calcareous pelagites overlain by siliceous pelagites reflecting later deposition in deeper water. If the segment of the oceanic basin bearing this doublet then crosses the equator, its transit may be marked by the deposition of a layer of equatorial calcareous pelagites. After the segment of the basin has moved away from the equator into the other hemisphere, it will then carry two calcareous-siliceous doublets. The two successive calcareous horizons, each overlain gradationally by siliceous sediment, record the times of positioning at the rise crest and equatorial transit, respectively. Whether details of this kind can ever be read clearly from the deformed oceanic facies of orogenic belts is a moot question at present, but avenues for inquiry are surely open.

A special set of oceanic facies is associated with islands and seamounts built as isolated mounds or in linear chains across oceanic regions. The thick volcanic piles themselves may be capped by reefoid sediment and flanked by archipelagic aprons of volcaniclastic turbidites derived locally. In certain instances, widespread and thick carbonate platforms like those in the

Bahamas may also be built within oceanic regions, probably on the quasioceanic crust of marginal fracture ridges (see below) or the submarine ridges of hotspot-generated island-seamount chains.

One of the most remarkable corollaries of plate-tectonic theory is the inference that all the old oceanic crust—igneous rocks and sediments alike older than the present ocean floors—has been placed into one of three non-oceanic repositories: (1) the mantle, into which crustal materials capable of pressure-induced inversion to suitably dense phases could be swept together with the bulk of the plates of lithosphere consumed through time by descent into or through the asthenosphere; (2) subduction complexes, into which crustal materials could be scraped from the tops of descending plates and thus welded by accretion to the flanks of continental and island-arc crust; or (3) magmatic arc structures, into the roots of which crustal materials melted from the upper levels of descending plates could be fed from below. Given the ages currently estimated for the present ocean floors, this inference means that the presumably immense bulk of all the turbidites in all the subsea fans and abyssal plains of all pre-Jurassic oceanic basins have met one of those fates, of which the second seems the most likely at present.

Rifted Continental Margins

Rifted continental margins form in pairs when continental separations occur along divergent plate junctures, and form singly when magmatic arc structures are rifted away from the margins of continental blocks by spreading behind the arcs. In the former case of simple continental separation, each rifted continental margin presents the juxtaposition of a high-standing continental block with sediment sources against a newly formed oceanic basin to serve as a sediment sink. The resulting sedimentation forms a characteristic sedimentary prism spanning the interface between continental and oceanic crust. Different components of the prism, here called *rifted-margin prism*, rest on continental, transitional, and oceanic crust. The prism thus contains strata of both miogeosynclinal and eugeosynclinal affinities. The nearshore assemblage of mainly paralic and shelf facies resting on continental basement has been aptly termed the miogeocline (Dietz and Holden, 1967) in recognition of the fact that these strata form, in transverse section, a wedge thickening seaward toward the shelf edge in existence at the time of deposition. Similarly, the offshore assemblage of turbidites and other deposits in deep water near the foot of the continental slope can be termed the eugeocline in analogous recognition of the asymmetric form of the thick accumulation, whose site is controlled by the position of the continental margin. However, as these latter deposits may grade imperceptibly into those of the broad oceanic basin nearby, the designation eugeocline is commonly less useful in practice than the term miogeocline.

The rifted-margin prism, when completed, includes a number of distinctive sedimentary phases within a complex assemblage of deposits. Each of the phases reflects either deposition in a particular plate-tectonic setting during the time when the rifted continental margin still lay within the zone of plate interaction along a divergent plate juncture, or else deposition during a particular stage in the growth of the prism during the time when the rifted margin was later in an intraplate setting. Variations arise within the total sedimentary assemblage as rates of spreading during different continental separations vary in relation to rates of sediment delivery to the rifted margins. In principle, the process of accumulation of a rifted-margin prism can also be terminated during any given phase of sedimentation by orogeny. Such orogeny may be related either to the activation of an arc-trench system along the continental margin, beneath which the offshore oceanic lithosphere thus begins to be consumed, or to crustal collision with an arc-trench system that approaches the continental margin by consuming the intervening oceanic lithosphere offshore (Dickinson, 1971b). By assuming that the sedimentation of a rifted-margin prism can be arrested at any stage in the growth of its successive depositional phases by several kinds of orogeny, a broad spectrum of individual geosynclinal developments can be accommodated within the same conceptual framework of plate tectonics. Important complications in the succession of depositional phases within rifted-margin prisms are introduced also by the presence locally of marginal offsets of the continental blocks involved and by the failed or aborted arms of triple junctions distributed along the trend of a rift belt. The marginal offsets may give rise to marginal fracture ridges and the triple junctions, to aulacogens.

The series of plate-tectonic settings that mark the successive stages of the evolution of rifted continental margins can be denoted loosely by the following five terms: pre-rift arch, rift valley, proto-oceanic gulf, narrow ocean, and open ocean (see also Schneider, 1972). The five gradational stages of structural evolution are associated with depositional phases whose strata are intercalated locally as contemporaneous facies. The successive phases of deposition may form markedly diachronous facies along any rifted-margin prism, for the geometry of plate tectonics requires most continental separations to proceed as wedge-like openings, rather than as instantaneous separations along the whole length of given rift belts (Dickinson, 1972).

Pre-rift arch.—During the thermal arching that precedes and accompanies incipient rifting, peralkaline volcanism is characteristic (fig. 6a). This activity is apparently not uniform along the rift belt but is concentrated near the crests of broad domal uplifts, from 250 to 1250 km across, that are spaced like beads with centers at intervals of roughly 1000 to 2000 km along the trend of the rift belt (LeBas, 1971). The balance between the rate of accumulation of such volcanics and the rate of erosion of the thermal arches that they crown is uncertain, but relations in Africa and South America adjacent to the South Atlantic suggest that erosion of the thermal arch is the dominant effect areally. In that region, uplifted terranes of Precambrian basement are prominent along both coasts between the ocean and extensive inland basins in which Paleozoic and Mesozoic strata are preserved on both continents (Burke and others, 1971).

Rift valley.—When sufficient crustal extension affects the arched region, rift valleys begin to form as grabens and half-grabens (fig. 6b). Probably these develop first within the domal uplifts, but later they extend as an essentially continuous branching network along the full trend of the rift belt. In the rift valleys, continental redbeds are intercalated with volcanics that continue to erupt through the growing system of crustal fractures (Scrutton, 1973). Broad regions to either side of the eventual zone of rupture between the separating continents apparently can be scarred by extensional faulting during this time. Large-scale extensional faulting has offset continental basement rocks across a belt 100 to 250 km wide west of the axial rift zone in the modern Red Sea (Hutchinson and Engels, 1972), and the Triassic basins of the Appalachian region lie as much as 250 to 500 km inland from the present continental slope, which can be taken as marking roughly the line of Jurassic continental separation.

Proto-oceanic gulf.—As continued crustal distension induces subsidence along the zone of incipient continental rupture despite continued thermal effects, the floors of the main rift valleys become partially or intermittently flooded to form proto-oceanic gulfs. Restricted conditions in these basins, which are probably still rimmed by uplifts that block delivery of clastic sediment, promote the deposition of evaporites in suitable climates (fig. 6c). Immense thicknesses, as much as 5 to 7.5 km, of evaporites are present in the subsurface beneath parts of the Red Sea (Lowell and Genik, 1972; Hutchinson and Engels, 1972). Buried salt layers that feed extensive diapir fields are present off many North Atlantic coasts (Pautot and others, 1970). Extensive evaporites are known also from coastal basins on both sides of the South Atlantic, where they are apparently correlative and represent dismembered portions of the same elongate and initially continuous evaporite basin (Reyment, 1972). The proto-oceanic evaporites are presumably deposited mainly on transitional crust, probably in most instances of the quasicontinental variety representing attenuated continental basement. Deposition on oceanic crust, or as part of the sedimentary component of quasioceanic crust, could conceivably occur if thermal uplift along the rift belt were sufficiently pronounced during and just after full continental rupture.

Narrow ocean.—Once new oceanic crust and lithosphere begin to form along the belt between two separated continental blocks, fully oceanic conditions are attained in the structural sense (fig. 7). In the geographic sense, however, a distinction can be drawn between narrow oceans and open oceans. In narrow oceans, sediment delivery to a given oceanic site from both bounding continental blocks could con-

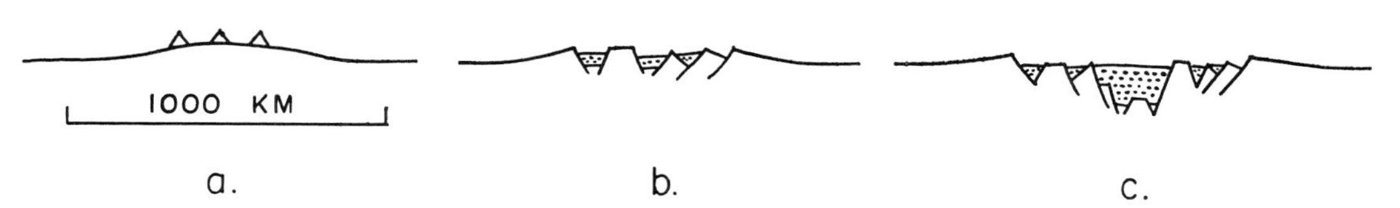

FIG. 6.—Sketches to illustrate successive pre-oceanic stages in evolution of rifted continental margins in cross-section (vertical exaggeration 25× except on dips of faults): *a,* pre-rift thermal arch shown about 750 km across with idealized volcanoes capping it; *b,* rift valley system with terrestrial sediments ponded locally across a belt about 500 km wide; *c,* proto-oceanic gulf with thick saline deposits shown within a belt about 250 km wide. Stipples indicate sediment.

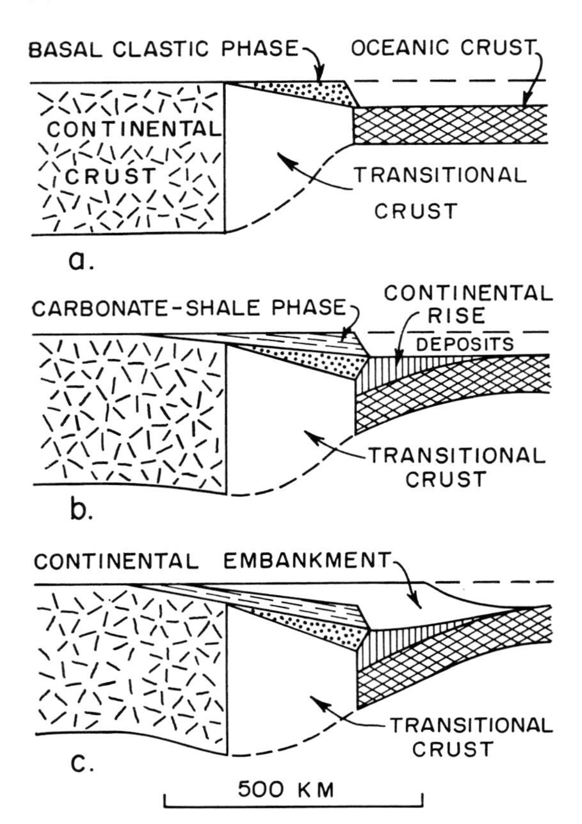

FIG. 7.—Idealized diagrams to illustrate successive depositional phases in evolution of rifted-margin prism along continent-ocean interface with sea level shown as dashed line (vertical exaggeration 10×): *a*, basal clastic phase of miogeocline deposited during thermal subsidence of transitional crust, within which earlier deposits of rift-valley redbeds, proto-oceanic evaporites, rift lavas, etc. are not differentiated; *b*, carbonate-shale phase of miogeocline deposited during shelf-slope-rise configuration of rifted continental margin; *c*, progradational continental embankment. See text for discussion.

ceivably occur, although the tendency of the spreading center to form an elevated midoceanic rise would tend to divide the oceanic basin into two halves with different sediment sources for beds deposited from bottom-hugging turbidity currents. If turbidity currents could reach or cross the actual spreading center, the net effect would be to continue forming transitional crust of the quasioceanic type.

More important in a narrow ocean is the fact that the transitional crust along and adjacent to the attenuated continental margins would continue to subside thermally as a closing stage of the plate interaction that formed the rifted continental margins. This period of subsidence probably persists along a rifted continental margin for perhaps 50 to 75 my, during which it proceeds independently of sedimentary loading. Its influence would tend to eliminate the uplifted belts that previously acted to bar sediment delivery to the rifted continental margin. This early thermal subsidence of the substratum beneath the belt of quasicontinental transitional crust is probably the factor that induces rather rapid accumulation of thick clastics as a basal phase of typical miogeoclines (fig. 7a). Such strata form a basal wedge, thickening seaward, as the oldest areally continuous deposits in the outer or oceanward parts of both the Appalachian and Cordilleran miogeoclines (King, 1969, p. 11-12; Stewart, 1972), in both of which the basal clastic phase is latest Precambrian and earlier Cambrian in age.

Open ocean.—When the strictly thermal subsidence of a rifted continental margin is complete, the margin is left in an intraplate setting and facing an open ocean. At this point, the drowned belt of transitional crust along the margin is already complex geologically. The attenuated continental basement rocks are faulted and covered locally by continental clastics, volcanics, and evaporites concentrated in varying degree within downfaulted blocks. Across this compound substratum, the basal clastic phase deposited during and closely following the main thermal subsidence is draped as a wedge of marine and paralic strata built upward to form an isostatically balanced continental terrace in the initial configuration of that feature. From this point onward, any further subsidence apparently is the result of sedimentary loading of crust offshore from the shelf break at the edge of the continental terrace.

The continued evolution of the rifted margin prism can be described using the terminology of Dietz (1963) for continental terrace, slope, rise, and embankment. The continental terrace, upon which shelf and paralic sedimentation dominates, extends to the slope break at the shelf edge, from which the continental slope leads down to deep water where the continental rise of turbidites accumulates along the edge of oceanic crust (fig. 7b). Bending of the lithosphere caused by the loading of the continental rise (Walcott, 1972) causes the continental terrace to tilt progressively seaward. This process enables the continental terrace to receive successive wedges of strata that thicken rather uniformly from a nearly common landward hinge zone toward the shelf edge (*e.g.*, Rona, 1973). The flexure at the hinge zone lies perhaps 100 to 250 km from the shelf edge. As subsidence of this kind is not linked directly to sedimentary loading of the continental terrace itself, erosional episodes on the shelf may produce disconformities within the shallow marine and shoreline deposits of the accumulating terrace wedge. These miogeoclinal strata, deposited more slowly than the underlying basal clastic phase, are probably represented by the succeeding carbonate-shale phase (*e.g.*, King, 1969) of the lower Paleozoic section in the Appalachian and Cordilleran miogeoclines. For the Appalachians, a carbonate platform marginal to the continent is recognized clearly for this interval (Rodgers, 1968), and similar strata appear in the Cordilleran case (Armstrong, 1968b).

The continental slope beyond the shelf break is largely a region of sediment bypass that serves for the transit of turbidity currents headed from shallower water toward the continental rise. Sediment thicknesses beneath the terrace-slope-rise association thus give an hourglass effect in section, with the pinched region of thin strata lying along the continental slope. Available data suggest that sediment thicknesses beneath the shelf break at the outer edge of the continental terrace, and also those beneath the continental rise, can reach at least 5 km.

If clastic sedimentation along a rifted continental margin is voluminous enough, upward construction of the continental rise and outward construction of the continental terrace lead to the development of a progradational continental embankment (fig. 7c). This type of feature is discussed here as a sequel to the stage of development represented by the terrace-slope-rise association, but appropriate relations among the details of structural development of transitional crust, the rate of thermal subsidence, and the timing and rate of sediment delivery could blur the distinction between the two stages of development in some instances. The continental slope on the front of a continental embankment becomes a constructional feature owing to wholesale progradation of the continental edge. Shelf break and slope thus advance seaward from the region of transitional crust until both reach a position above fully oceanic crust, as Dietz (1963) suggests for the Texas coast. The top of the embankment receives mainly fluvial and paralic sediments while the frontal slope and toe receive mainly turbidites. Immense thicknesses of sediment are possible for continental embankments; at least 12.5 km of sediment are present beneath the Texas coast and Walcott (1972) suggests that thicknesses of 17.5 km could be attained by simple isostatic subsidence of oceanic crust and lithosphere. By analogy with the deep-water Niger delta, the structure of the continental embankment in the region beyond the edge of the pre-existing continental terrace can be inferred to include three main depositional phases (Burke, 1972): a basal phase of sandy turbidites deposited near the toe of the embankment, a middle phase of mainly shaly rocks deposited on the advancing frontal slope of the embankment, and an upper phase of largely sandy paralic strata deposited along the prograding outer edge of the top of the embankment.

Marginal offsets.—On the floor of the modern Atlantic Ocean, the major transforms that offset the crest of the midoceanic rise are extensions of fracture zones whose extremities at the flanks of the ocean appear to coincide with abrupt offsets of the adjacent continental margins (Le Pichon and Hayes, 1971; Le Pichon and Fox, 1971). To some extent, therefore, the gross shape of the rectilinear trellis of rise segments and connecting transforms in the ocean is inherited from the shape of the initial rupture formed by continental separation. The marginal offsets were transform fault zones, rather than extensional rift zones, during continental separation. The edges of the continental blocks along the trends of the marginal offsets thus underwent a different early evolution than the edges that face toward the center of the ocean and are masked now by rifted-margin prisms of the type just discussed (fig. 8). Strike-slip along the marginal offsets during continental separation would disrupt and displace segments of the earlier phases of any sedimentary accumulations that might form along those parts of the continental edges. More important, however, is the fact that the structural character of the transitional crust along the marginal offsets is likely to be different in kind (Francheteau and Le Pichon, 1972).

During continental separation, continental margins along the marginal-offset transforms are swept by the butt ends of incipient midoceanic rise segments (see fig. 8). Although the full thermal and petrologic effects of this process are unclear, the lateral transit of the end of a rise segment along a marginal offset apparently leads to the formation of a distinctive feature termed a marginal fracture ridge (Le Pichon and Hayes, 1971; Le Pichon and Fox, 1971). Where fully developed, marginal fracture ridges extend along the marginal offset itself and at least that far again out to sea along the same trend. They are formed probably in part by crumpling and shearing along the slip zone, and in part by overthickened quasi-oceanic crust formed by exceptional leakage of igneous materials from the regions near the butt

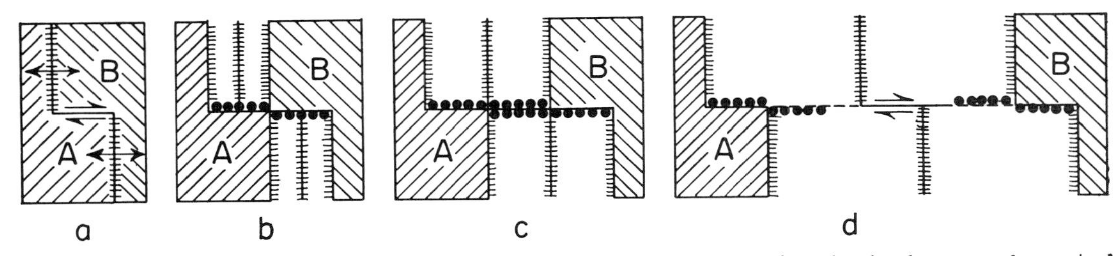

FIG. 8.—Diagram after LePichon and Hayes (1971) showing in plan view the development of marginal fracture ridges during continental separation of continental fragments A and B shown joined prior to separation in *a* and progressively farther apart in *b, c, d*. Barred lines are spreading centers linked by a transform and hachured lines show positions of normal rifted-margin prisms. Heavy dots are marginal fracture ridges along marginal offsets and offshore. Dashed lines in *d* are fracture zones along same trend.

ends of the migrating rise segments. Thick sedimentation along marginal offsets is probably delayed by prolonged thermal uplift, and clastic sedimentation may be inhibited locally by high-standing marginal fracture ridges. However, the marginal offsets are subject to prolonged thermal uplift and marginal fracture ridges may later actually promote abnormally thick sedimentation of biogenic sediments in the oceanic realm along offshore trends in line with marginal offsets. The Bahama platform, an elongate accumulation of 5 km of mainly shallow-water carbonates above perhaps 15 km of quasioceanic crust (Dietz and others, 1970) may reflect such a phenomenon, although other means of generating the quasioceanic crust beneath the carbonate platform have also been suggested.

Aulacogens.—The term aulacogen is applied here in the usage of Hoffman (and others, 1974) as adapted from the Russian literature, in which the term was devised for a class of features initiated mainly in the later Precambrian (Salop and Scheinmann, 1969). Aulacogens are elongate sedimentary basins that extend, as gradually narrowing wedges or pie slices in plan view, from the margins of cratons toward the interiors of cratons. The sedimentary sequences of aulacogens are mainly similar in general nature to facies equivalents in platformal sequences of the cratons adjacent on both sides, but are much thicker. Aulacogens are thought to evolve from semioceanic gashes formed at re-entrants in rifted continental margins during continental separations (fig. 9).

The overall geometry of RRR and RRF triple junctions (McKenzie and Morgan, 1969) is attractive as an explanation for the tectonic setting of aulacogens. Such a triple junction has two or three spreading centers as arms. The Afar region linking the Red Sea, Gulf of Aden, and East African rift systems is a modern example. Burke and others (1971) and Grant (1971) argue that the Benue Trough, which extends into Africa from the head of the Gulf of Guinea, was temporarily one spreading arm of a triple junction in the Cretaceous. When continuation of motion along the other two arms opened the South Atlantic, the Benue arm was aborted in an incipient stage of development. Cretaceous and younger sediments beneath the trough are more than 5 km thick for at least 500 km along its axis. Their accumulation was probably accommodated by the subsidence of transitional crust beneath the trough. Thermal subsidence following the failure of the Benue spreading arm to continue into a fully oceanic configuration probably served as a trigger to initiate sedimentation, which then forced further isostatic subsidence under sedimentary loading. The location of a long-lived Benue depression also apparently controlled the course of the Niger River and the position of its delta. The delta itself is evidently a local continental embankment containing sediment some 10 km thick built into deep water beyond the initial continental margin off the seaward end of the aulacogen.

In North America, the best example of an aulacogen containing Phanerozoic strata is the Anadarko-Ardmore basin, which extends inland nearly 500 km from the southern margin of the Paleozoic continent. Ham and Wilson (1967) describe the section in the elongate basin as 10 to 12 km of Paleozoic strata overlying at least 2 km of Cambrian volcanics. Late Paleozoic deformation and coarse clastic sedimentation within the basin was greatest toward its open end, and was probably related to the Ouachita orogeny along the nearby continental margin.

Arc-Trench Systems

Arc-trench systems are the characteristic geologic expression of convergent plate junctures (Dickinson, 1970). As recognized plainly by Kay (1951), the volcaniclastic rocks of volcanic island chains built along magmatic arcs are prominent within many eugeosynclinal terranes. Eugeosynclinal terranes also include the subduction complexes associated with trenches, where oceanic strata are mingled tectonically as they are detached from the tops of slabs of lithosphere descending beneath the flanks of arc-trench systems. Sedimentary sequences that accumulate on the flanks of magmatic arcs, which stand as positive topographic features during arc activity, receive a variety of geosynclinal designations locally, depending on details of their relationships to various types of substratum and also upon the nature of the strata themselves. A full discussion of the evolution of arc-trench systems requires an essential focus on magmatism and metamorphism, but the emphasis here is solely upon the facets of behavior that affect the associated sedimentary basins (Dickinson, 1974).

Arc-trench systems include the following five major morphotectonic elements (*e.g.,* Dickinson, 1973): (1) the *trench,* a bathymetric deep floored by oceanic crust; (2) the *subduction zone* beneath the inner wall of the trench and the trench slope break marking the top of the inner wall; (3) the arc-trench gap, a belt within which a *forearc basin* may occur between the trench slope break and the magmatic arc; (4) the magmatic arc, within which *intra-arc basins* may occur; and (5) the backarc area, within which may lie either an *interarc basin* floored by oceanic crust and separated from the rear of the arc by a normal fault system, or a *retroarc basin* floored by continental basement and separated from the rear of the arc by a thrust fault system.

Sedimentation in the various types of basins noted for the different elements of arc-trench systems is contemporaneous with both volcanism and plutonism along the magmatic arc, and with metamorphism both in the cool subduction zone and in the hot roots of the magmatic arc. Faulting and other deformation in the subduction zone, within the magmatic arc, and in the backarc area is also contemporaneous with sedimentation. Although sequential phases of sedimentation can doubtless be recognized for each of the kinds of sedimentary basins noted in the various morphotectonic settings, the areal contrast in facies among the various kinds of sedimentary sequences forms the most important and regular genetic pattern. This pattern of distinctive and parallel sedimentary terranes, coupled with their igneous and metamorphic associates, can be used as a means to identify the petrotectonic assemblages that form the geologic record of past arc-trench systems.

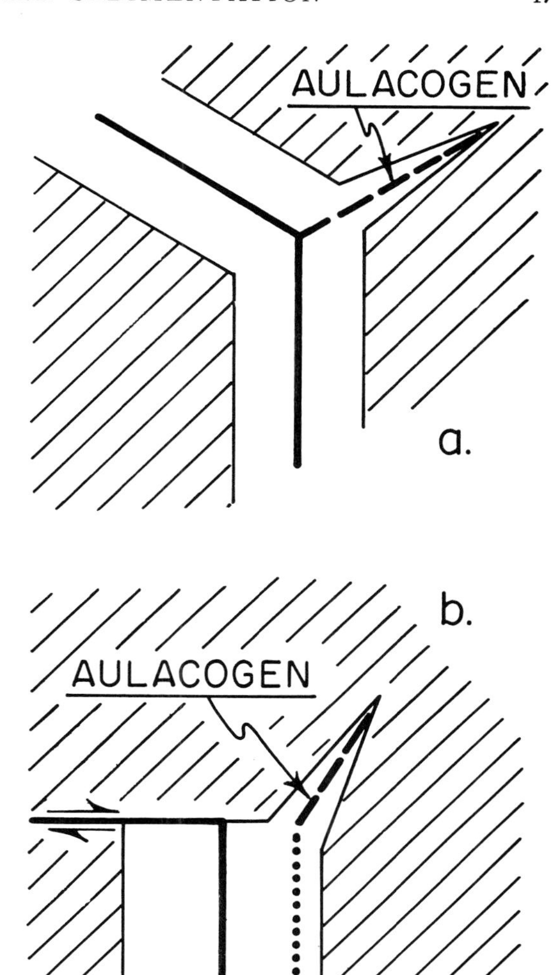

FIG. 9.—Diagrams showing in plan view alternate mechanisms for the development of aulacogens at re-entrants in rifted continental margins (continental blocks shaded, plate junctures that continue active shown as solid heavy lines, failed spreading centers along axis of aulacogen shown as dashed heavy line): *a,* aulacogen as failed arm of formerly stable RRR triple junction (spreading directions changed along the two arms that continued spreading when motion stopped along the failed arm); *b,* aulacogen as failed arm of inherently unstable RRF triple junction (after Grant, 1971).

Thick sedimentation associated with arc-trench systems is best discussed, therefore, in relation to subduction zones, forearc basins, intra-arc basins, interarc basins, and retroarc basins (fig. 10). The progressive development of these features presumably will continue until the plate consumption that fosters an arc-trench

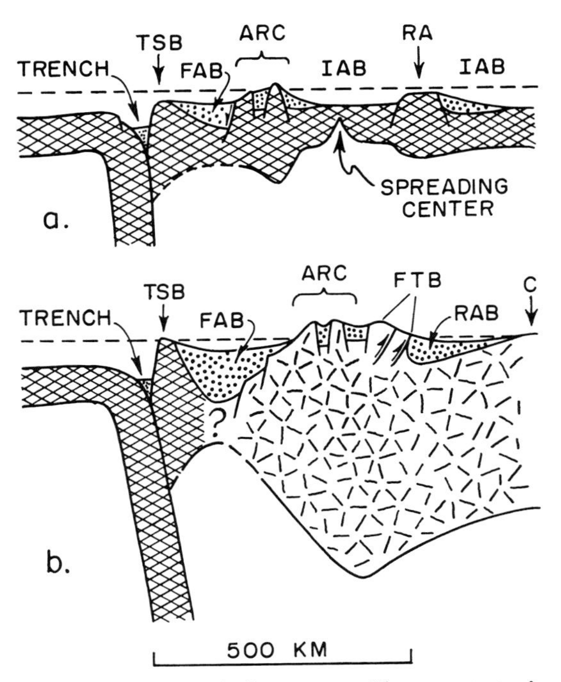

FIG. 10.—Idealized diagrams to illustrate tectonic settings of sedimentary basins (stippled) associated with arc-trench systems. Dashed line is sea level. Vertical exaggeration is 10× ; note apparent steep angles of descent of plates beneath trench although true angles depicted are 60 degrees (*a*) and 30 degrees (*b*). Oceanic and paraoceanic crust is crosshatched; continental and paracontinental crust is jackstrawed. Trench slope breaks (TSB) lie above subduction zone complexes at thresholds of arc-trench gaps within which forearc basins (FAB) are shown. For intraoceanic arc (*a*), active or frontal island arc (ARC) is shown with a marine intra-arc basin, and remnant arc (RA) stands between two interarc basins (IAB), one active (left) and one inactive (right) but both with volcaniclastic wedges along one flank. For continental margin arc (*b*), volcanic highlands (ARC) are shown with a terrestrial intra-arc basin, a flanking intermontane lowland, and a foreland fold-thrust belt (FTB) above zone of partial crustal subduction lies between magmatic arc and retroarc basin (RAB) on depressed crust of pericratonic foreland adjacent to margin of interior craton (C).

system is terminated, commonly by crustal collision to form a suture belt.

Subduction zones.—Seaward from the trench in typical arc-trench systems is a broad upwarp of the ocean floor marking the flexure of the lithosphere as it bends to descend beneath the arc-trench system. The inner slope of this outer arch is the gentle outer slope of the trench, and is scarred in places by normal faults reflecting local extensional deformation of the ophiolite sequence represented by the igneous oceanic crust and its sediment cover. The trench is a bathymetric deep immediately adjacent to the tectonic front of the subduction zone, which begins at the base of the steep inner wall of the trench. On the trench floor, variable thicknesses of turbidites are ponded above the sediment layers rafted tectonically into the trench from the open ocean floor. Transport by turbidity currents within a trench is mainly longitudinal along the trench axis (von Huene, 1972; Marlow and others, 1973), although the initial entry of sediment into the trench may occur at intervals along the inner wall as well as from the ends of the trench (Ross, 1971; J. C. Moore, 1973). Where the rate of sediment delivery to a trench is high enough in relation to the rate of plate consumption, the trench may be filled with sediment and the trench site covered by subsea fans that mask the position of the tectonic front of the subduction zone (Silver, 1969). It is fair to infer that the volume of locally deposited turbidite sediment incorporated within the nearby subduction complex is thus in some measure inversely proportional to the bathymetric depth of the associated trench. An empty trench leaves little evidence in the geologic record.

Recent data leave little doubt that the steep inner wall of the trench is underlain almost directly by deformed and uplifted oceanic strata with only a local cover of undeformed sediment (Karig, in press). This material is interpreted here as a subduction complex of mélanges and crumpled beds sliced by thrusts and including ophiolitic scraps. The tectonic top of the subduction complex is assumed to lie at or just beneath the sea bottom at the trench slope break, a bathymetric transition point located at the top of the inner wall of the trench. The mass of the subduction complex is inferred to grow by the accretion of successive increments of oceanic crustal materials that either are jammed against its seaward flank at the trench axis or scraped into its basal levels from the top of the slab of lithosphere that descends beneath the subduction zone. It is important to note that these materials thus added to the subduction complex include not only indigenous trench turbidites deposited nearby, but potentially also include samplings of all the turbidites deposited over extensive areas of the ocean floor from sources wholly exotic to the arc-trench system into whose flank they are incorporated.

As oceanic materials are stacked tectonically within subduction zones, net uplift of the subduction complex must occur even while subduction continues, and should be dramatic when subduction ceases for any reason. The condition of a subduction complex where exposed to view on land is thus never the initial condition. Always there is the overprint of deformation during uplift, which must amount to a minimum of 5 to 10 km if the depths of modern trenches are representative. Mass movement of material off the steep inner wall of the trench may in time recycle some materials back through the process of subduction. We are still at a loss to understand fully the complex structures of mélanges and thrust-bounded slabs in subduction complexes (Suppe, 1972). It seems clear, however, that their great apparent thicknesses are tectonic, rather than stratigraphic.

Forearc basins.—The topographic and bathymetric configuration within the arc-trench gap between the trench slope break and the volcanic front is highly varied. The elevation of the threshold at the trench slope break is evidently controlled by the elevation of the top of the subduction complex, which may be emergent as islands or may lie at depths as great as 2 to 3 km. Different arc-trench gaps contain, singly or in combination, such diverse geographic elements as mountainous uplifts, longitudinal troughs, transverse submarine slopes, shallow shelves, deep-marine terraces or plains, and terrestrial plains or valleys. In a number of modern arc-trench gaps, thick sequences of largely undeformed sediments attest to progressive subsidence to develop forearc basins as the term is used here, and sequences interpreted as forearc basins in the geologic record attain thicknesses of 5 to 12 km. Subsidence may be related to the descent of a dense slab of lithosphere beneath the arc-trench gap.

The sedimentary sequences of forearc basins rest on a substratum of variable and partly uncertain character. On the arc flank of the arc-trench gap the substratum may include eroded igneous rocks, both plutonic and volcanic, of the magmatic arc. On the trench flank of the arc-trench gap the substratum may include parts of the subduction complex. Beneath the center of a forearc basin the substratum may be paraoceanic crust made of previously accreted elements of a subduction complex that broadens by growing seaward with time, or may be oceanic or transitional crust that existed before the arc-trench system was activated (*e.g.*, Grow, 1973). Thus, in general, forearc basins are commonly successor basins (King, 1969) in the sense that they overlie older, deformed elements or orogenic belts.

Forearc basins receive sediment mainly from the extensive nearby arc structures, where not only volcanic rocks but also plutonic and metamorphic rocks exposed by uplift and erosion may serve as sources. Sources may also include local uplands along the trench slope break or within the arc-trench gap itself. Facies gradations may presumably occur between strata of forearc basins and volcaniclastic beds of the volcanic arcs, but prominent normal fault zones commonly bound the basins on the arc side (Karig, in press). On the trench side, tectonic gradation into the disrupted strata of the subduction zone presumably occurs locally. In several instances, however, nearly intact ophiolite sequences that underlie continuous sedimentary sequences of inferred forearc basins are in sharp fault contact with the adjacent subduction complexes. This circumstance implies little or no transfer of material into the subduction zone from the part of the forearc basin now preserved. Instead, the forearc basins appear to have wholly overridden the subduction zones. This relation holds for the late Mesozoic Great Valley sequence of California where faulted against the coeval Franciscan complex (Bailey and others, 1971), for the late Paleozoic and early Mesozoic western marginal facies of New Zealand where faulted against the coeval eastern axial facies or Torlesse Group (Landis and Bishop, 1972; Blake and Landis, 1973), and for the early Tertiary succession of the central Burmese lowland where faulted against the Indoburman flysch terrane (Brunnschweiler, 1966).

By inference from the bathymetry of modern forearc basins, and from the sedimentology of older sequences inferred to have been deposited in similar settings, forearc basins may contain a variety of facies. Shelf and deltaic or terrestrial sediments, as well as turbidites with either transverse or longitudinal paleocurrents, may occur in different examples. The local bathymetry is presumably controlled by the elevation of the trench slope break, the rate of sediment delivery to the forearc basin, and the rate of basin subsidence acting in combination. Various facies patterns and successive depositional phases probably can occur in different cases.

Intra-arc basins.—Magmatic arcs include both intraoceanic and continental margin types (Dickinson, 1974). Intraoceanic arcs include those with only paraoceanic crust built by magmatic additions to oceanic crust and those underlain at depth by a sliver of continental crust detached as part of a migratory arc structure. Continental margin arcs include island arcs backed by shallow epicontinental seas as well as those standing along the edges of landmasses. Fault-bounded extensional basins that occur within many magmatic arcs may be related to local volcano-tectonic subsidence, or to arching that accompanies uplift of paracontinental crust, or to the development of an incipient interarc basin. Volcaniclastic strata are characteristic, but may include a range of types from terrestrial redbeds to turbidites, as well as various intermediary facies. Local sources of sediment within the arc structure are typical, but low-standing island arcs located near continental blocks may accumulate clastic strata from external sources as well; these have been termed basinal arcs (Berg and others, 1972).

Backarc areas.—The distinction made here between interarc and retroarc basins in the backarc area reflects the existence of two distinct variants of arc-trench systems (see fig. 10). Both types of basins are related indirectly to the convergent plate junctures to which the trenches and their associated subduction zones are related directly. The contrast between the tectonic settings of interarc and retroarc basins seemingly stems, therefore, from influences other than simple plate interactions at convergent plate junctures. The key control is apparently the relative motion of the plate of lithosphere in the backarc area with respect to the underlying asthenosphere (Coney, 1971; Dickinson, 1972; Hyndman, 1972; Wilson and Burke, 1972; Wilson, 1973).

The lithosphere is apparently not wholly intact across the region beneath the magmatic arc owing to thermal softening from the high heat flux. The narrow belt of lithosphere beneath the arc-trench gap can thus be viewed as a separate narrow plate. Where the lithosphere behind the arc has a component of motion, relative to underlying asthenosphere, away from the magmatic arc, then the arc structure may split. An interarc basin underlain by newly formed oceanic crust built by a backarc spreading center then opens between the active or frontal arc and a remnant arc (Karig, 1972), which may be of either intraoceanic or continental margin type. This mode of behavior is characteristic of eastward-facing island arcs in the western Pacific region (Karig, 1970, 1971a). Where the lithosphere behind the arc has a component of motion, relative to underlying asthenosphere, toward the magmatic arc, then partial subduction of continental lithosphere beneath the rear of the arc structure is assumed here to occur (e.g., Coney, 1972). A fold-thrust belt thus develops in the backarc area as cover rocks are stripped off descending basement. The resulting highlands shed debris into a downbowed retroarc basin along a belt that can be termed pericratonic between the continental margin arc and the craton. This mode of behavior is characteristic of the westward-facing Andean arc, which is flanked on the east by the Subandean fold-thrust belt, beyond which are the Subandean sedimentary basins that lie between the Andes and the craton (Ham and Herrerra, 1963; Sonnenberg, 1963).

Conceivably, the contrasting behavior of eastward-facing and westward-facing arc-trench systems may reflect the different tectonic regimes, respectively extensional and contractional, induced in barkarc areas by the postulated net westward drift of lithosphere with respect to asthenosphere as a result of tidal influences (G. W. Moore, 1973). By implication, arc-trench systems with a roughly east-west orientation might experience no marked deformation in backarc areas, and hence might display neither interarc nor retroarc basins.

Interarc basins.—The sedimentary record of interarc basins is not well documented, but their global abundance at present suggests that eugeosynclinal terranes of the past probably contain numerous examples. It must be inferred that some ophiolitic sequences of orogenic belts represent oceanic crust formed as the floors of interarc basins, rather than in open oceans. If there are significant differences between the igneous rocks of the two kinds of ophiolitic sequences, the distinction is not yet established.

The sedimentary strata in modern interarc basins include distinctive turbidite aprons of volcaniclastic beds shed backward from the rifted rear sides of migratory frontal arcs (Karig, 1970, 1971a, 1972). These turbidite wedges appear to rest almost directly on the igneous oceanic crust with little or no intervening pelagites present. Beyond the interarc spreading centers sedimentation varies markedly. Where a given interarc basin is bounded on the side away from the arc-trench system by a submerged remnant arc, no effective source of clastic sediment is present and oceanic pelagites accumulate. Where successive remnant arcs with paraoceanic crust are calved in succession from migratory frontal arcs, a broad oceanic region is formed in which the only thick sedimentary accumulations are turbidite wedges stranded behind each submerged remnant arc.

On the other hand, where an interarc basin forms by disruption of a continental margin arc, one side of the interarc basin is a form of rifted continental margin along which some variant of a rifted-margin prism can be formed (Mitchell and Reading, 1969) beside a marginal sea (Karig, 1971b; Packham and Falvey, 1971; Moberly, 1972). It may be argued that the pattern of parallel facies belts associated with such a continental margin fringed by migratory intraoceanic arcs lying offshore faithfully reproduces the classic miogeosyncline-eugeosyncline couple. If so, extreme horizontal motions of lithosphere may be unnecessary assumptions to explain the juxtaposition of diverse terranes within orogenic belts. The rifted continental margin on the inner side of the interarc basin is interpreted then as the miogeosynclinal belt, whereas the adjacent interarc basin, the offshore island arc, and the open ocean beyond together represent the complex tectonic elements of the eugeosynclinal belt. Although attractive, the analogy harbors a potential fallacy. Only if the substratum beneath the supposed miogeoclinal wedge includes igneous rocks representing part of the geologic record of the earlier stages of arc evolution prior to arc migration can the analogy be defended in detail. As most miogeoclinal wedges appear to rest on truncated continental basement considerably older than the base of the miogeoclinal wedge, this logical requirement of the analogy does not appear to be met in typical orogenic belts.

Retroarc basins.—The sedimentary record of retroarc basins includes fluvial, deltaic, and marine strata as much as 5 km thick deposited in terrestrial lowlands and epicontinental seas along elongate pericratonic belts between continental margin arcs and cratons. Where the magmatic arcs stand along continental margins that have grown seaward by tectonic accretion, some retroarc basins may be successor basins in the sense of resting upon previously deformed terranes. Sediment dispersal into and across retroarc basins is mainly tranverse, in a gross sense, from highlands on the side toward the magmatic arc, although contributions from the craton are also present. The deposits of retroarc basins are thus exogeosynclinal in the sense that debris is shed toward the craton from sources within marginal orogenic belts.

The sources of sediment may include the magmatic arc itself, but commonly the principal sources are uplifted strata in the fold-thrust belt formed by partial subduction behind the arc. Such was the case for the Cretaceous retroarc basin of the interior and Rocky Mountain region of North America (Weimer, 1970). The main highland sources were uplands of folded and faulted pre-Mesozoic strata lying just west of the retroarc basin, but still east of the batholith belt that marks the position of the associated magmatic arc (Hamilton, 1969). Part of the subsidence in retroarc basins is probably in response to flexure of the lithosphere or other isostatic adjustments induced by the tectonic load of thrust sheets in the adjacent foreland fold-thrust belt (Price and Mountjoy, 1971). As the retroarc basin evolved, contractional deformation disrupted piedmont facies along the highland flank of the basin, and ultimately crumpled the flank of the basin fill within the fold-thrust belt (Armstrong, 1968a). Where the main sources of sediment are thus in the fold-thrust belt behind the arc, rather than within the magmatic arc, the nature of the sources depends upon the previous history of the continental margin. Where the magmatic arc arises following the initiation of plate consumption along a previously inactive continental margin draped with a rifted-margin prism, the sources are apt to be uplifted miogeoclinal strata.

The fold-thrust belts that parallel the orogenic margin of retroarc basins thus may be described commonly as foreland thrust belts (Coney, 1973). In this sense, the foreland is simply the cratonal or platformal interior of the continent, and the foreland basin is a retroarc basin. However, foreland basins in this same setting with respect to the continental interior may form as a result of crustal collisions in which a rifted continental margin with its rifted-margin prism encounters the main subduction zone associated with the trench of an arc-trench system. The designation of the foreland can thus be ambiguous with respect to the polarity of the arc-trench system responsible for the orogenic belt. So long as parts of a rifted-margin prism are thrust back toward the continental interior, and a pericratonic fringe of continental basement is drawn down by partial subduction to form an elongate basin parallel to the fold-thrust belt, the concept of a foreland to the orogenic belt is appropriate. Foreland basins formed by partial subduction of continental margins during crustal collisions are here termed peripheral basins as discussed in the next section. Designation of a given foreland basin as either a retroarc basin or a peripheral basin thus depends upon a knowledge of the sequence and timing of tectonic events in the adjacent orogen.

Suture Belts

The term suture belt is used here for the complexly deformed joins along which crustal blocks are welded together by the crustal collisions that occur when lithosphere bearing thick crustal blocks reaches a subduction zone along a convergent plate juncture where oceanic lithosphere was previously being consumed (fig. 11). Crustal collisions include a variety of types involving both intraoceanic and continental margin arc-trench systems (Dickinson, 1971c). In all cases, crustal collision involves juxtaposition of the tectonic elements of an arc-trench system, together with its variety of sedimentary basins, against other crustal blocks across the

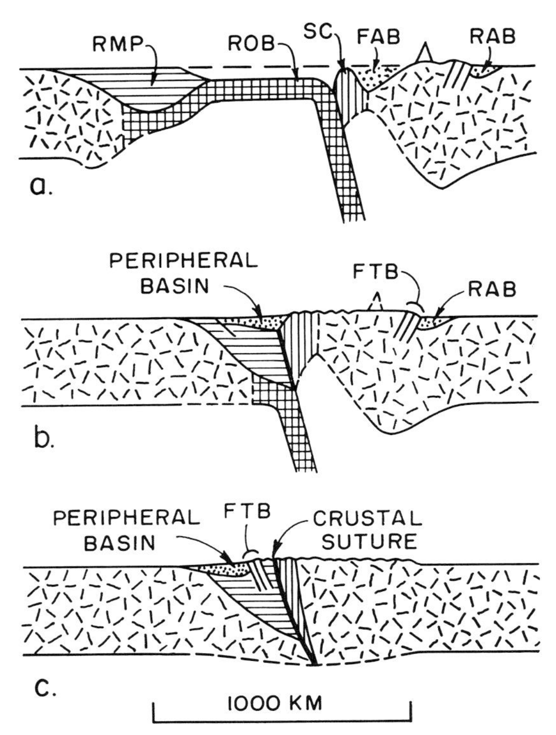

FIG. 11.—Idealized diagrams to illustrate hypothetical sequence of events, and associated sedimentary basins (stippled), during crustal collision between rifted continental margin (see fig. 7) on left and continental margin arc (see fig. 10b) on right: *a,* prior to collision; *b,* initial collision; *c,* final suturing (vertical exaggeration is 10×). Oceanic and quasioceanic crust is crosshatched; continental, quasicontinental, and paracontinental substratum is jackstrawed. Symbols: RMP, rifted-margin prism (horizontal rules); ROB, remnant ocean basin (sea level as dashed line); SC, subduction complex (vertical rules); FAB, forearc basin; RAB, retroarc basin; FTB, foreland foldthrust belts.

suture belt. Along a sutured join, deformed sedimentary sequences that were deposited on the ophiolitic basement of an open oceanic basin or an interarc basin can be caught between the sutured crustal blocks. Such sequences commonly appear to view after suturing as tectonically scrambled mélanges of ophiolitic scraps and oceanic facies. Such suture-belt mélanges may not be visible within a suture belt if the extent of subduction during crustal collision was sufficient to hide them beneath rocks of the overriding plate. Clear evidence of the sutured join may also be absent if contractional deformation during collision is sufficient to squeeze the suture-belt mélanges upward to tectonic levels that are later removed by erosion. These two cases of obscured suture belts can be described as hidden sutures where telescoping thrust sheets cover the suture and as cryptic sutures where the materials caught in the suture are pressed out tectonically and lost by later erosion (e.g., Dewey and Burke, 1973).

Suture belts contain deformed examples of all the various types of sedimentary sequences discussed in connection with oceanic basins, rifted-margin prisms, and arc-trench systems. In addition, sedimentary basins of a unique type here termed ***peripheral basins*** with strata as thick as 5 km are also formed by processes related to collision (see fig. 11). As a continental crustal block is drawn toward a subduction zone just prior to crustal collision, bending of the lithosphere probably first causes extensional faulting analogous to that seen on the oceanic outer arch seaward of arc-trench systems. These faults might offset the strata of a rifted-margin prism in a sense similar to that of the earlier faults that were associated with continental rifting or with growth faulting during deposition of the prism. Later in the progress of crustal collision, the edge of the continental block is depressed by partial subduction to form a pericratonic or foreland basin peripheral to the suture belt on the plate being partly consumed. As the process of subduction is braked by crustal collision, the subsidence in this peripheral basin may be succeeded by marked uplift. As the peripheral basin is drained, a phase of evaporite deposition could conceivably ensue. The well-known sabhka deposition in the Persian Gulf may be an example of such deposition in a restricted seaway remaining along a belt parallel to the suture belt of the Zagros Crush Zone in Iran (Wells, 1969).

Perhaps the most characteristic deposits of peripheral basins are exogeosynclinal clastic wedges spread toward the craton as fluvial and deltaic strata shed from a suture belt involving the continental margin (Graham and others, in press). If the peripheral basin is deep enough, however, these deposits may be preceded by turbidites deposited on depressed continental or transitional crust, rather than oceanic crust (W. M. Neill, personal commun., 1973). Paleocurrent trends in the clastic wedges may be dominantly transverse to the orogenic trend, whereas those in the turbidites may be dominantly longitudinal to the orogenic trend. Clastic wedges of peripheral basins, as well as any clastic wedges shed toward the other side of highlands along the suture belt, may thus be termed molasse in many cases. The turbidites of peripheral basins, as well as the turbidites of oceanic basins or forearc basins caught within the suture belt, may thus be termed flysch in many cases.

The evolution of suture belts forms an attractive explanation, though not the only one, for the tectonic relations of flysch and molasse (Graham and others, in press). In general, any completed suture belt will represent the end result of a sequential closure of a remnant ocean basin (Dickinson, 1972). Only if the shapes of colliding continental margins are mirror images of one another, and the vector of the relative plate motion causing crustal collision is exactly as required, can crustal collisions be synchronous along their whole length. In the general case, extensive suture belts must be diachronous in development as successive adjustments in plate motions and boundaries allow progressive welding of crustal blocks to proceed. A tectonic transition point between the segment already sutured (fig. 11c) and the segment yet to be sutured (fig. 11a) will migrate along the developing suture belt with time. Behind the transition point, orogenic highlands, clastic wedges and filled peripheral basins are characteristic. Ahead of the transition point, remnant ocean floor and incipient peripheral basins are present. As the drainage of orogenic highlands is commonly longitudinal, much of the sediment derived from the collision orogen will not be shed transversely as clastic wedges, but will be shed longitudinally into the remnant oceanic basin and deepening peripheral basins along tectonic strike. Many of the deposits that reflect erosion of the collision orogen thus will be incorporated later into the same orogenic belt as the tectonic transition point migrates along the growing suture belt. In this fashion, synorogenic flysch of turbidites with mainly longitudinal paleocurrents and postorogenic molasse of clastic wedges with largely transverse paleocurrents may be seen as the natural result of crustal collision to form suture belts.

Intracontinental Basins

Intracontinental basins are the most difficult type to treat constructively in terms of plate tectonics if basins related to apparently intracontinental orogenic belts like the Urals are excluded. Provided such orogenic terranes are interpreted as suture belts (Hamilton, 1970), the associated basins can be interpreted variously in terms of former oceanic basins, rifted continental margins, arc-trench systems, and collision orogens. Basins related to these kinds of features include foreland basins of both retroarc and peripheral types where the basin fill is supracontinental in the sense of resting on continental crust or on older rifted-margin prisms.

Basins bounded on all sides by anorogenic terranes forming a basement that is uniformly older than the basin fill are the intracontinental ones difficult to explain using principles of plate tectonics. For infracontinental basins (see above) the basement does not extend unmodified beneath the floor. Partial attenuation of continental basement along an aborted rift that never advanced beyond an incipient stage could lead to conditions permitting marked crustal subsidence locally, especially under sedimentary loading. Unfortunately, detection of transitional crust hidden beneath an infracontinental basin depends upon geophysical observations, for the basin fill permanently masks the substratum.

Presumably, infracontinental basins would tend to be elongate in many cases, but not necessarily in all. If antecedent or contemporaneous domal uplifts were distributed at intervals along the belt of partial crustal attenuation, as appears to be the pattern for early stages of continental rifting, then crustal thinning by stretching and erosion might be concentrated within relatively equant areas. As a result, the infracontinental basins that developed after thermal tumescence gave way to thermal decay might appear as apparently isolated and more or less round features distributed apparently at random across a continental block. The only clue to their essentially common origin might be a rough contemporaneity of development. The initial stages of a major continental separation may well involve extensive gashing of the continental block, while still joined, over a broad region that would lapse eventually into quiescence except where the rifting was fully established along a single trend. Dispersed infracontinental basins might then remain as a record of the widespread extent of incipient rifting.

Alternatively, a fundamentally elongate belt of incipient continental separation might be marked by a chain of isolated infracontinental basins linked only by intracontinental transforms. If the transform segments of the integrated tectonic system were masked by cover rocks or later deformation, the fundamental pattern might be difficult to detect by any means. There seems an especially strong possibility that successor basins might form along recently completed suture belts in this fashion as residual plate motions were resolved into translation along transforms roughly parallel to the suture belt.

None of these speculations touch upon the possibility of long-lived supracontinental basins with underpinnings of normal continental crust. The motion of a plate of lithosphere over a bumpy asthenosphere accounts well only for reversible epeirogenic warping and temporary subsidence. Note that this effect might affect local areas distributed in unpredictable fashion over a continental block, but that any local and temporary subsidence would occur as part of a wave of shbsidence. The passage of the lithosphere over a bump or depression on top of the asthenosphere thus might leave a sort of subtle track in the stratigraphic record of any epicontinental seas covering a continental block. Mechanisms for permanent subsidence of supracontinental basins on a large scale in truly intraplate settings are not apparent from plate-tectonic theory.

Summary

The preceding tentative classification of sedimentary basins in a plate-tectonic framework indicates that satisfactory alternatives to the geosynclinal terminology can be devised, and that points of correspondence between the two schemes of nomenclature can be appreciated. The discussion also indicates that direct equivalency between individual terms in the two sets cannot be expected. For example, eugeosynclines apparently contain the strata of oceanic basins, intra-arc basins, and interarc basins as modified by deformation in subduction complexes, magmatic arcs, and suture belts. On the other hand, rifted-margin prisms may include the superimposed strata of taphrogeosynclines, miogeosynclines, and paraliageosynclines. Exogeosynclinal foreland basins may be either retroarc basins or peripheral basins in the terminology suggested. Forearc basins and aulacogens have been described by some as epieugeosynclines and zeugogeosynclines, respectively, but others have applied different terms to analogous features and the same terms to different kinds of features. Such discordances in terminology are to be expected, given the dramatic change in frame of reference. Whatever terminology is used, progress in applying plate-tectonic theory to problems of sedimentation can come easily only if sedimentary basins are classified and discussed in a manner that is congruent with concepts of plate tectonics. In this paper, I have tried simply to find phraseology that would convey meaning now, without prejudice to either past or future usage.

REFERENCES CITED

ARMSTRONG, R. L., 1968a, Sevier orogenic belt in Nevada and Utah: Geol. Soc. America Bull., v. 79, p. 429–458.

———, 1968b, The Cordilleran miogeosyncline in Nevada and Utah: Utah Geol. and Mineralog. Survey Bull. 78, 58 p.

ATWATER, TANYA, 1970, Implications of plate tectonics for the Cenozoic tectonic evolution of western North America: Geol. Soc. America Bull., v. 81, p. 3513–3536.

BAILEY, E. H., BLAKE, M. C., JR., AND JONES, D. L., On-land Mesozoic crust in California Coast Ranges: U.S. Geol. Survey Prof. Paper 700-C, p. 70–81.

BAKER, B. H., MOHR, P. A., AND WILLIAMS, L. A. J., 1972, Geology of the eastern rift system of Africa: Geol. Soc. America Special Paper 136, 67 p.

BELOUSSOV, V. V., 1970, Against the hypothesis of sea-floor spreading: Tectonophysics, v. 9, p. 489–511.

BERG, H. C., JONES, D. L., AND RICHTER, D. H., 1972, Gravina-Nutzotin belt—tectonic significance of an upper Mesozoic sedimentary and volcanic sequence in southern and southeastern Alaska: U.S. Geol. Survey Prof. Paper 800-D, p. 1–24.

BERGER, W. H., 1973, Cenozoic sedimentation in the eastern tropical Pacific: Geol. Soc. America Bull., v. 84, p. 1941–1954.

BLAKE, M. C., JR., AND LANDIS, C. A., 1973, The Dun Mountain ultramafic belt—Permian oceanic crust and upper mantle in New Zealand: Jour. Research U.S. Geol. Survey, v. 1, p. 529–534.

BOSTROM, R. C., 1971, Westward displacement of lithosphere: Nature, v. 234, p. 536–538.

BRUNNSCHWEILER, R. O., 1966, On the geology of the Indoburman Ranges (Arakan Coast and Yoma, Chin Hills, Naga Hills): Jour. Geol. Soc. Australia, v. 13, p. 137–194.

BURKE, K. C. A., 1972, Longshore drift, submarine canyons, and submarine fans in development of Niger delta: Am. Assoc. Petroleum Geologists Bull., v. 56, p. 1975–1983.

———, AND DEWEY, J. F., 1972, Permobile, metastable, and plate-tectonic regimes in the Precambrian: Geol. Soc. America Abstracts with Programs, v. 4, p. 462.

———, DESSAUVAGIE, T. F. J., AND WHITEMAN, A. J., 1971, Opening of the Gulf of Guinea and geological history of the Benue Depression and Niger Delta: Nature Phys. Sci., v. 233, p. 51–55.

———, KIDD, W. S. F., AND WILSON, J. T., 1973, Relative and latitudinal motion of Atlantic hot spots: Nature, v. 245, p. 133–137.

———, AND DEWEY, J. F., 1973, An outline of Precambrian plate development, *in* Tarling, D. H., and Runcorn, S. K., Implications of continental drift to the earth sciences: Academic Press, N.Y., p. 1035–1045.

CHASE, C. G., 1972, The *N* plate problem of plate tectonics: Geophys. Jour. Roy. Astron. Soc., v. 29, p. 117–122.

CONEY, P. J., 1970, Geotectonic cycle and the new global tectonics: Geol. Soc. America Bull., v. 81, p. 739–748.

———, 1971, Cordilleran tectonic transitions and motion of the North America plate: Nature, v. 233, p. 462–465.

———, 1972, Cordilleran tectonics and North American plate motion: Am. Jour. Sci., v. 272, p. 603–655.
———, 1973, Plate tectonics of marginal foreland thrust-fold belts: Geology, v. 1, p. 131–134.
Crowell, J. C., and Frakes, L. A., 1970, Phanerozoic glaciation and the causes of ice ages: Am. Jour. Sci., v. 268, p. 193–224.
Dewey, J. F., 1972, Plate tectonics: Sci. American, May, p. 56–68.
———, and Bird, J. M., 1970a, Mountain belts and the new global tectonics: Jour. Geophys. Research, v. 75, p. 2625–2647.
———, and Bird, J. M., 1970b, Plate tectonics and geosynclines: Tectonophysics, v. 10, p. 625–638.
———, and Burke, K. C. A., 1973, Tibetan, Variscan, and Precambrian basement reactivation: products of continental collision: Jour. Geology, v. 81, p. 683–692.
Dickinson, W. R., 1970, Relations of andesites, granites, and derivative sandstones to arc-trench tectonics: Rev. Geophysics and Space Physics, v. 8, p. 813–862.
———, 1971a, Plate tectonic models of geosynclines: Earth and Planetary Sci. Lettrs., v. 10, p. 165–174.
———, 1971b, Plate tectonic models for orogeny at continental margins: Nature, v. 232, p. 41–42.
———, 1971c, Plate tectonics in geologic history: Science, v. 174, p. 107–113.
———, 1972, Evidence for plate tectonic regimes in the rock record: Am. Jour. Sci., v. 272, p. 551–576.
———, 1973, Widths of modern arc-trench gaps proportional to past duration of igneous activity in associated magmatic arcs: Jour. Geophys. Research, v. 78, p. 3376–3389.
———, 1974, Sedimentation within and beside ancient and modern magmatic arcs, *in* Dott, R. H., Jr., and Shaver, R. H. (eds.): Soc. Econ. Paleontologists and Mineralogists Special Pub. 19, p. 230–239.
Dietz, R. S., 1963, Wave-base, marine profile of equilibrium, and wave-built terraces: a critical appraisal: Geol. Soc. America Bull., v. 74, p. 971–990.
———, and Holden, J. C., 1966, Miogeoclines in space and time: Jour. Geology, v. 74, p. 566–583.
———, and Holden, J. C., 1970, Reconstruction of Pangaea, breakup and dispersion of continents, Permian to present: Jour. Geophys. Research, v. 75, p. 4939–4956.
———, Holden, J. C., and Sproll, W. P., 1970, Geotectonic evolution and subsidence of Bahama platform: Geol. Soc. America Bull., v. 81, p. 1915–1928.
Doyle, H. A., 1971, Australian seismicity: Nature Phys. Sci., v. 234, p. 174–175.
Elsasser, W. M., 1971, Sea-floor spreading as thermal convection: Jour. Geophys. Research, v. 76, p. 1101–1112.
Ernst, W. G., 1970, Tectonic contact between the Franciscan melange and the Great Valley sequence, crustal expression of a late Mesozoic Benioff zone: *ibid.*, v. 75, p. 886–902.
———, 1972, Occurrence and mineralogic evolution of blueschist belts with time: Am. Jour. Sci., v. 272, p. 657–668.
Frakes, L. A., and Kemp, E. M., 1972, Generation of sedimentary facies on a spreading ocean ridge: Nature, v. 236, p. 114–117.
Francheteau, Jean, and Lepichon, Xavier, 1972, Marginal fracture zones as structural framework of continental margins in South Atlantic Ocean: Am. Assoc. Petroleum Geologists Bull., v. 56, p. 991–1007.
Graham, S. A., Dickinson, W. R., and Ingersoll, R. V., in press, Himalayan-Bengal model for flysch dispersal in Appalachian-Ouachita system: Geol. Soc. America Bull.
Grant, N. K., 1971, South Atlantic, Benue trough, and Gulf of Guinea Cretaceous triple junction: Geol. Soc. America Bull., v. 82, p. 2295–2298.
Grow, J. A., 1973, Crustal and upper mantle structure of the central Aleutian arc: *ibid.*, v. 84, p. 2169–2192.
Ham, C. K., and Herrerra, L. J., Jr., 1963, Role of Subandean fault system in tectonics of eastern Peru and Ecuador, *in* Childs, O. E. and Beebe, B. W. (eds.), Backbone of the Americas: Am. Assoc. Petroleum Geologists Mem. 2, p. 47–61.
Ham, W. E., and Wilson, J. L., 1967, Paleozoic epeirogeny and orogeny in the central United States: Am. Jour. Sci., v. 265, p. 332–407.
Hamilton, Warren, 1969, Mesozoic California and the underflow of Pacific mantle: Geol. Soc. America Bull., v. 80, p. 2409–2430.
———, 1970, The Uralides and the motion of the Russian and Siberian platforms: *ibid.*, v. 81, p. 2553–2576.
———, 1973, Tectonics of the Indonesian region: Geol. Soc. Malaysia Bull. 6, p. 3–10.
———, and Myers, W. B., 1966, Cenozoic tectonics of the western United States: Rev. Geophysics, v. 4, p. 509–549.
Harland, W. B., 1971, Tectonic transpression in Caledonian Spitzbergen: Geol. Mag., v. 108, p. 27–42.
Heezen, B. C., and 11 co-authors, 1973, Diachronous deposits, a kinematic interpretation of the post-Jurassic sedimentary sequence on the Pacific plate: Nature, v. 241, p. 25–32.
Hoffman, Paul, Burke, K. C. A., and Dewey, J. F., 1974, Aulacogens and their genetic relation to geosynclines, with a Proterozoic example from Great Slave Lake, Canada, *in* Dott, R. H., Jr., and Shaver, R. H. (eds.): Soc. Econ. Paleontologists and Mineralogists Special Pub. 19, p. 38–55.
Hsu, K. J., 1968, Principles of melanges and their bearing on the Franciscan-Knoxville problem: Geol. Soc. America Bull., v. 79, p. 1063–1074.
———, 1972, The concept of the geosyncline, yesterday and today: Leicester Lit. Philos. Soc. Trans., v. 66, p. 26–48.
Hutchinson, R. W., and Engels, G. G., 1972, Tectonic evolution in the southern Red Sea and its possible significance to older rifted continental margins: Geol. Soc. America Bull., v. 83, p. 2989–3002.
Hyndman, R. D., 1972, Plate motions relative to the deep mantle and the development of subduction zones: Nature, v. 238, p. 263–265.
Isacks, Bryan, Oliver, Jack, and Sykes, L. R., 1968, Seismology and the new global tectonics: Jour. Geophys. Research, v. 73, p. 5855–5899.
James, D. E., 1971, Plate tectonic model for the evolution of the central Andes: Geol. Soc. America Bull., v. 82, p. 3325–3346.

Johnson, J. G., 1971, Timing and coordination of orogenic, epeirogenic, and eustatic events: *ibid.*, v. 82, p. 3263–3298.
Karig, D. E., 1970, Ridges and basins of the Tonga-Kermadec island arc system: Jour. Geophys. Research, v. 75, p. 239–254.
———, 1971a, Structural history of the Mariana island arc system: Geol. Soc. America Bull., v. 82, p. 323–344.
———, 1971b, Origin and development of marginal basins in the western Pacific: Jour. Geophys. Research, v. 76, p. 2542–2561.
———, 1972, Remnant arcs: Geol. Soc. America Bull., v. 83, p. 1057–1068.
———, in press, Crustal accretion in subduction zones.
Kay, Marshall, 1951, North American geosynclines: Geol. Soc. America Mem. 48, 143 p.
King, P. B., 1969, The tectonics of North America—a discussion to accompany the tectonic map of North America: U.S. Geol. Survey Prof. Paper 628, 94 p.
Knopoff, Leon, and Leeds, A., 1972, Lithospheric momenta and the deceleration of the earth: Nature, v. 237, p. 93–95.
Lachenbruch, A. H., and Thompson, G. A., 1972, Oceanic ridges and transform faults: their intersection angles and resistance to plate motion: Earth and Planetary Sci. Lettrs., v. 15, p. 116–122.
Landis, C. A., and Bishop, D. G., 1972, Plate tectonics and regional stratigraphic-metamorphic relations in the southern part of the New Zealand geosyncline: Geol. Soc. America Bull., v. 83, p. 2267–2284.
Le Bas, M. J., 1971, Per-alkaline volcanism, crustal swelling and rifting: Nature Phys. Sci., v. 230, p. 85–86.
Le Pichon, Xavier, 1968, Sea-floor spreading and continental drift: Jour. Geophys. Research, v. 73, p. 3661–3698.
———, and Fox, P. J., 1971, Marginal offsets, fracture zones, and the early opening of the North Atlantic: *ibid.*, v. 76, p. 6294–6308.
———, and Hayes, D. E., 1971, Marginal offsets, fracture zones, and the early openings of the South Atlantic: *ibid.*, v. 76, p. 6283–6293.
Lowell, J. D., and Genik, G. J., 1972, Sea-Floor spreading and structural evolution of southern Red Sea: Am. Assoc. Petroleum Geologists Bull., v. 56, p. 247–259.
Markhinin, E. K., 1968, Volcanism as an agent of formation of the earth's crust, *in* Knopoff, Leon, Drake, C. L., and Hart, P. J. (eds.), The crust and upper mantle of the Pacific area: Am. Geophys. Union Mon. 12, p. 413–423.
Marlow, M. S., Scholl, D. W., Buffington, E. D., and Alpha, T. R., 1973, Tectonic history of the central Aleutian arc: Geol. Soc. America Bull., v. 84, p. 1555–1574.
McElhinny, M. W., 1973, Palaeomagnetism and plate tectonics: Cambridge Univ. Press, 358 p.
McKenzie, D. P., 1969, Speculations on the consequences and causes of plate motions: Geophys. Jour. Roy. Astron. Soc., v. 18, p. 1–32.
———, 1970, Plate tectonics of the Mediterranean region: Nature, v. 226, p. 239–243.
———, 1972a, Plate tectonics, *in* Robertson, E. C. (ed.), The nature of the solid earth: McGraw-Hill, N.Y., p. 323–360.
———, 1972b, Active tectonics of the Mediterranean region: Geophys. Jour. Roy. Astron. Soc., v. 30, p. 109–185.
———, and Morgan, W. J., 1969, Evolution of triple junctions: Nature, v. 224, p. 125–133.
———, and Parker, R. L., 1967, The North Pacific, an example of tectonics on a sphere: Science, v. 216, p. 1276–1280.
Menard, H. W., 1973a, Depth anomalies and the bobbing motion of drifting islands: Jour. Geophys. Research, v. 78, p. 5128–5138.
———, 1973b, Epeirogeny and plate tectonics: Am. Geophys. Union Trans. (EOS), v. 54, p. 1244–1255.
Meyerhoff, A. A., and Meyerhoff, H. A., 1972, "The new global tectonics": major inconsistencies: Am. Assoc. Petroleum Geologists Bull., v. 56, p. 269–336.
Mitchell, A. H., and Reading, H. G., 1969, Continental margins, geosynclines, and ocean-floor spreading: Jour. Geology, v. 77, p. 629–646.
Moberly, Ralph, 1972, Origin of lithosphere behind island arcs, with reference to the western Pacific: Geol. Soc. America Mem. 132, p. 35–55.
Molnar, Peter, and Atwater, Tanya, 1973, Relative motion of hot spots in the mantle: Nature, v. 246, p. 288–291.
Moore, D. G., 1973, Plate-edge deformation and crustal growth, Gulf of California structural province: Geol. Soc. America Bull., v. 84, p. 1883–1906.
Moore, G. W., 1973, Westward tidal lag as the driving force of plate tectonics: Geology, v. 1, p. 99–100.
Moore, J. C., 1973, Cretaceous continental margin sedimentation, southwestern Alaska: Geol. Soc. America Bull., v. 84, p. 595–614.
Morgan, W. J., 1968, Rises, trenches, great faults, and crustal blocks: Jour. Geophys. Research, v. 73, p. 1959–1982.
———, 1972, Plate motions and deep mantle plumes: Geol. Soc. America Mem. 132, p. 7–22.
Murauchi, S., and 12 Co-authors, 1968, Crustal structure of the Philippine Sea: Jour. Geophys. Res., v. 73, p. 3143–3172.
Packham, G. H., and Falvey, D. A., 1971, An hypothesis for the formation of marginal seas in the western Pacific: Tectonophysics, v. 11, p. 79–109.
Pautot, Guy, Auzende, J-M., and Lepichon, Xavier, 1970, Continuous deep sea salt layer along North Atlantic margins related to early phase of rifting: Nature, v. 227, p. 351–354.
Price, R. A., and Mounjoy, E. W., 1971, The Cordilleran foreland thrust and fold belt in the southern Canadian Rockies: Geol. Soc. America Abstracts with Programs, v. 3, p. 404–405.
Reyment, R. A., 1972, The age of the Niger delta (West Africa): 24th Internat. Geol. Cong. Rept., Sec. 6, p. 11–13.

RODGERS, JOHN, 1968, The eastern edge of the North American continent during the Cambrian and Early Ordovician, *in* Zen, E-An, and White, W. S. (eds.), Studies of Appalachian geology, northern and maritime: Wiley Interscience, N.Y., p. 141–149.

ROMAN, CONSTANTIN, 1973a, Buffer plates where continents collide: New Scientist, 25 Jan, p. 180–181.

———, 1973b, Rigid plates, buffer plates, and subplates: Geophys. Jour. Roy. Astron. Soc., v. 33, p. 369–373.

RONA, P. A., 1973, Relations between rates of sediment accumulation on continental shelves, sea-floor spreading, and eustacy inferred from the central North Atlantic: Geol. Soc. America Bull., v. 84, p. 2851–2871.

ROSS, D. A., 1971, Sediments of the northern Middle America trench: *ibid.,* v. 82, p. 303–322.

SALOP, L. I., AND SCHEINMANN, Y. M., 1969, Tectonic history and structures of platforms and shields: Tectonophysics, v. 7, p. 565–597.

SCHNEIDER, E. D., 1972, Sedimentary evolution of rifted continental margins: Geol. Soc. America Mem. 132, p. 109–118.

SCLATER, J. G., AND FRANCHETEAU, JEAN, 1970, The implications of terrestrial heat flow observations on current tectonic and geochemical models of the crust and upper mantle of the earth: Geophys. Jour. Roy. Astron. Soc., v. 20, p. 509–542.

SCLATER, J. G., ANDERSON, R. N., AND BELL, M. L., 1971, Elevation of ridges and evolution of the central Pacific: Jour. Geophys. Research, v. 76, p. 7888–7915.

SCRUTTON, R. A., 1973, The age relationship of igneous activity and continental break-up: Geol. Mag., v. 110, p. 227–234.

SHOR, G. G., JR., KIRK, H. K., AND MENARD, H. W., 1971, Crustal structure of the Melanesian area: Jour. Geophys. Research, v. 76, p. 2562–2586.

SILVER, E. A., 1969, Late Cenozoic underthrusting of the continental margin off northernmost California: Science, v. 166, p. 1265–1266.

SLEEP, N. H., 1971, Thermal effects on the formation of Atlantic continental margins by continental breakup: Geophys. Jour. Roy. Astron. Soc., v. 24, p. 325–350.

SONNENBERG, F. P., 1963, Bolivia and the Andes, *in* Childs, O. E., and Beebe, B. W. (eds.), Backbone of the Americas: Am. Assoc. Petroleum Geologists Mem. 2, p. 36–46.

STEWART, J. H., 1972, Initial deposits in the Cordilleran geosyncline, evidence of a late Precambrian continental separation: Geol. Soc. America Bull., v. 83, p. 1345–1360.

SUPPE, JOHN, 1972, Interrelationships of high-pressure metamorphism, deformation, and sedimentation in Franciscan tectonics, U.S.A.: 24th Internat. Geol. Cong. Rept., Sec. 3, p. 552–559.

SYKES, L. R., 1970, Seismicity of the Indian Ocean and a possible nascent island arc between Ceylon and Australia: Jour Geophys. Research, v. 75, p. 5041–5055.

TALWANI, MANIK, AND ELDHOLM, OLAV, 1972, Continental margin off Norway; a geophysical study: Geol. Soc. America Bull., v. 83, p. 2575–3606.

TURCOTTE, D. L., AND OXBURGH, E. R., 1973, Mid-plate tectonics: Nature, v. 244, p. 337–339.

VALENTINE, J. W., AND MOORES, E. M., 1972, Global tectonics and the fossil record: Jour. Geology, v. 80, p. 167–184.

VINE, F. J., AND MOORES, E. M., 1972, A model for the gross structure, petrology, and magnetic properties of oceanic crust: Geol. Soc. America Mem. 132, p. 195–205.

VON HUENE, ROLAND, 1972, Structure of the continental margin and tectonism at the eastern Aleutian trench: Geol. Soc. America Bull., v. 83, p. 3613–3626.

WALCOTT, R. I., 1972, Gravity, flexure, and the growth of sedimentary basins at a continental edge: *ibid.,* v. 83, p. 1845–1848.

WANLESS, H. R., AND SHEPARD, F. P., 1936, Sea level and climatic changes related to late Paleozoic cycles: *ibid.,* v. 47, p. 1177–1206.

WELLS, A. J., 1969, The crush zone of the Iranian Zagros Mountains, and its implications: Geol. Mag., v. 106, p. 385–394.

WEIMER, R. J., 1970, Rates of deltaic sedimentation and intrabasin deformation, Upper Cretaceous of Rocky Mountain region, *in* Morgan, J. P. (ed.), Deltaic sedimentation, modern and ancient: Soc. Econ. Paleontologists and Mineralogists Special Pub. 15, p. 270–292.

WHITE, D. A., ROEDER, D. H., NELSON, T. H., AND CROWELL, J. C., 1970, Subduction: Geol. Soc. America Bull., v. 81, p 3431–3432.

WILSON, J. T., 1965, A new class of faults and their bearing on continental drift: Nature, v. 207, p. 343–347.

———, 1973, Mantle plumes and plate motions: Tectonophysics, v. 19, p. 149–164.

———, AND BURKE, K. C. A., 1972, Two types of mountain building: Nature, v. 239, p. 448–449.

WISE, D. U., 1972, Freeboard of continents through time: Geol. Soc. America Mem. 132, p. 87–100.

LOWER PALEOZOIC AND UPPERMOST PRECAMBRIAN CORDILLERAN MIOGEOCLINE, GREAT BASIN, WESTERN UNITED STATES[1]

J. H. STEWART AND F. G. POOLE
U.S. Geological Survey, Menlo Park, California; U.S. Geological Survey, Denver, Colorado

ABSTRACT

Shallow-marine intertidal and supratidal detrital and carbonate strata of latest Precambrian (<850 my) and early Paleozoic (>345 my) age thicken from a few hundred meters in cratonic areas east of the Great Basin to nearly 10,000 m (c. 30,000 ft) in the central Great Basin 350 to 450 km (c. 250 mi) to the west. Coeval rocks in the western Great Basin are deep-water strata characterized by shale and radiolarian chert associated with mafic pillow lavas. Strata deposited at moderate depths between the shallow- and deep-water facies have a limited distribution that suggests a relatively abrupt transition from shelf to deep water. The thick accumulation of shallow-water deposits in the Great Basin is similar to deposits along present-day stable continental margins. Such accumulations have been termed miogeoclines, rather than miogeosynclines, because they are bordered outward by open ocean, or a marginal sea, and are not synclinal in form.

The continental margin along which the early Paleozoic and latest Precambrian miogeocline was constructed apparently developed by rifting in late Precambrian time (<850 my). Extensional faulting and flowage related to this rifting extended well into the continent and may have caused major crustal thinning as far east as the Wasatch line, across which the rate of westward thickening of uppermost Precambrian and Paleozoic strata increases markedly. A persistent positive belt, perhaps analogous to the buried ridge beneath the outer edge of the present-day Atlantic continental shelf, may account for regional thinning and local erosional truncation of lower Paleozoic strata along the western margin of the Cordilleran miogeocline.

INTRODUCTION

The extensively exposed lower Paleozoic and uppermost Precambrian strata in the Great Basin are an excellent record of ancient sedimentation along a continental margin. The region described extends from the Wasatch Mountains in central Utah to the Sierra Nevada in westernmost Nevada and eastern California and from the Snake River Plain in southern Idaho to the Mojave Desert in California. The total area is about 5 × 10^5 km^2. The completeness of the record within the region contrasts with that in adjoining areas where upper Precambrian and lower Paleozoic rocks either are less extensively exposed, are covered by younger rocks, or occur in structurally complex settings where stratigraphic details are difficult to interpret. The purpose of this paper is to summarize the upper Precambrian and lower Paleozoic stratigraphy in the Great Basin, with particular attention to characteristics that indicate environment of deposition and tectonic setting.

The strata described range in age from latest Precambrian[2] (<850 my) to Late Devonian (>345 my) and were deposited during a time of relative tectonic stability following a marked change in the tectonic framework of the region in latest Precambrian time. Deposition preceded the Antler orogeny, a major deformational event of Late Devonian and Early Mississippian time (Poole, this volume). The deposits lie mostly within the Cordilleran geosyncline (fig. 1), the dominating structural feature of western North America during late Precambrian, Paleozoic, and early Mesozoic time.

Various classifications have been proposed for major belts of Paleozoic sediments in the Cordilleran geosyncline in the Great Basin. Geologists have long recognized that the eastern part of the geosyncline contains carbonate and quartzite rocks, and the western part chert, shale, and volcanic rocks. The terms "eastern" and "western" facies or assemblages have been applied to these belts of rocks (Merriam and Anderson, 1942, p. 1704; Nolan and others, 1956, p. 6, 23, 34; Roberts and others, 1958, p. 2816-2817). These same facies were also noted by Kay (1951), who recognized an eastern miogeosynclinal (Millard) belt and a western eugeosynclinal (Fraser) belt of the Cordilleran geosyncline. Roberts and others

[1] Publication authorized by the Director, U.S. Geological Survey.

[2] Strata referred to here as latest or uppermost Precambrian correspond in large part or entirely to the Precambrian Z age (800 to 570 my) of James (1972).

(1958) described rocks of an intermediate facies that crop out in areas between the two major assemblages and referred to these rocks as the "transitional" assemblage. More recently the terms "carbonate" and "siliceous and volcanic" (or simply "siliceous" have replaced the terms "eastern" and "western" in the nomenclature (Roberts, 1964, p. A8, 1968a, 1972; Gilluly and Gates, 1965; Smith and Ketner, 1968). Roberts (1968b), in describing Silurian and Devonian strata, further divided each of the major assemblages into two subfacies.

This report uses the threefold facies nomenclature of common usage—carbonate, transitional, and siliceous assemblages; we further divide the siliceous assemblage in the Great Basin into two subassemblages, the inner and outer belts (table 1). Assignment of a particular rock unit to one assemblage or another is not always clear; some rocks called transitional here are part of the carbonate assemblage of other geologists.

The eugeosynclinal (Fraser) belt, as defined by Kay (1951), includes lower Paleozoic rocks that outcrop in the northern and western Sierra Nevada and in the Klamath Mountains in California as well as in the Great Basin. In this report, only those rocks of this belt that outcrop in the Great Basin are described.

No entirely satisfactory nomenclature for geosynclines is available. Dietz and Holden (1967) and Dietz (1972) have pointed out that a modern analogy of a geosyncline is a deposit along a stable continental margin where a seaward-thickening wedge of sediment underlies the continental shelf and is limited seaward by the continental slope. Farther oceanward, sediment in this system consists of continental-rise and ocean-basin deposits. This system does not have the classic form of a geosynclinal trough bounded on both sides by landmasses, or bounded on one side by a landmass and on the other by an island arc system. Dietz and Holden (1966) and Dietz (1972) proposed the term miogeocline to describe the wedge of sediment underlying the continental shelf and slope, and the term eugeocline to describe continental rise deposits. They used these terms instead of miogeosyncline or eugeosyncline in order to indicate the nonsynformal shape of the deposit.

In this report we use the term miogeocline to describe the wedge-shaped deposit formed by lower Paleozoic and uppermost Precambrian carbonate and transitional assemblage rocks in the Great Basin. We do not intend, however, that this practice should prohibit use of the term miogeosyncline in a more general sense to describe thick sequences of nonvolcanic sedi-

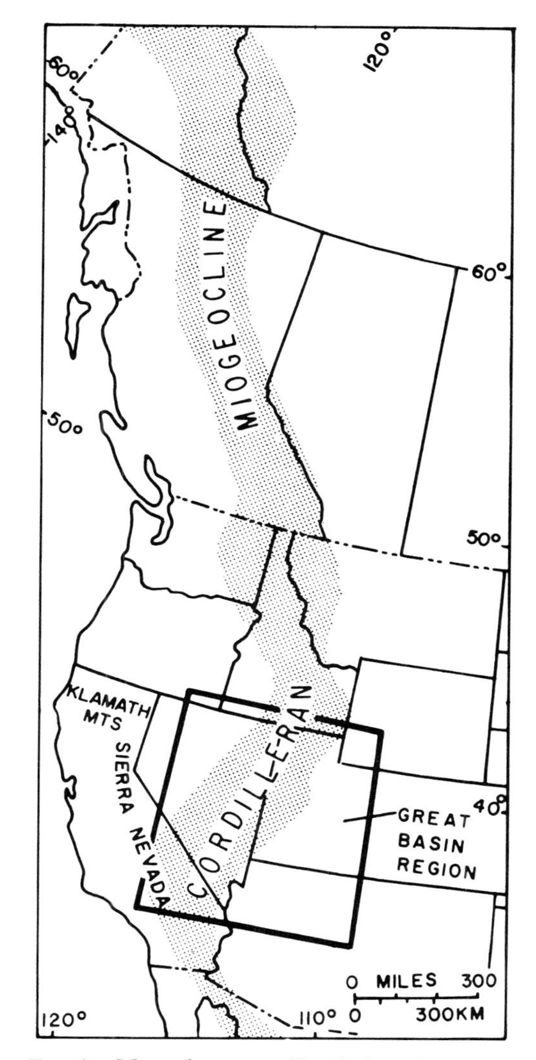

FIG. 1.—Map of western North America showing location of Cordilleran miogeocline and Great Basin region.

ments in geosynclines of different character. The term eugeocline is awkward etymologically, meaning "true earth slope," but is tentatively used in the Great Basin to describe lower Paleozoic siliceous assemblage rocks that may be continental rise deposits. The term eugeosyncline is used in a general sense to describe the volcanic part of a geosyncline.

This paper is based on our own studies as well as on published reports by hundreds of geologists. Where possible we have acknowledged the work of others, but we have not indicated the sources of all the data on the isopach maps. The excellent summary of the Cordilleran

TABLE 1.—LOWER PALEOZOIC AND UPPERMOST PRECAMBRIAN ASSEMBLAGES AND FACIES IN THE GREAT BASIN.

AGE	Eugeocline		Miogeocline	
	Siliceous assemblage		*Transitional assemblage* Limestone and shale, including quartzite in upper Precambrian and Lower Cambrian	*Carbonate assemblage* Includes largely detrital rocks in upper Precambrian and Lower Cambrian
	Outer belt Chert, shale, quartzite, greenstone, arkose, and feldspathic sandstone	*Inner belt* Shale and chert		
DEVONIAN	Chert facies	Shale and chert facies	Limestone and shale shale	Carbonate and quartzite facies
SILURIAN	Feldspathic sandstone facies	Chert and shale facies	Laminated limestone facies	Dolomite facies
ORDOVICIAN	Siliceous and volcanic facies	Shale and chert facies	Shale and limestone facies	Carbonate and quartzite facies
LATE AND MIDDLE CAMBRIAN	Arkosic facies	—	Limestone and shale facies, includes abundant chert in Emigrant Formation	Carbonate facies
LATEST PRECAMBRIAN AND EARLY CAMBRIAN	Siliceous and volcanic facies	—	Siltstone, carbonate, and quartzite facies	Quartzite and siltstone facies

miogeosyncline in eastern Nevada and western Utah by Armstrong (1968a) is of great value both for the information and ideas it contains and as a guide to methods of presenting data. Our palinspastic base maps are similar to those prepared by Armstrong. We are indebted for discussions with many colleagues, some of whom have supplied unpublished data. The report has benefited from thoughtful comments of Michael Churkin, Jr., M. D. Crittenden, Jr., W. R. Dickinson, H. Gabrielse, R. K. Hose, K. B. Ketner, F. J. Kleinhampl, and T. E. Mullens.

REGIONAL STRUCTURE AND PALINSPASTIC BASE MAP

Original geographic positions of upper Precambrian and lower Paleozoic strata in the Great Basin have been severely disrupted by major displacements on thrust and strike-slip faults (fig. 2) and distorted by oroflexural bending. To compensate for this disruption and distortion, thickness and facies data in this report are plotted on a palinspastic base map (fig. 3).

The oldest structure shown on the palinspastic base map is the Roberts Mountains thrust along which siliceous and some transitional assemblage rocks have been transported about 145 km (90 mi) to the east primarily over carbonate and transitional assemblage rocks during the Late Devonian and Early Mississippian Antler orogeny (Roberts and others, 1958; Roberts, 1968a, p. 106; 1972, p. 1995; Smith and Ketner, 1968). On the palinspastic base, the reconstructed map position for upper plate (siliceous assemblage) rocks is an area where standard county line symbols are used and is west of an area, shown with dotted county lines, where lower plate rocks occur. In westernmost Nevada, only rocks in northern Esmeralda County and southern Mineral County are considered here to be allochthonous and part of the upper plate of the Roberts Mountains thrust; McKee (1968), on the other hand, has suggested that Ordovician rocks in southern Esmeralda County may also be allochthonous and part of the upper plate, an interpretation not followed here.

The western Great Basin is an area of 130 to 195 km (80 to 120 mi) of cumulative right-lateral distortion resulting from fault slip and more pervasive large-scale drag from oroflexural bending (Albers, 1967; Stewart, 1967; Poole and others, 1967; Stewart and others, 1968), although other geologists (Wright and

Troxel, 1966, 1967, 1970) have questioned this concept. Our palinspastic base was constructed on the assumption of about 160 km (100 mi) of right-lateral displacement in a general northwest direction. More than half of the displacement is considered to be by oroflexural bending, which can be detected on the palinspastic base by curving of originally straight county and state lines; the remainder, by fault displacement on two major right-lateral strike-slip fault zones, the Las Vegas Valley shear zone and the Death Valley-Furnace Creek fault zone.

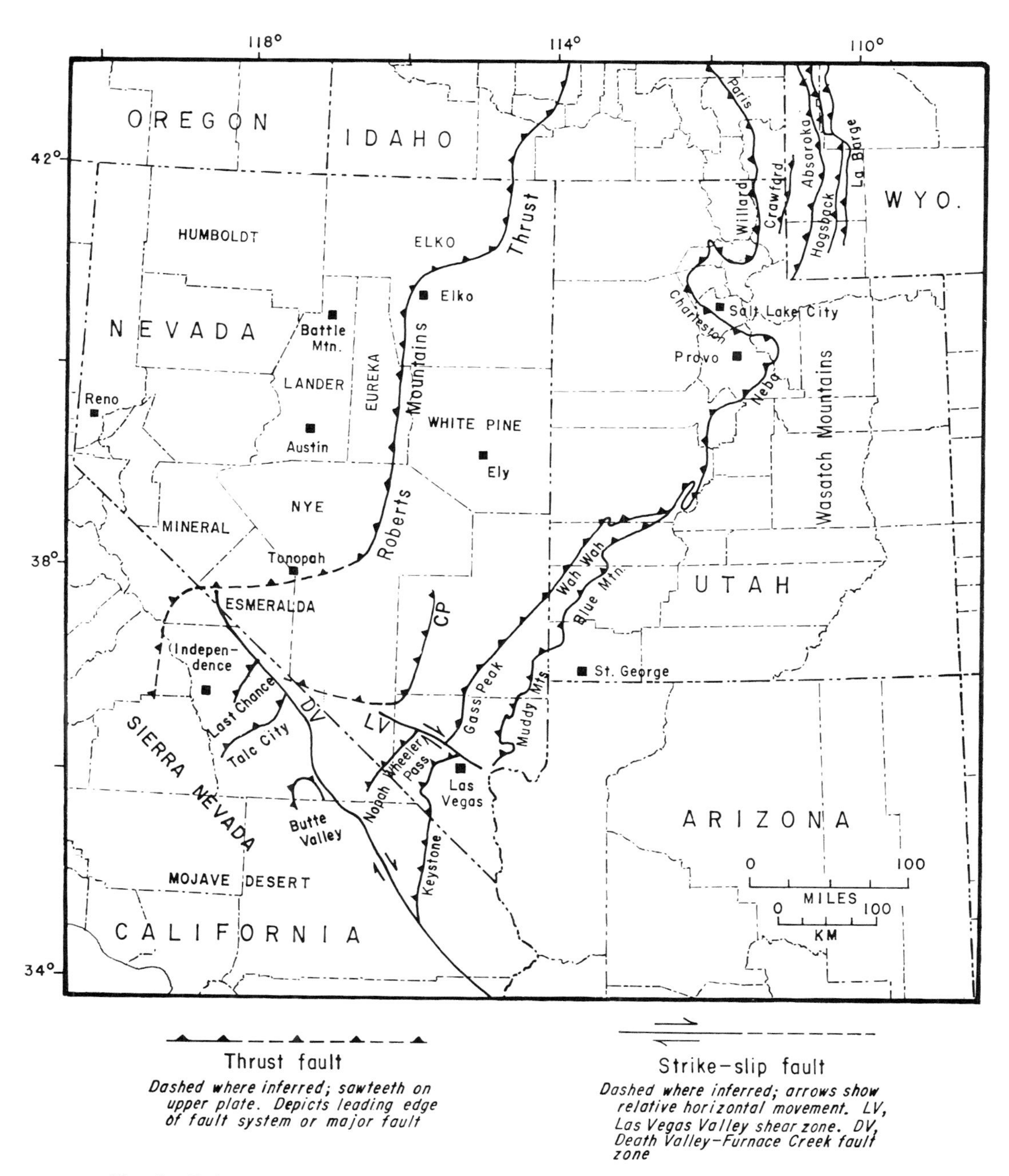

FIG. 2.—Index map of Great Basin region showing major thrust and strike-slip faults. Only counties referred to in text are labeled.

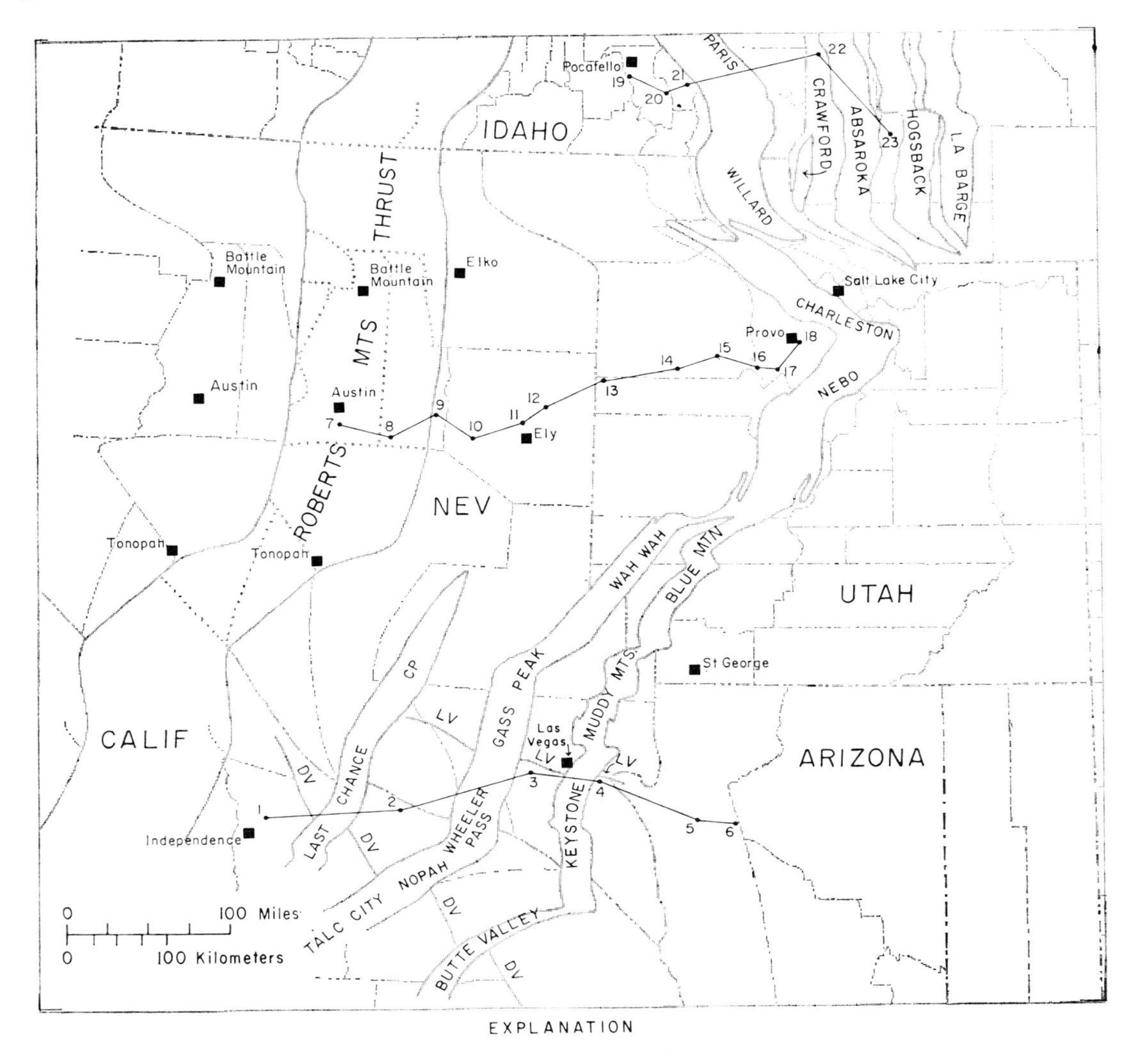

FIG. 3.—Palinspastic base map showing location of cross sections shown in figures 5, 6, 7 and 8. Dotted county lines indicate location of lower plate of Roberts Mountains thrust.

The displacement on the Las Vegas Valley shear zone occurred in late Tertiary time (Ekren and others, 1968; Fleck, 1970a; Anderson and others, 1972, p. 284), and perhaps much of the other right-lateral displacement in the western Great Basin is of this age.

Another oroflexural feature may occur in northeastern Nevada, where lower and upper Paleozoic facies and thickness trends and the leading edge of the Roberts Mountains thrust curve northeastward at latitude 41° for 120 km (75 mi) before regaining a more northerly direction. The curving of these trends has generally been assumed to mark an original curve in the Cordilleran geosyncline, although Thorman (1970, p. 2441) has suggested that displacement along major northwest-trending right-lateral strike-slip faults may account for anomalous facies trends in the area. We prefer the interpretation that the curving of structural and stratigraphic features is due to oroflexural bending. Several regionally anomalous east-trending fold structures (Hope, 1970; oral commun., 1973) near latitude 41° N. are compatible with the presumed oroflexural feature.

A system of thrust faults extends along the

eastern part of the Great Basin and the adjoining part of the Rocky Mountains to the east. In Wyoming, this belt consists of several major thrusts along each of which tectonic transport may have been about 15–25 km (10–15 mi) subject to a factor of uncertainty of about two, either way, according to Rubey and Hubbert (1959, p. 187). S. S. Oriel (oral commun., 1973) considers Rubey and Hubbert's estimate of displacement to be conservative. Our reconstruction is based on an assumed total tectonic transport of 120 km (75 mi) across the thrust belt in Wyoming. In Idaho and Utah, we used 65 km (40 mi) as the amount of transport on the Paris, Willard, Charleston, and Nebo thrust system; this figure appears to be a minimum based on the estimates of Crittenden (1961), Rigo (1968, p. 64), and Armstrong (1968b, p. 440–441). Estimates of displacement on faults farther south in the thrust belt are less certain. A minimum of 30–35 km (c. 20 mi) is required on the combined Muddy Mountains and Glendale thrusts in southern Nevada (Longwell, 1961, as quoted in Armstrong, 1968b, p. 440). Fleck (1970b) has suggested a total displacement of 30 to 50 km (18 to 32 mi) on the Keystone, Wheeler Pass, and associated minor thrusts. A minimum of 30-35 km (c. 20 mi) has been suggested on the Last Chance thrust (Stewart and others, 1966), although Stevens and Olson (1972) have a somewhat different interpretation of the geometry of this thrust. Tectonic transport on the CP thrust has been inferred to be about 55 km (35 mi; Barnes and Poole, 1968). We show a total of about 120 km (75 mi) telescoping on thrust faults in southern Nevada and adjacent California.

The correlation of thrust faults in the southern Great Basin follows the interpretation of Poole and others (1967, p. 891–892). Burchfiel and Davis (1972) have presented a somewhat different picture and interpretation of thrusting in this region.

Crustal shortening due to folding and to telescoping on seemingly minor thrust faults has not been incorporated on the palinspastic map, although we realize that some of these structures could be of major significance. Nor have we included late Tertiary crustal extension, which resulted in the development of basin and range faulting and is commonly considered to be 50–100 km (30–60 mi) across the entire Great Basin (see summary by Stewart, 1971, p. 1035–1036) or even as much as 160 km (100 mi; Proffett, 1971). The factors of crustal shortening and extension that were not evaluated on the palinspastic base are compensating and perhaps could be nearly balancing.

PREGEOSYNCLINAL ROCKS

Three major divisions of Precambrian are recognized in the Great Basin and adjoining regions: metasedimentary (schist and paragneiss) and plutonic rocks older than about 1400 my; unmetamorphosed sedimentary rocks approximately comparable in age to the Belt Supergroup, 850 to 1250 my; and unmetamorphosed uppermost Precambrian sedimentary rocks probably less than about 850 my old.

A distinctive diamictite[3] unit (Crittenden and others, 1972; Stewart, 1972) at or near the base of the third (uppermost Precambrian) group of these rocks occurs at scattered localities throughout westernmost North America from Alaska to California and is the oldest unit that clearly has a depositional pattern related to the Cordilleran geosyncline. A major change in the tectonic pattern of North America appears to have taken place shortly before the deposition of the diamictite, and this change is inferred to mark the beginning of the geosyncline (Stewart, 1972). Pregeosynclinal rocks in the Great Basin and adjoining regions (fig. 4), therefore, consist of the two older groups of Precambrian rocks.

Metasedimentary and plutonic rocks older than 1400 my crop out or are known in the subsurface over a large area of Wyoming, Utah, Arizona, and California east of the Wasatch line[4] (fig. 4) but are sparsely exposed west of the Wasatch line. They are as old as 2300–2400 my (Hansen, 1965, p. 31; Armstrong and Hills, 1967, p. 1300), although most are 1500–1800 my old (King, 1969, p. 41). West of the Wasatch line, rocks clearly older than 1400 my occur in the Death Valley-Mojave Desert region of southeastern California (Wasserburg and others, 1959; Lanphere and others, 1963), in the Albion and Raft River Ranges in southern Idaho and northern Utah (locs. 1 and 2, fig. 4; Armstrong and Hills, 1967; Compton, 1972), and in the upper plate of the Willard thrust in the Wasatch Mountains (loc. 5; Crittenden, McKee, and Peterman, 1971). Rocks in the Ruby (loc. 3), East Humboldt (loc. 4), and Snake Ranges (loc. 8) in eastern Nevada commonly have been considered

[3] Diamictite: A nonsorted sedimentary rock consisting of sand and(or) larger particles in a muddy matrix (Crittenden, Schaeffer, Trimble, and Woodward, 1971, *modified from* Flint and others, 1960).

[4] The Wasatch line (Kay, 1951, p. 14) is the hinge line along the eastern border of the Cordilleran geosyncline across which the rate of westward thickening of upper Precambrian, Paleozoic, and Mesozoic strata increases greatly.

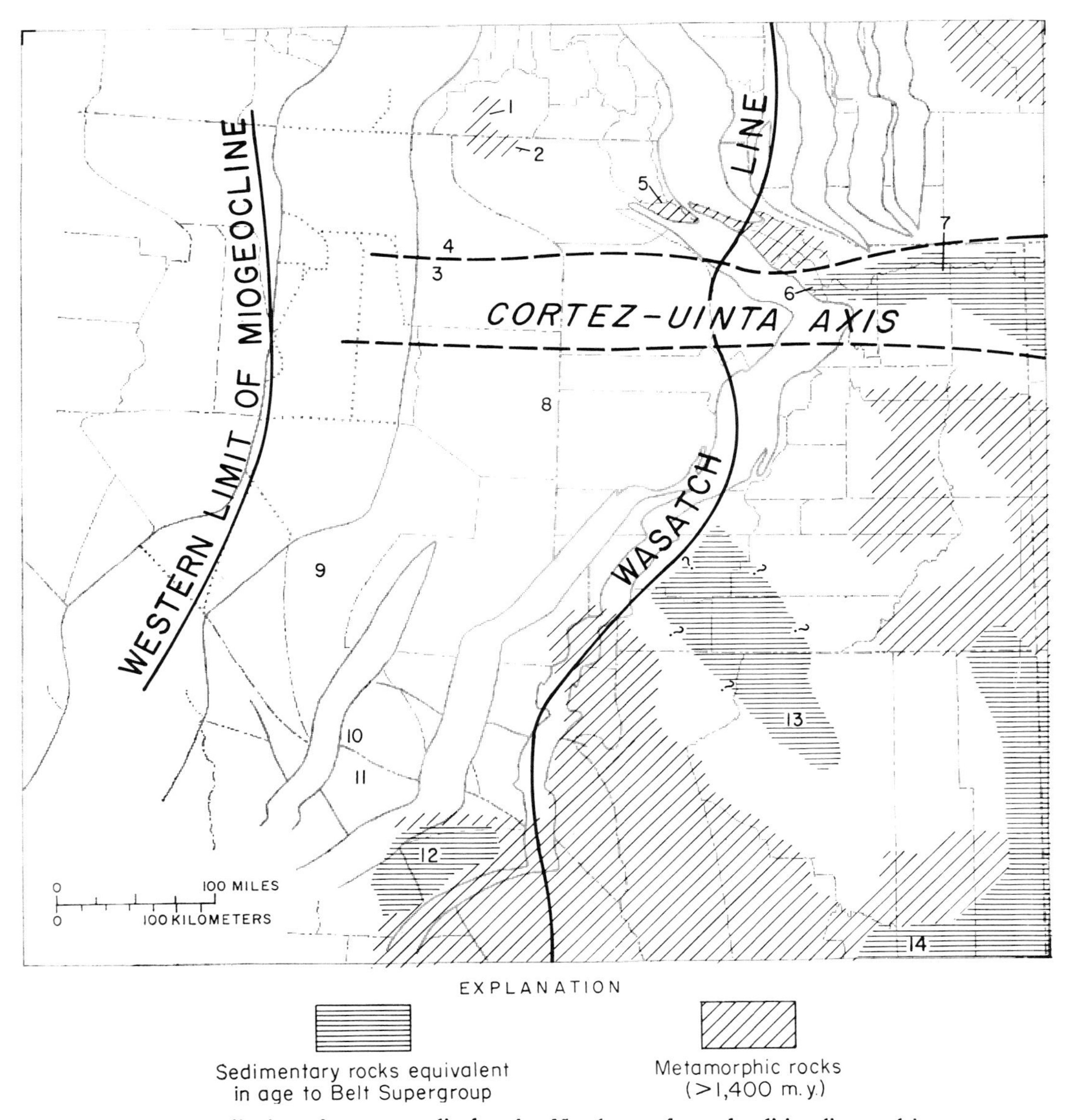

FIG. 4.—Distribution of pregeosynclinal rocks. Numbers refer to localities discussed in text.

to be Precambrian in age (Bayley and Muehlberger, 1968), but recent studies (Howard, 1971; Hose and Blake, 1970) indicate that they are probably metamorphosed Paleozoic strata and that the only Precambrian present is very young Precambrian (Howard, 1971, p. 260; Hose and Blake, 1970) that is a part of the geosynclinal sequence of this paper. Supposed older Precambrian rocks that crop out in southern Nevada (locs. 9 and 10; Ekren and others, 1971; Cornwall and Kleinhampl, 1964), may also be metamorphosed strata of latest Precambrian or Paleozoic age.

Unmetamorphosed to slightly metamorphosed pregeosynclinal sedimentary rocks equivalent in age to the Belt Supergroup are inferred to include (1) the Uinta Mountain Group (loc. 7) and the Big Cottonwood Formation (loc. 6) in northern Utah (Wallace and Crittenden, 1969; Wallace, 1972; Crittenden and others, 1952), (2) the Crystal Spring Formation and Beck Spring Dolomite in California (loc. 12, and possibly at loc. 11; Hewett, 1956; Wright and Troxel, 1967; Troxel and Wright, 1968), (3) the Grand Canyon Supergroup in northern Arizona, loc. 13; Walcott, 1894; Maxson, 1961; Ford and Breed, 1973), and (4) the Apache Group and Troy Quartzite in southern Arizona (loc. 14; Shride, 1967). More data on the correlation of these strata, including possible alternative correlations, were given recently by Crittenden and others (1972).

The Uinta Mountain Group appears to be at least 7350 m (24,000 ft) thick (Hansen, 1965, p. 33); the Big Cottonwood Formation is 4900 m (16,000 ft) thick (Crittenden and others, 1952). These rocks crop out in an east-west belt that seems to represent at least in part an original depositional trough (Wallace and Crittenden, 1969, p. 140). Deep linear troughs extending into a platform or cratonic area at high angles to the trend of a bordering geosyncline have been termed aulacogens in the USSR (Salop and Scheinmann, 1969, p. 586; Hoffman, 1971) and Paul Hoffman (oral commun. to M. D. Crittenden, Jr., 1971) has suggested that the Uinta Mountain trough may be such a feature. This aulacogen could extend west of the Wasatch and Uinta Mountains and underlie the Cordilleran geosyncline in the Great Basin. Such a postulated buried aulacogen may coincide with the Cortez-Uinta axis (fig. 4) of Roberts and others (1965, p. 1928), along which thickness trends of lower Paleozoic rocks are disrupted. Whether or not the aulacogen was connected to a geosyncline to the west is not known. C. A. Wallace (Crittenden and others, 1972, p. 337) has suggested that the Uinta Mountain Group may be correlative with post-diamictite strata of the Cordilleran geosyncline, but this correlation is not accepted here. If his correlation is correct, the Uinta Mountains aulacogen would be an offshoot of the Cordilleran geosyncline, which seems unlikely to us.

The Grand Canyon Supergroup is about 3650 m (12,000 ft) thick (Walcott, 1894) and may have been deposited in another aulacogen. The Apache Group and Troy Quartzite, on the other hand, have a combined thickness of less than about 765 m (2500 ft; Shride, 1967) and the Crystal Spring Formation and Beck Spring Dolomite have a combined thickness of 900 to 1500 m (3000 to 5000 ft; Hewett, 1956; Wright and Troxel, 1967), suggesting that these formations may have been deposited in a platform area rather than in an aulacogen.

GEOSYNCLINAL ROCKS

Geosynclinal rocks described here range in age from latest Precambrian to Late Devonian and consist of miogeoclinal carbonate and transitional assemblage rocks and of eugeoclinal siliceous assemblage rocks in the Great Basin. In the eastern and central Great Basin, miogeoclinal rocks consist of lithologically distinctive and persistent units exposed in relatively simple structural blocks where accurate measurements of thickness can be made, whereas in the western Great Basin, eugeoclinal rocks consist of lithologically monotonous units exposed in highly complex structural blocks where only crude estimates of thickness are possible. On the isopach maps, therefore, thicknesses are shown only in the eastern and central Great Basin.

A summary of the stratigraphy of the geosynclinal rocks for each of five major groups of strata is presented below. The assemblages and constituent facies described here have been summarized in table 1.

Uppermost Precambrian and Lower Cambrian Rocks

At least half of the thickness of the lower Paleozoic and uppermost Precambrian Cordilleran miogeocline is represented by uppermost Precambrian and Lower Cambrian rocks consisting of quartzite and siltstone, and, to a lesser extent, of carbonate rock and conglomerate (figs. 5–8). These strata are less than 150 m (500 ft) thick along the eastern edge of the Great Basin and thicken to over 6000 m (20,000 ft) in areas 240–320 km to the west (fig. 9). Regional descriptions of these rocks have been given by Misch and Hazzard (1962), Woodward (1963, 1965, 1967, 1968), Stewart (1970), Crittenden, Schaeffer, Trimble and Woodward (1971), and Oriel and Armstrong (1971).

Three facies of uppermost Precambrian and Lower Cambrian rocks are recognized: (1) a quartzite and siltstone facies in the eastern Great Basin; (2) a siltstone, carbonate, and quartzite facies in the central Great Basin; and (3) a siliceous and volcanic facies (one locality) in the western Great Basin (fig. 9).

The quartzite and siltstone facies typically consists of cliff-forming fine- to medium-grained quartzite in units that are 25 to 1225 m (100 to 4000 ft) thick and are separated by units of siltstone and fine- to very fine-grained quartzite from 15 to 300 m (50 to 1000 ft) thick. The quartzite ranges in composition from arkose to highly pure (97 percent SiO_2) orthoquartzite (Stewart, 1970). Cross strata of both trough and tabular planar types are common and indicate westerly transport (fig. 9; Seeland, 1968, 1969; Stewart, 1970, p. 11 and 12). Conglomerate that contains pebbles of quartz and

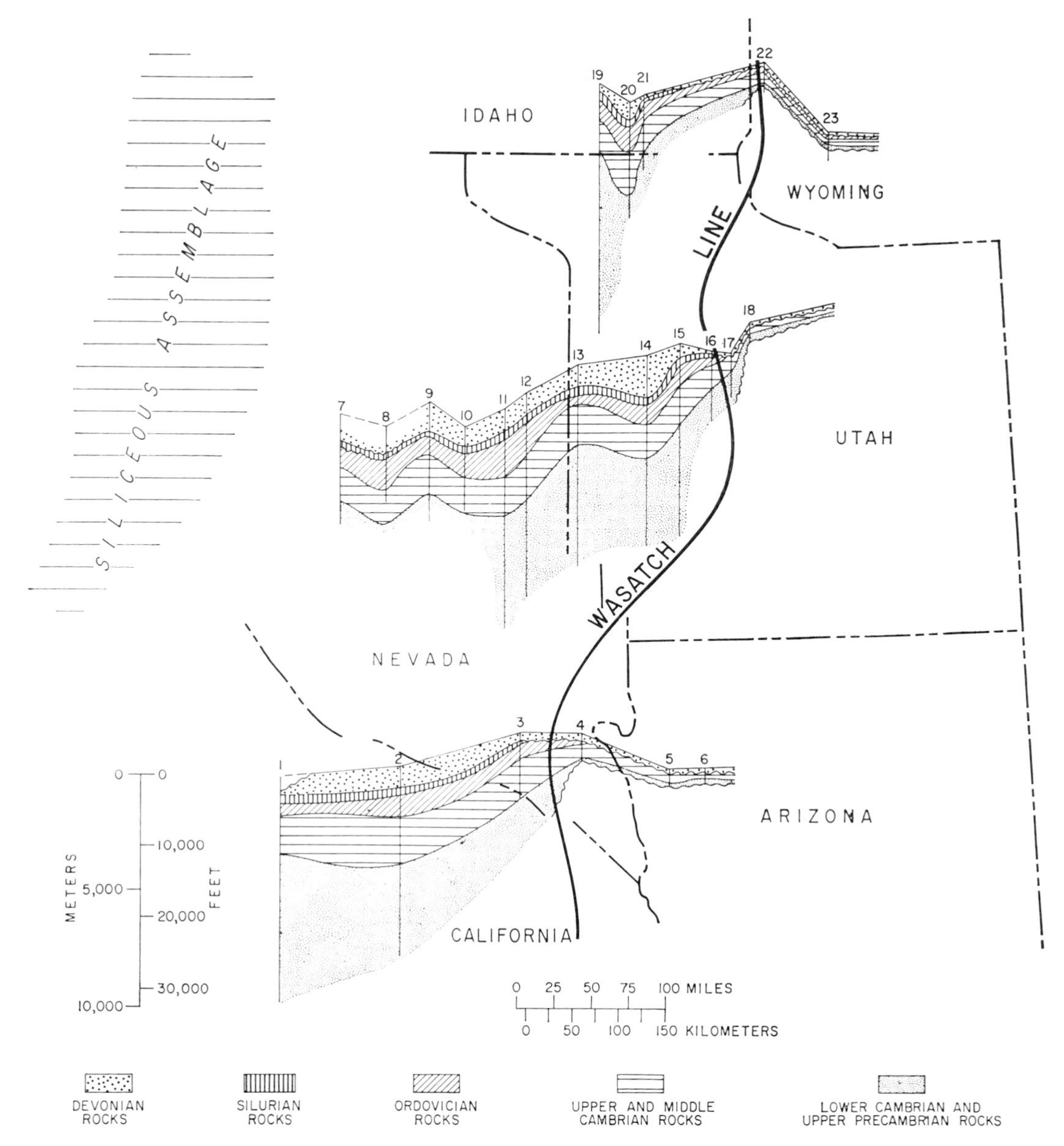

FIG. 5.—Stratigraphic diagram showing the miogeoclinal carbonate and transitional assemblages and the eugeoclinal siliceous assemblage in the Great Basin. Top of column is location of section. Column numbers are same as those on figures 6, 7, and 8.

quartzite is sparsely distributed in the quartzite. Limestone and dolomite are present in layers less than a foot to several hundred feet thick. Individual units of quartzite, siltstone, or carbonate persist for hundreds of miles along the trend of the miogeocline (Stewart, 1970, p. 64; Crittenden, Schaeffer, Trimble, and Woodward, 1971).

As noted above, a poorly sorted diamictite consisting of rounded to subangular pebbles to boulders of diverse rock types in a sandy or argillaceous matrix is recognized at or near the base of the geosynclinal sequence at several localities in Utah and California (Stewart, 1972, table 1), and appears to be of glacial origin (Troxel, 1967; Crittenden, Schaeffer,

Trimble, and Woodward, 1971; Crittenden and others, 1972; Stewart, 1972).

Volcanic rocks occur in the quartzite and siltstone facies in the eastern Great Basin (Stewart, 1972, table 2) and are unknown in other Paleozoic rocks in this part of the miogeocline. The thickest unit is the Bannock Volcanic Member of the Pocatello Formation consisting of porphyritic, vesicular, or amygdaloidal volcanic flows and breccias about 300 m (1000 ft) thick (Crittenden, Schaeffer, Trimble, and Woodward, 1971) occurring low in the uppermost Precambrian and Lower Cambrian sequence in Idaho (fig. 8, column 19). Vesicular basalt flows from less than a hundred to a few hundred feet thick occur within quartzite units (Tintic Quartzite and Prospect Mountain Quartzite) high in the sequence in Utah and Nevada (Abbott, 1951; Morris and Lovering, 1961, p. 15; Kellogg, 1963, p. 687–688).

The siltstone, carbonate, and quartzite facies of the uppermost Precambrian and Lower Cambrian sequence in central Nevada and southeastern California (fig. 8) is thick and fossil-rich. It contains large amounts of siltstone (or phyllitic siltstone), thin to thick units of limestone and dolomite, and fine- to very fine-grained quartzite (Nelson, 1962, Stewart, 1970; Albers and Stewart, 1973). The fine- to medium-grained type of quartzite characteristic of uppermost Precambrian and Lower Cambrian strata in the eastern Great Basin is largely absent. Trilobites, archeocyathids, pelecypods, echinoderms, pelmatozoan debris, *Hyolithes, Salterella, Scolithus* and algae (Palmer, 1971; Stewart, 1970), are locally abundant.

The siliceous facies is represented by only one formation, the Lower or Middle Cambrian

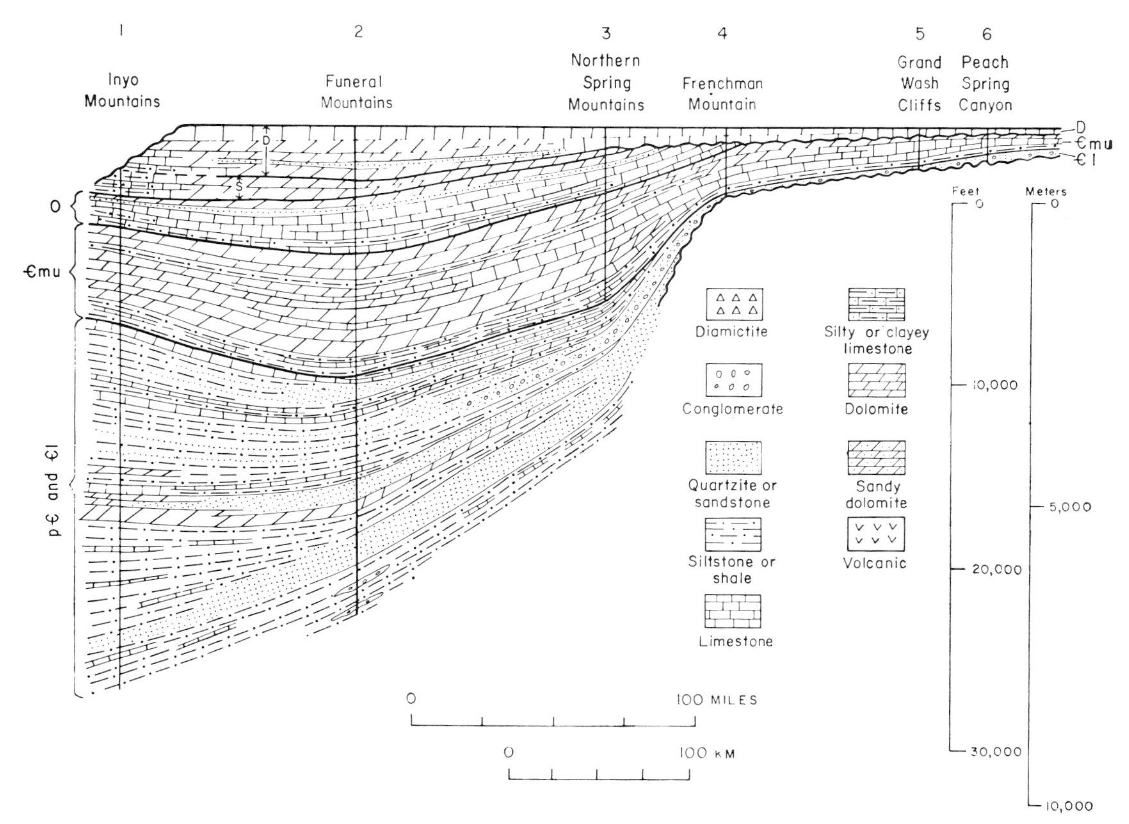

FIG. 6.—Cross section of miogeoclinal strata in southern Great Basin. Sources of data: 1, Ross (1965), Nelson (1962), Stewart (1970); 2, J. F. McAllister (written commun., 1972), Stewart (1970); 3, Fleck (1967); 4, McNair (1952); 5, McNair (1951); 6, McNair (1951). Symbols: p€, Precambrian; €l, Lower Cambrian; €mu, Middle and Upper Cambrian; O, Ordovician; S, Silurian; D, Devonian. Explanation includes some rock types shown only on figures 7 and 8.

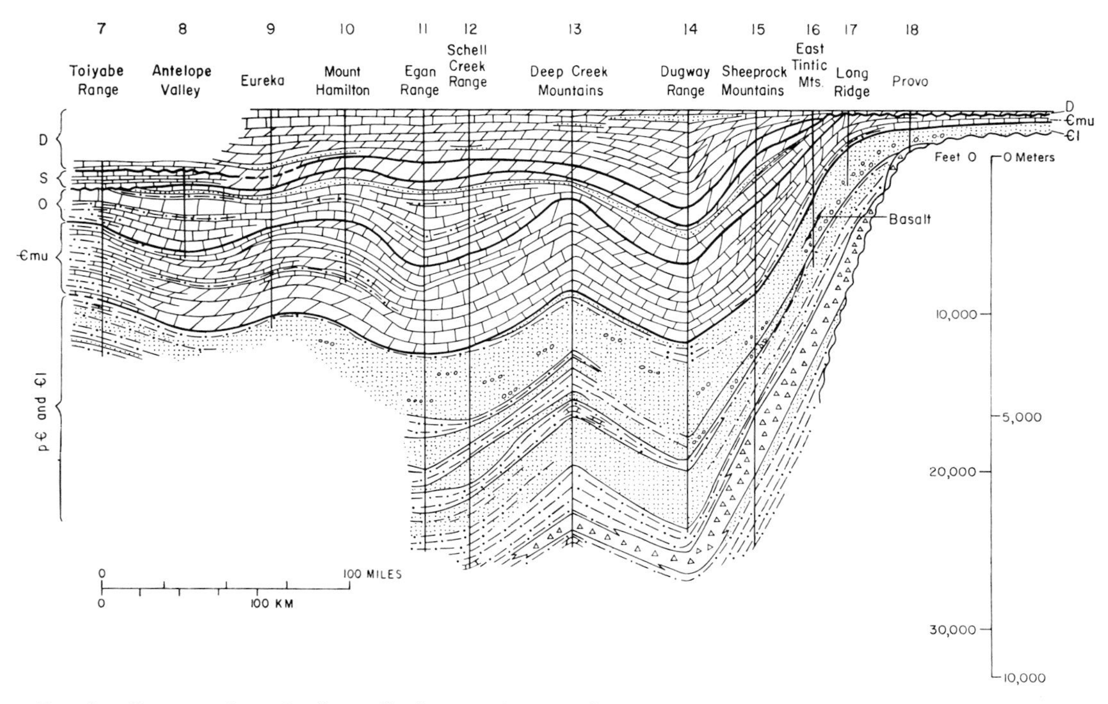

FIG. 7.—Cross section of miogeoclinal strata in central Great Basin. Sources of data: 7, J. H. Stewart and T. E. Mullens (unpub. data), Stewart and Palmer (1967); 8, Merriam (1963); 9, Nolan and others (1956); 10, Humphrey (1960); 11, Woodward (1962); 12, Young (1960) and R. K. Hose (unpub. data); 13, Bick (1966); 4, Staatz and Carr (1964); 15, Cohenour (1959); 16, Morris and Lovering (1961); 17, Brady (1965); 18, Baker (1947). Rock and letter symbols same as on figure 6.

Scott Canyon Formation[5] of north-central Nevada (Roberts, 1964; p. A14–A17). It consists of many thousands of feet of chert, argillite, and greenstone (in part pillow lavas) and minor amounts of sandstone, quartzite, and limestone (Roberts, 1964; Theodore and Roberts, 1971). Limestone lenses contain algae, sponges, archeocyathids, and trilobites.

Rocks of the quartzite and siltstone facies and the siltstone, carbonate, and quartzite facies are considered to have been deposited in shallow water (Stewart, 1970, p. 66–68; Klein, 1972), on the basis of the local abundance of such presumed shallow-water fossils as algae and archeocyathids and the abundance of cross strata produced by strong, presumably shallow-water currents.

Two different hypotheses can be considered for the origin of the widespread quartzite units: (1) they were laid down in a nearshore environment during repeated transgressions and regressions of the sea across the region; (2) they were laid down in an open-ocean environment far removed from a shoreline. Klein (1972) has suggested that they are deposits formed in a prograding tidal flat coastline, a concept that fits the first hypothesis. Stewart (1970, p. 66–68), however, has suggested that nearshore environment is difficult to reconcile with the lithologic uniformity of units over large areas, and with the absence, or at least scarcity, of deposits that can be interpreted as those of bars, beaches, inland bays, or distributary channels. The fact that current directions, as indicated by cross-strata studies, are unidirectional (fig. 9), and not bidirectional as in a tidal regime, and that the upper and lower boundaries of quartzite units are probably time conformable (Stewart, 1970, p. 64–66), rather than time transgressive as would be expected in a prograding system, argue against a nearshore origin for the sands. If an open-ocean environ-

[5] The presence of archeocyathids in the Scott Canyon Formation suggests that it may be entirely Early Cambrian, and not Middle Cambrian in age. Elsewhere in the Great Basin, archeocyathids occur only in Lower Cambrian strata (Palmer, 1971). In any case, the age span of the Scott Canyon Formation is uncertain because fossil material is sparse.

ment is a feasible alternative, strong ocean currents would be required to move the bedload sand. Perhaps storm-induced currents, such as the current velocity of 70 cm/sec registered in 80 m of water on the continental shelf of Washington (Smith and Hopkins, 1971), would be sufficient to move the sand.

The Scott Canyon Formation, by analogy with other siliceous assemblage rocks, is considered to be a relatively deep-water oceanic deposit. Fossil material such as algae and archeocyathids that suggest shallow-water conditions might occur in blocks that were transported into deep water by slumping, or they could indicate shallow-water conditions on the flanks of oceanic volcanoes.

Middle and Upper Cambrian Rocks

The maximum thickness of Middle and Upper Cambrian rocks in the Great Basin is about one-third that of uppermost Precambrian and Lower Cambrian strata. Middle and Upper Cambrian strata thicken (fig. 10) from 300–600 m (1000–2000 ft) in the eastern part of the

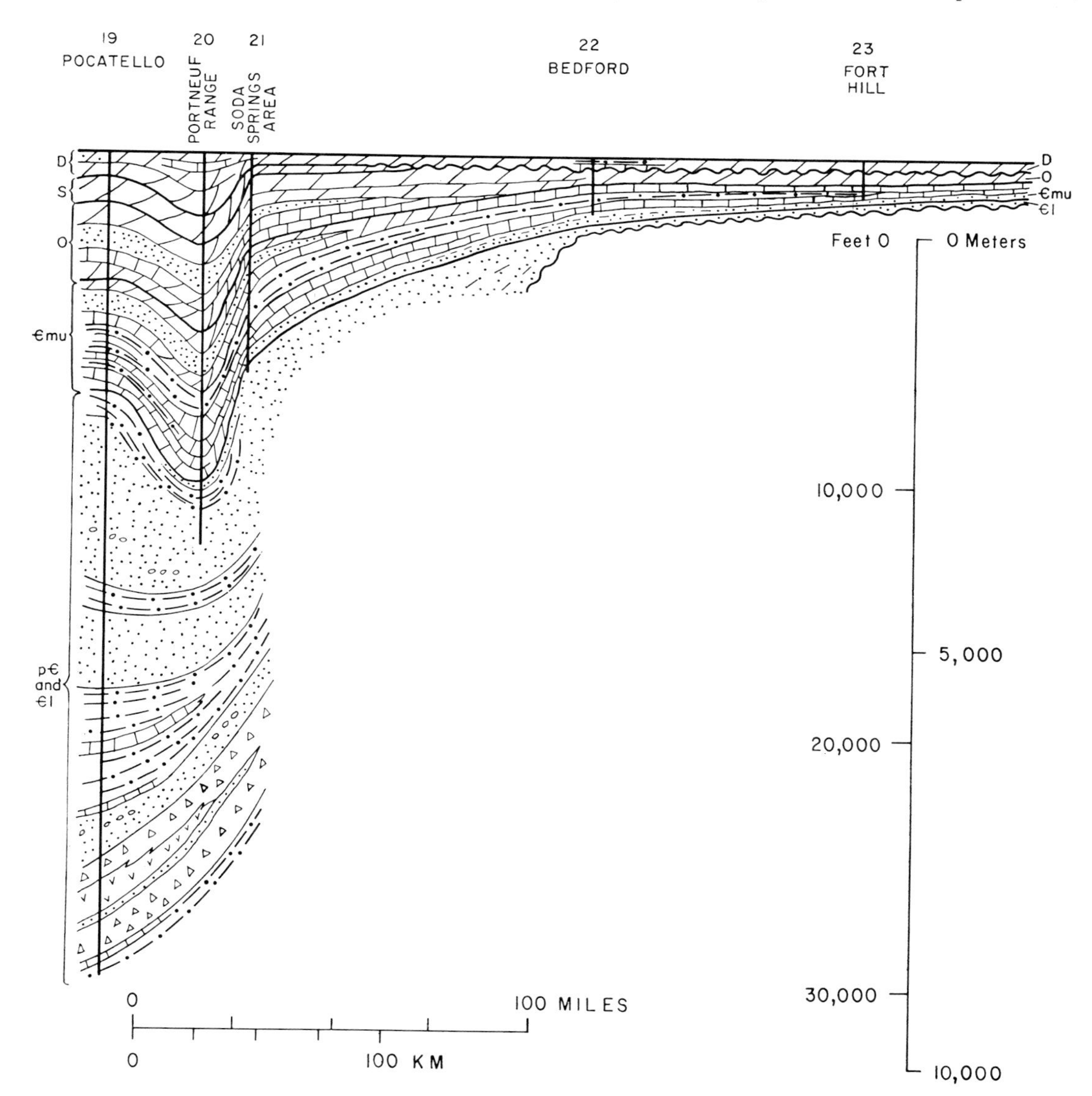

FIG. 8.—Cross section of miogeoclinal strata in northern Great Basin. Sources of data: 19, Trimble and Carr (1962) and Crittenden, Schaeffer, Trimble, and Woodward (1971); 20, Oriel (1965); 21, Armstrong (1969); 22, Rubey (1958); 23, Oriel (1969). Rock and letter symbols same as on figure 6.

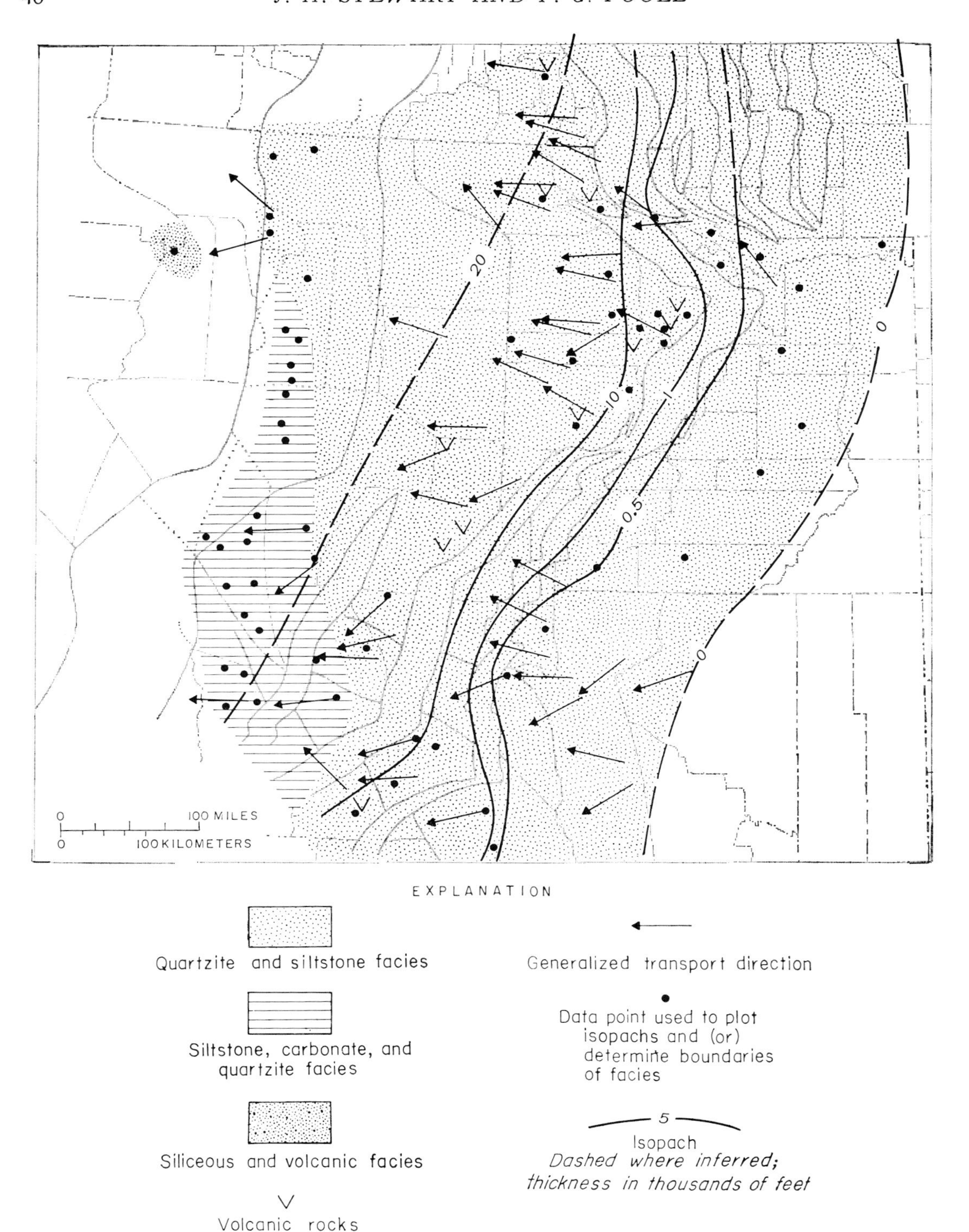

FIG. 9.—Isopach and facies map of uppermost Precambrian and Lower Cambrian strata showing generalized current directions (after Seeland, 1968, 1969; Stewart, 1970) and distribution of volcanic rocks in miogeoclinal sequence.

Great Basin to more than 2500 m (8000 ft) in the "Ibex basin"—a basin in western Utah and eastern Nevada (Webb, 1958; Hintze, 1959)—and to more than 1750 m (6000 ft) in the "northern Utah basin"—a basin in northwestern Utah (Webb, 1958). During much of Paleozoic time, these two basins were separated by the Tooele arch (Hintze, 1954, 1959; Webb, 1958),

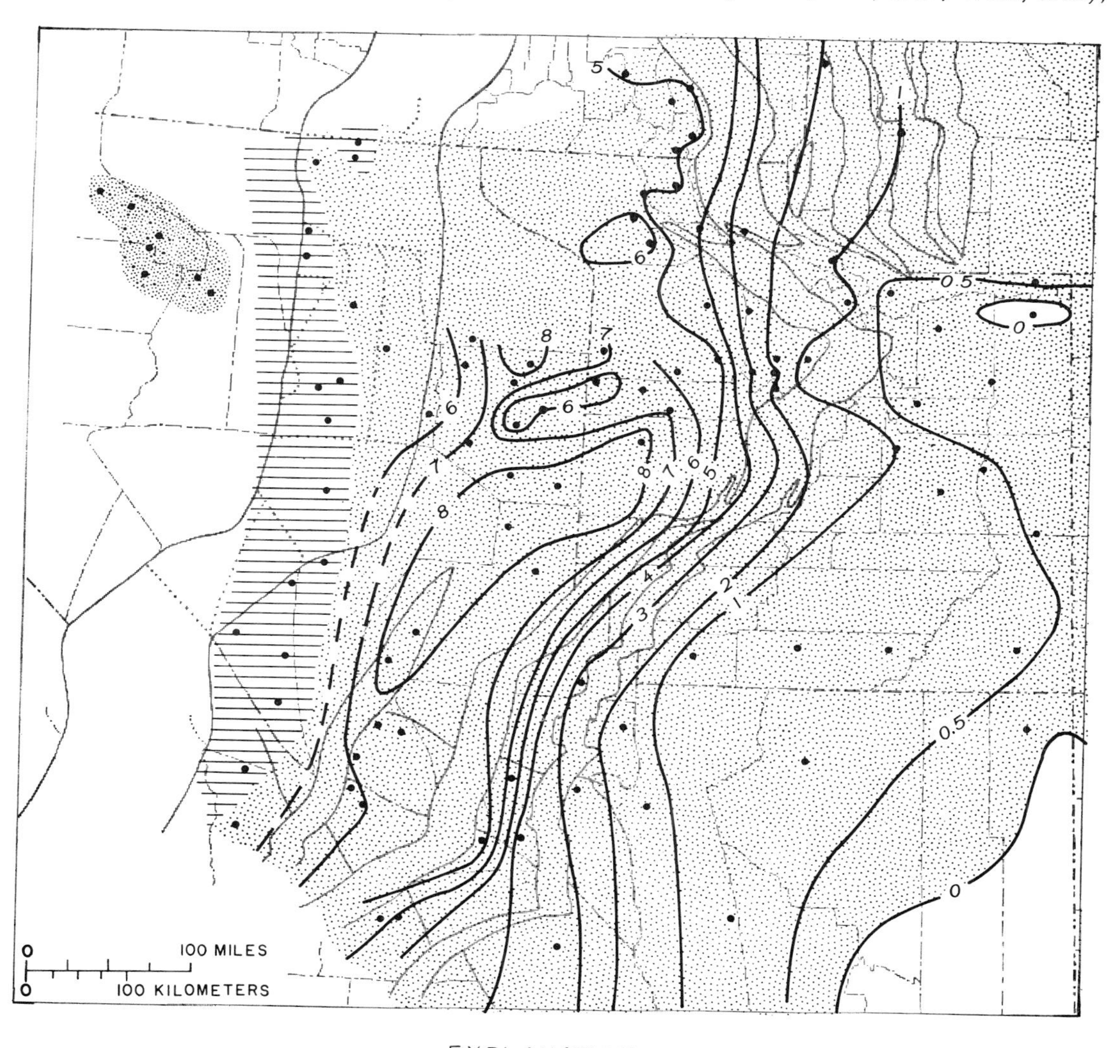

FIG. 10.—Isopach and facies map of Middle and Upper Cambrian strata (partly restored).

the Cortez-Uinta axis of Roberts and others (1965), but the Tooele arch was not well defined during Middle and Late Cambrian time.

An excellent summary of Cambrian stratigraphy in the Great Basin is given by Palmer (1971). Other important papers include those of Robison (1960, 1964), Rigo (1968), and Kepper (1972).

Three major facies of Middle and Upper Cambrian rocks are recognized in the Great Basin: (1) a carbonate facies in the eastern part, (2) a limestone and shale facies in the central part, and (3) an arkosic facies (Paradise Valley Chert and Harmony Formation) in the western part.

The carbonate facies consists typically of several thousand feet of massive- to well-bedded, light- to dark-gray dolomite and limestone. In detail, the stratigraphy is moderately complex; shale, laminated limestone, chert, and intraformational conglomerate are major units locally; some are regionally persistent. An unusual quartzite unit, the Worm Creek Quartzite Member of the St. Charles Limestone (Rigo, 1968; Oriel, 1965), occurs in the carbonate facies in southern Idaho. The lower few tens or hundreds of meters of the carbonate facies are shale-rich and gradational downward into the quartzite and siltstone of the Lower Cambrian.

Rocks of the limestone and shale facies consist of light- to dark-gray, evenly laminated to very thin-bedded limestone and dark-gray to olive-gray, extremely fine-textured shale. The proportion of limestone to shale in this facies is variable; some sections are predominantly limestone, others predominantly shale. The Emigrant Formation, a formation of this facies in western Nevada (Albers and Stewart, 1973), contains abundant beds of chert in addition to limestone and shale. Other strata in the limestone and shale facies include the Swarbrick, Tybo, and Hales formations in the Tybo area, Nye County, Nevada (Ferguson, 1933), the Crane Canyon sequence of Means (1962), part of the Broad Canyon sequence or Formation of Means (1962) and Washburn (1970), an unnamed sequence in the Mount Callaghan area, Lander County, Nevada (Stewart and Palmer, 1967), the Preble Formation (Hotz and Willden, 1964; Gilluly, 1967), and part of the Tennessee Mountain Formation of Bushnell (1967). Rocks lithologically similar to those of the limestone and shale facies also occur locally within the carbonate facies, and the boundary between the limestone and shale facies and the carbonate facies is irregular and difficult to define. The only volcanic rocks reported in the Middle and Upper Cambrian sequence anywhere in the Great Basin are in the Shwin Formation (Gilluly and Gates, 1965), a unit mainly of limestone and shale, which may represent that facies.

Strata of the arkosic facies consist of the Paradise Valley Chert (Hotz and Willden, 1964), an Upper Cambrian unit composed predominantly of chert and at least 100 m thick, and the depositionally overlying Harmony Formation, a late Late Cambrian unit composed predominantly of medium- to coarse-grained arkosic sandstone, perhaps 1000 to 1500 m thick. The Harmony contains minor amounts of shale, sparse limestone, and rare coarse-grained "gritty" beds containing granules and pebbles of quartz and feldspar. The sandstone is composed of quartz, orthoclase, microcline, plagioclase, and mica. Zircon from a sandstone bed in the Harmony has been dated as 958 my old (Jaffe and others, 1959, p. 130). Graded bedding is characteristic of the Harmony.

Carbonate facies rocks of the eastern Great Basin formed in shallow inner shelf lagoons and on shoals, whereas the limestone and shale facies formed in open deeper water of outer shelf environments to the west (Kepper, 1972). The shoals, which consisted of an interlacing pattern of tidal algal mudbanks and shallow subtidal basins, at times separated shelf lagoons on the east from the open water on the west (Kepper, 1972).

The graded bedding in the Harmony Formation suggests deposition by turbidity currents, probably in a deep-water continental rise environment. The presence of microline indicates a plutonic source rock, and the isotopic age of the zircons indicates a Precambrian terrane, but the source area's location (as discussed under "Sedimentary and tectonic models") is uncertain.

Ordovician System

Strata of the Ordovician System thicken from 0 near the eastern edge of the Great Basin to more than 1500 m (5000 ft) in the Ibex and northern Utah basins and are thin across the Tooele arch (fig. 11).

Ordovician strata have been studied in more detail than any other system in the Great Basin. Important regional studies or summaries include those of Webb (1958), Lowell (1958, 1960), Hintze (1959, 1963), Ross (1964a and b, 1970), Ross and Berry (1963), Kay and Crawford (1964) and Ketner (1966, 1968, 1969).

Four facies of Ordovician strata recognized in the Great Basin are, from east to west: (1) carbonate and quartzite; (2) shale and limestone; (3) shale and chert; and (4) siliceous and volcanic.

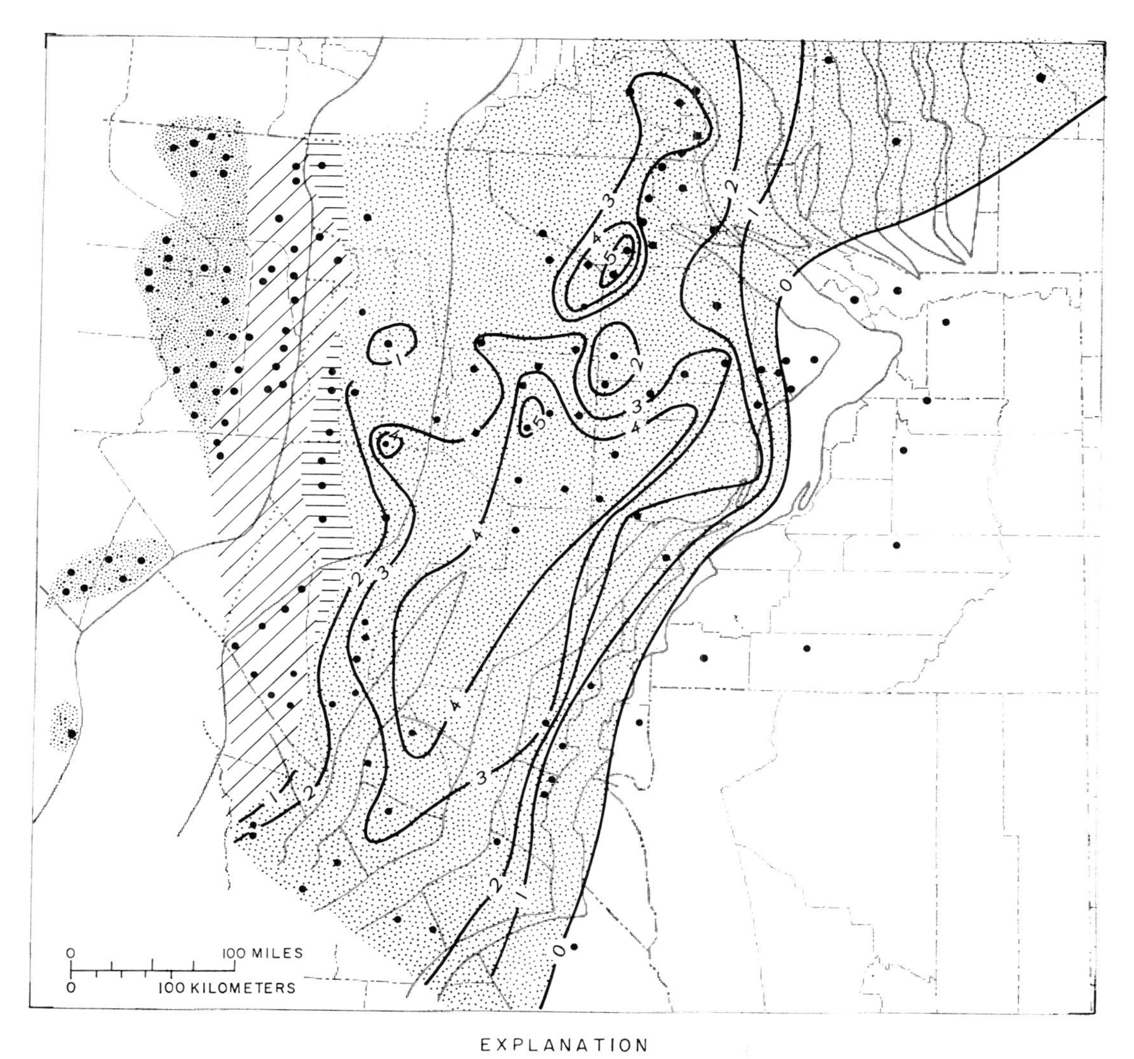

EXPLANATION

Carbonate and quartzite facies

Shale and limestone facies

Shale and chert facies

Siliceous and volcanic facies

Data point used to plot isopachs and (or) determine boundaries of facies

5
Isopach
Thickness in thousands of feet

FIG. 11.—Isopach and facies map of Ordovician strata.

The carbonate and quartzite facies has a typical threefold division: the Pogonip Group or Garden City Formation below (limestone with minor shale), the Eureka Quartzite and(or) Swan Peak Quartzite in the middle, and the Fish Haven Dolomite and equivalent units above. The Eureka and Swan Peak quartzites are distinctive and widespread; they are absent over the Tooele arch (Webb, 1958) where Upper Ordovician dolomite overlies various parts of the Lower Ordovician sequence. The Eureka wedges out along the western or outer edge of the miogeocline where Ordovician, Silurian, and Devonian strata are characterized by local or regional facies changes as well as by internal disconformities or unconformities (Webb, 1958, figs. 7, 8, 11, and 13; Lowell, 1960, fig. 2; Kay and Crawford, 1964, fig. 4; Merriam, 1963, fig. 7; McKee and Ross, 1969, fig. 5; Ketner, 1968, fig. 3; Ross, 1966, p. 24 and pl. 3).

The shale and limestone facies is relatively limited in distribution and consists of dark shale and siltstone and well-bedded argillaceous limestone. It contains both graptolite and shelly faunas. This facies includes the Aura Formation of Decker (1962) in northwest Elko County, Nevada, the Perkins Canyon Formation of Kay (1960), the Zanzibar Limestone and Toquima Formation (Ferguson, 1924), all in northern Nye County, Nevada, and some related rocks (Lowell, 1958; 1960) in southern Lander County, Nevada.

The shale and chert facies is widespread and thick. It consists of dark-gray or commonly varicolored graptolite-rich shale, several thick units of thin-bedded chert, minor amounts of limestone, quartzite, and volcanic rock, and very minor amounts of bedded barite. Rocks of this facies occur in both the lower and upper plates of the Roberts Mountains thrust. Rocks in the lower plate include the Comus Formation (Hotz and Willden, 1964) and most of the Palmetto Formation (Albers and Stewart, 1973). Rocks in the upper plate include the Vinini Formation (Merriam and Anderson, 1942; Gilluly and Mazursky, 1965), the Clipper Canyon sequence of Kay and Crawford (1964), the Valder Formation and Agort Chert of Riva (1970), and the Basco Formation of Lovejoy (1959). Volcanic rocks, mostly mafic lava flows that include pillow lavas, are sparse in this facies (Merriam and Anderson, 1942; Kay and Crawford, 1964, fig. 2; Riva, 1970). Radiolaria have been reported in some beds (Merriam and Anderson, 1942; Kay and Crawford, 1964, p. 437, Riva, 1970, p. 2692 and 2697; Ketner, 1969).

Rocks of the siliceous and volcanic facies are mainly assigned to the Valmy Formation, a widespread formation in north-central Nevada (Gilluly and Gates, 1965, p. 23–34; Roberts, 1964, p. 17–22; Churkin and Kay, 1967). Also included in this facies are Ordovician strata in southern Mineral County and the northern part of Esmeralda County, Nevada and in the Mount Morrison roof pendant in the Sierra Nevada (Rinehart and Ross, 1964). The Valmy Formation consists of thousands of meters of chert, quartzite, shale, siltstone, greenstone (including pillow lava), sandstone, and very minor limestone and bedded barite. It contains much more greenstone and quartzite, and less shale, than the shale and chert facies. The quartzite is highly vitreous and quartz-rich (98 percent SiO_2; Roberts, 1964, p. A19; Gilluly and Gates, 1965, table 2; Ketner, 1966, table 1), and forms units that are commonly fault-bounded and roughly 100 m thick. The quartzite in the Valmy is distinctly different from that in miogeoclinal rocks (Eureka and Swan Peak Quartzites); it is darker gray and coarser, not so well sorted, and typically has a conspicuous seriate texture ranging from fine to coarse (Ketner, 1966). The Valmy Formation is highly faulted and thickness estimates are uncertain. Gilluly and Gates (1965, p. 23) suggest a thickness of 6000 to 7500 m (20,000–25,000 ft), although Churkin (1973) indicates that it is much thinner, generally only perhaps 1000 m thick. A few radiolaria have been reported in some chert beds (Gilluly and Masursky, 1965, p. 42; Ketner, 1969; Rinehart and Ross, 1964, pl. 2).

Ordovician carbonate rocks probably were deposited in shallow shelf lagoons and on shoals. Conspicuous bioherms occur locally near the western limit of carbonate facies (Ross and Cornwall, 1961). The dolomite in the upper part of the Ordovician System may be mainly dolomitized limestone related to the development of magnesium-rich water in shallow shelf lagoons and local supratidal areas. Although the Eureka and Swan Peak Quartzites are of uncertain origin, they are generally considered to be shallow-water marine deposits (Ketner, 1968).

Rocks of the shale and limestone, shale and chert, and siliceous and volcanic facies appear to have been deposited in progressively deeper water from east to west. The presence of radiolarian chert, volcanic rocks, and bedded barite (Poole and others, 1968) in the shale and chert facies and the siliceous and volcanic facies supports an interpretation of a relatively deepwater oceanic environment. Depths greater than 500 m (1500 ft) are suggested by the results of a preliminary study made by Churkin (1974) on the vesicularity of Ordovician pillow basalts as compared with modern pillow lavas. The silica-rich quartzite, abundant in the siliceous and volcanic facies, is anomalous for such an environment, and the origin of these rocks is not clearly understood. As discussed under "Sedimentary and tectonic models," the source for the sand making up these quartzites is not clear.

Silurian System

The Silurian System is the most limited areally and the thinnest of any lower Paleozoic system in the Great Basin (fig. 12). Silurian strata thicken westward from an erosional edge in the eastern Great Basin to more than 300 m (1000 ft) in the Ibex and northern Utah basins and to at least 600 m (2000 ft) along the western edge of the miogeocline in central Nevada. Regional studies of Silurian rocks of the Great Basin include those of Winterer and Murphy (1960), Merriam (1963), and Berry and Boucot (1970). T. E. Mullens of the U.S. Geological Survey has recently completed a comprehensive study of the Silurian and Devonian Roberts Mountains Formation of central Nevada. Poole is completing a regional study of the dolomite facies in the eastern and southern Great Basin.

Four facies of Silurian rocks recognized in the Great Basin are, from east to west: (1) dolomite, (2) laminated limestone, (3) chert and shale, and (4) feldspathic sandstone.

The dolomite facies (Laketown Dolomite and equivalent strata) consists almost entirely of gray thin- to thick-bedded dolomite but contains some dark-gray, locally cherty units. Westward, in the central Great Basin, the dolomite changes rather abruptly into platy weathering laminated limestone (the laminated limestone facies) that is black to very dark gray on fresh surfaces and contains abundant detrital silt of quartz and feldspar, common carbonaceous material and pyrite, and locally abundant graptolites. The laminated limestone facies includes part of the Roberts Mountains Formation (Merriam, 1940; Winterer and Murphy, 1960) in north-central Nevada, the Chellis and Storff Formations of Decker (1962) in northwest Elko County, Nevada, the Noh Formation of Riva (1970) in the upper plate of the Roberts Mountains allochthon in northeast Elko County, the Roberts Mountains Formation of southern Nevada (Cornwall and Kleinhampl, 1964), and the basal part of the Sunday Canyon Formation in California (Ross, 1966).

Rocks of the chert and shale facies include the Fourmile Canyon Formation (Gilluly and Masursky, 1965) and unnamed Silurian rocks mapped by Lovejoy (1959), Kerr (1962), and Gardner and Peterson (1968), Evans (1972), and Evans and Cress (1972). These strata consist of chert, argillite, shale, and siltstone; the Fourmile Canyon Formation contains a few thin beds of fine-grained sandstone. Dark volcanic flow rocks with pillow structures have been identified by Kerr (1962, p. 449) in rocks that he assigned to the Silurian, but these rocks may be misidentified Ordovician units, as no other volcanic flow rocks have been identified in the Silurian of the Great Basin. Although thicknesses are uncertain, the Fourmile Canyon could be 1200–1800 m (4000–6000 ft), or even thicker (Gilluly and Masursky, 1965).

The feldspathic sandstone facies consists of the Elder Sandstone. The Elder is an unusual siliceous assemblage formation composed predominantly of light-colored, fine-grained, commonly silty sandstone containing quartz (70–80 percent), potassium feldspar (15–25 percent), muscovite (5 percent), and a little albite (Gilluly and Gates, 1965, p. 35). Siltstone, shale, and chert are present in minor amounts. Some siltstone beds contain ghosts of pumice shards and some sandstone beds contain grains that appear to be devitrified volcanic glass (Gilluly and Gates, 1965, p. 35–36; Gilluly and Masursky, 1965, p. 58). The thickness of the Elder is at least 600 m (2000 ft) and probably 1200 m (4000 ft) (Gilluly and Gates, 1965, p. 36).

Silurian dolomite may have originally been limestone formed on a broad shallow-water shelf containing many shoals and lagoons, and dolomitization may have been a secondary process related to the development of magnesium-rich waters in lagoons and on shoals. Possible reefs have been identified (Winterer and Murphy, 1960) along the western edge of the dolomite facies in north-central Nevada; the fossil-rich Vaughn Gulch Limestone (Ross, 1966, p. 30) of southeastern California could represent a similar, though undolomitized, reef complex. The laminated limestone probably represents moderately deep-water deposition on the outer shelf. Rocks of the chert and shale facies may have been deposited in relatively deep water, perhaps on the continental rise, although Gilluly and Masursky (1965 p. 55) have described abundant current bedding, which they believe indicates no unusual depth. The depositional environment of the Elder is uncertain, although the presence of algal fragments (Gilluly and Gates, 1965, p. 36), if not of detrital origin, would indicate at least local shallow-water deposition, whereas the presence of chert suggests deeper water deposition. The occurrence of tuffaceous beds in the Elder suggests that much of the feldspar in the formation was volcanically derived.

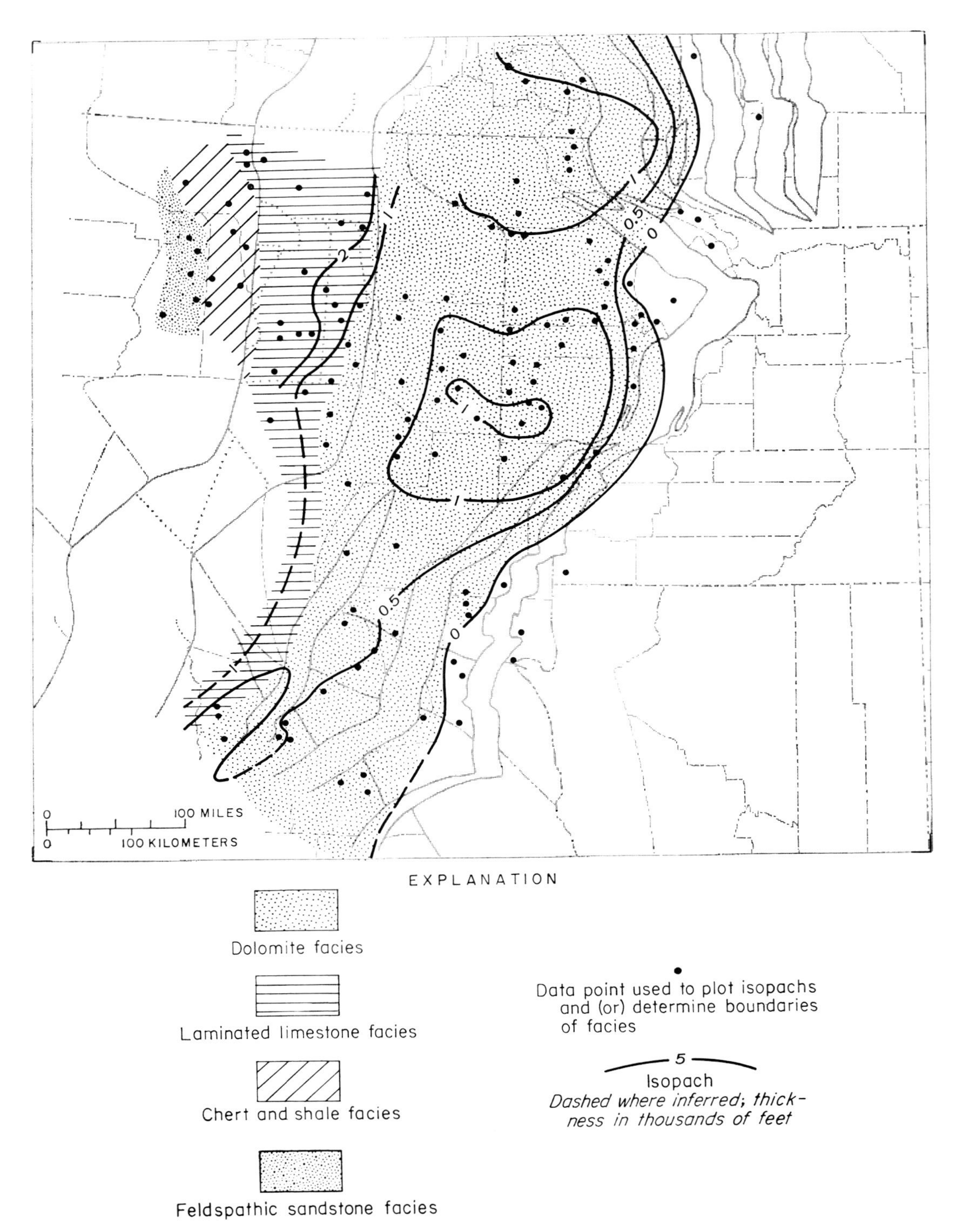

FIG. 12.—Isopach and facies map of Silurian strata.

Devonian System

Devonian rocks thicken from about 300 m (1000 ft) in the eastern Great Basin to more than 1800 m (6000 ft) in the Ibex basin, more than 1200 m (4000 ft) in the northern Utah basin, and locally more than 1500 m (5000 ft) in central Nevada (fig. 13). Uppermost Devonian strata in the eastern part of the Great Basin were deposited synchronously with the initial development of the Antler orogeny in the central part of the Great Basin (Poole, this volume). Some Devonian thickness trends may be slightly affected by the initial movement of the Antler orogeny during the Devonian, but the isopach map has been modified to restore any changes caused by post-Devonian erosion. The most comprehensive summary of Devonian stratigraphy in the Great Basin is that by Poole and others (1967). Other important regional studies include those of Osmond (1954, 1962), Merriam (1940, 1963), Carlisle and others (1957), Langenheim and others (1960), and Johnson (1965, 1971).

Four facies of Devonian strata recognized in the Great Basin are, from east to west: (1) carbonate and quartzite, (2) limestone and shale, (3) shale and chert, and (4) chert.

The carbonate and quartzite facies consists of cliff-forming thin-to thick-bedded limestone and dolomite, some of which contain sandy or silty units, and interbedded sandstone or quartzite. In north-central and northern Nevada, thin- to thick-bedded limestone and dolomite of the carbonate facies grade westward into units such as the Wenban Limestone (Gilluly and Masursky, 1965), Rabbit Hill Limestone (Merriam, 1963), and Van Duzer Limestone (Decker, 1962) of laminated limestone, silty limestone, calcareous shale, and clastic limestone. These formations and others are included in the limestone and shale facies. Farther east in Nevada (Elko and northern Eureka Counties), somewhat different Devonian strata are included with the limestone and shale facies. These include unnamed rocks composed of limestone, shale, and locally chert in (1) the Windermere Hills (Oversby, 1972), (2) the Snake Mountains (Gardner and Peterson, 1968), (3) the Pinon Range (Smith and Ketner, 1968, p. I1–I7), and (4) near Marys Mountain (Evans, 1972; Evans and Cress, 1972). Also included is the Coal Creek sequence of Lovejoy (1959) in western Elko County. This sequence consists of a lower plate of limestone, argillaceous limestone, and greenstone—the only Devonian volcanic rocks that have been reported in the Great Basin—and an upper plate of limestone, quartz-sandy limestone, siltstone, shale, and chert. In places the transitional Devonian strata in Elko and northern Eureka Counties occur as thrust slices within the upper plate of the Roberts Mountains thrust; in other places, they may be para-autochthonous. Their reconstructed position on the palinspastic base is not everywhere certain. Transitional Devonian rocks, composed of limestone and shale, also occur in California, where they constitute the upper part of the Sunday Canyon Formation (Ross, 1966).

The shale and chert facies includes the Woodruff Formation (Smith and Ketner, 1968), mostly dark-gray to black siliceous mudstone and radiolarian chert; the Cockalorum Wash Formation (Merriam, 1973), dark-gray mudstone and sparse coralline limestone; and siliceous mudstone, radiolarian chert, and sparse coralline limestone near Warm Springs in central Nye County, Nevada (Kleinhampl and Ziony, 1967).

The chert facies consists of the Slaven Chert (Gilluly and Gates, 1965; Gilluly and Masursky, 1965; Stewart and Palmer, 1967), composed predominantly of thin- to thick-bedded black chert, with very minor amounts of sandstone, shale, feldspathic siltstone, bedded barite, and limestone. Radiolaria occur in the chert (Gilluly and Masursky, 1965, p. 42). Thickness of the Slaven is uncertain but Gilluly and Gates (1965, p. 36) indicate that 1200 m (4000 ft) is probably not an excessive figure.

Rocks of the Devonian carbonate facies probably consist of shallow-water, subtidal, intertidal, and supratidal deposits formed on a broad inner shelf. Sand probably was derived from erosion of Ordovician sandstone on the shelf or from basal Cambrian and Precambrian sandstone on the craton (Osmond, 1962). Rocks of the limestone and shale facies are somewhat similar to those of the laminated limestone facies of the Silurian and probably were deposited in moderately deeper water, near the outer edge of the shelf. The radiolarian chert and the bedded barite of the shale and chert and the chert facies suggests deep water, probably oceanic conditions.

SEDIMENTARY AND TECTONIC MODELS

Two models of the Cordilleran geosyncline have been proposed (fig. 14). In one, deposition is visualized as occurring in a marginal sea bounded on the west by an island arc system (Eardley, 1947; Kay, 1951; Roberts, 1968a; Burchfiel and Davis, 1972, fig. 2; Moores, 1970; Churkin, 1974). In the other, deposition is visualized as taking place along a stable continental margin (Stewart, 1972; Burchfiel and Davis, 1972, fig. 3). Each model has its own particular

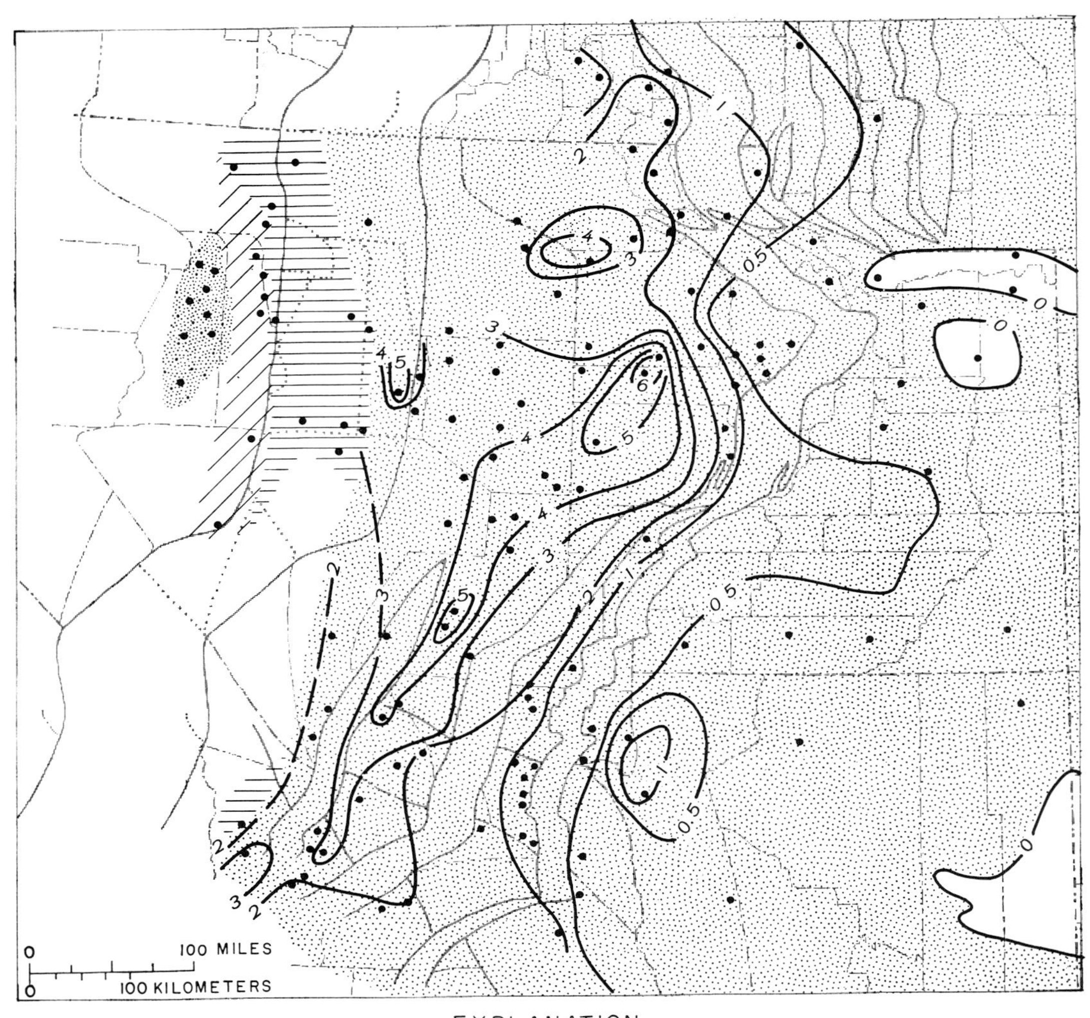

EXPLANATION

FIG. 13.—Isopach and facies map of Devonian strata (partly restored).

appealing attributes, as described below, and neither is necessarily unrelated to the other—theoretically a marginal sea-island arc system could change to a stable continental margin by the dying out of the subduction zone below the arc-trench system, or a stable continental margin could change to a marginal sea-island arc system by the development of a subduction zone.

In the marginal sea model, which has been described in detail by Churkin (1974), the miogeocline is along the inner side of the marginal basin at the edge of the main continental mass. Siliceous assemblage rocks are deep-water deposits within the basin; basalts formed by eruptions on the ocean floor. Small bodies of alpine-type serpentinite (Poole and Desborough, 1973) tectonically interleaved with strongly deformed allochthonous upper and lower Paleozoic oceanic rocks in western Nevada may in part represent lower Paleozoic mantle beneath the marginal sea. In this model, rocks farther west, including the lower Paleozoic rocks of the Klamath Mountains in California (Irwin, 1966), are inferred to be island-arc assemblages. Possible interpretations within the framework of this model are that a basement of continental crust in the island arc system could have supplied the plutonic rock debris in the Harmony Formation of north-central Nevada; that associated supracrustal sandstone could be a source of the mature sand-forming quartzite of the Ordovician Valmy Formation; and that island arc volcanic rocks could be a source for the probable abundant volcanic debris in the Silurian Elder Sandstone. A western source for the Valmy sand has been suggested by Ketner (1966) because quartzite of the Valmy is coarser and less well sorted than Ordovician sand on the shelf. A further indication of a western source is the greater abundance of quartzite in the outer belt of the siliceous assemblage rocks in the Great Basin than in the inner belt. Hopson (1973) has indicated, however, that Ordovician to Permian rocks in the Klamath Mountains rest on Ordovician oceanic crust, and Churkin (1974) also indicates that the island arc assemblages did not develop on continental crust. Conceivably, continental crust could have been present elsewhere in the island arc system and now be buried under younger rocks, or such continental crust could have been subsequently rifted away.

An alternative interpretation within the framework of the marginal sea model is that the Harmony and Valmy detritus had a source on the North American continent. Churkin (1974) suggests that the sand of the Valmy and sand of comparable age on the shelf in central Idaho may have had a common source—on the craton—and that the Valmy sand was carried into deeper parts of the basin and transported southward into the Great Basin, although he does not explain why basin sands are coarser than shelf sands. Another possibility is that the sand was derived from slumping or erosion of older (upper Precambrian and Lower Cambrian) mature sand deposits along the steep westward front (continental slope) of the mio-

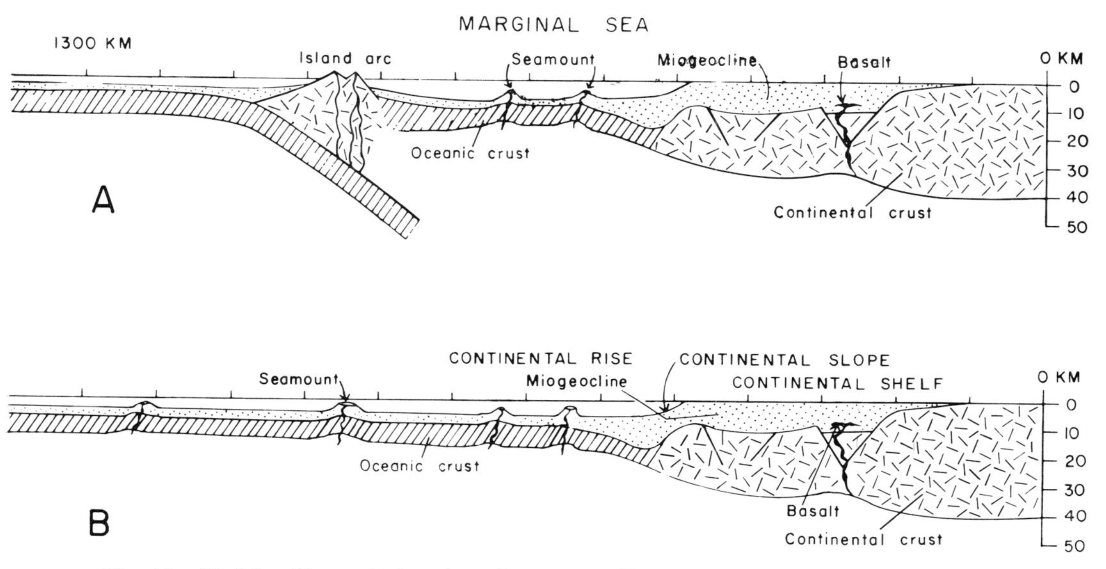

FIG. 14.—Models of lower Paleozoic and uppermost Precambrian Cordilleran geosyncline.

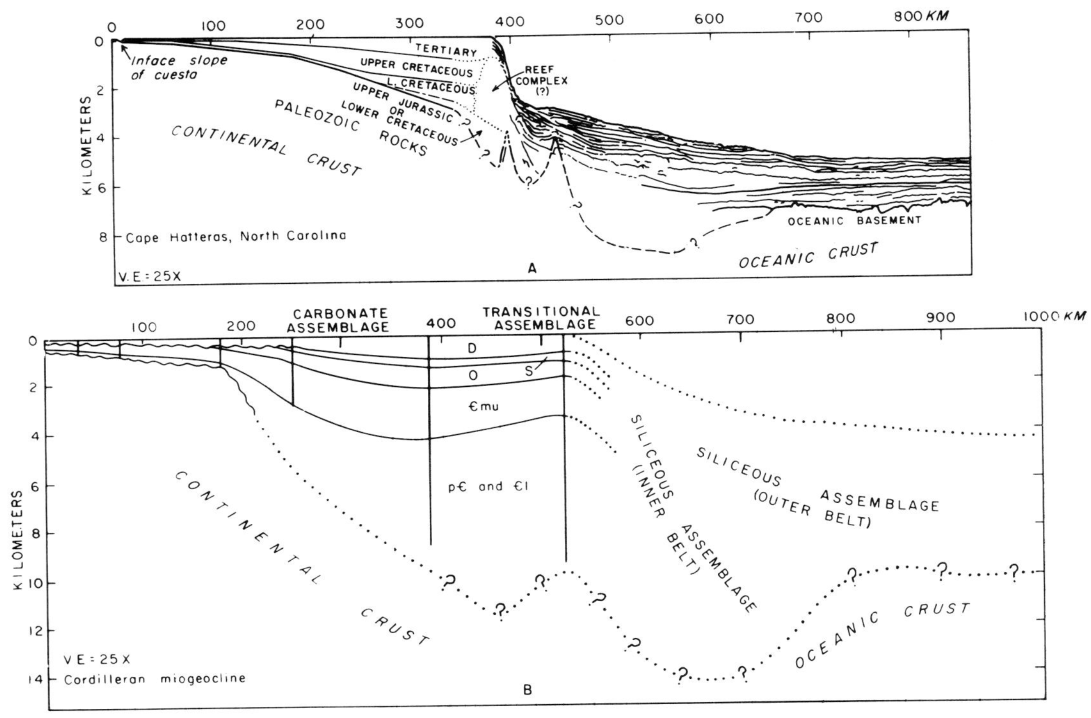

FIG. 15.—Cross sections comparing Cordilleran miogeocline with present-day miogeocline along continental margin of eastern North America. *A,* Section at Cape Hatteras, North Carolina, from Emery and others (1970, fig. 38). *B,* Cordilleran miogeocline showing distribution of assemblages. Based on figure 6 (orientation reversed). Letter symbols same as on figure 6. Dotted lines are hypothetical.

geocline, although uppermost Precambrian and Lower Cambrian quartzite generally contains some feldspar and a few quartz granules and pebbles whereas quartzite of the Valmy does not. The arkosic debris of the Harmony could have been derived from upfaulting and erosion of Precambrian crystalline rocks along the front of the miogeocline, second cycle from older (upper Precambrian?) arkosic deposits on the front of the miogeocline, or from an unknown Precambrian crystalline source, perhaps a westward extension of the craton lying either north or south of the Great Basin.

In the stable continental margin model (Atlantic type), the Cordilleran miogeocline is considered to be a continental terrace deposit with a broad continental shelf and a relatively narrow continental slope bordered on the west by a continental rise and, in turn, by an oceanic basin. In this model, rocks of the carbonate and transitional assemblages are continental shelf deposits, rocks of the transitional and siliceous assemblages represent progressively deeper water deposits to the west on the outer shelf, on the continental slope and rise, and in the ocean basin. The model is patterned after the Mesozoic and Cenozoic miogeocline along the eastern border of the North American continent (fig. 15). This feature is similar in size and shape and contains somewhat similar deposits. The stable continental margin model does not allow for an island arc system to the west. Accordingly, lower Paleozoic assemblages in the Klamath Mountains either are not true island arc deposits, as has been suggested by Hopson (1973), or, if they are, have been tectonically emplaced subsequent to the development of the miogeocline described here. Hamilton (1969), for example, has suggested that assemblages in the Klamath Mountains were tectonically emplaced by underflow of the Pacific in Mesozoic or Cenozoic time. Burchfiel and Davis (1972, fig. 3) speculate that the Klamath Mountains assemblages were related to an island arc system far removed from North America in early Paleozoic time, although they believe that this model is less likely than a marginal sea model.

The arkosic deposits of the Harmony and the mature quartzite of the Valmy present problems in the stable margin model, as in the marginal sea model. Either deposit could be derived from uplift and erosion of blocks along the front of the miogeocline or by slumping along the front, as was suggested in the marginal sea model, or, as was suggested by Churkin (1974), sand of the Valmy may have been derived from the shelf. Alternatively, the source area may have been a microcontinent rifted away from North America (an analogous situation today would be Madagascar), but evidence of such rifting is lacking. Still another possibility is that units like the Harmony and Valmy are not related to deposition on or near the North American continent but rather were deposited on the margin of another continent or an island arc system unrelated to North America and were tectonically emplaced (obducted) on the North American continent during the Late Devonian and Early Mississippian Antler orogeny or subsequent orogenies. This hypothesis is difficult to evaluate but is discounted because of the absence of a major serpentine belt that would mark the suture line at a continent-continent (or a continent-island arc) join, although small scattered bodies of Alpine-type serpentinite (Poole and Desborough, 1973) do occur. A further argument against such tectonic emplacement is that reconstructed facies changes, particularly in Ordovician rocks, seem gradual, indicating a progressive change in lithologic type within a single basin of deposition rather than an abrupt change related to tectonic telescoping of two unrelated facies.

The strongest argument in favor of the stable margin model is the similarity of the Cordilleran miogeocline to the Atlantic coast miogeocline (fig. 15), known to be an accumulation along a tectonically stable margin subsequent to a time of continental separation (Dietz and Holden, 1970). The remarkable lateral extent and lithologic uniformity of lower Paleozoic and uppermost Precambrian strata of the Cordilleran miogeocline indicate a long interval (perhaps as much as 500 my) of tectonic stability in western North America. Such stability is difficult to understand if a subduction zone and island arc system existed to the west. Perhaps the major tectonic activity was confined to the island arc region and North America, which lay landward of the marginal sea, was protected from direct interaction with an oceanic plate farther west. Nonetheless, we would expect that some energy, during part of the 500 my interval, would be transferred to the continent, either by collapse of the marginal sea or by the formation of a subsidiary subduction zone along the edge of the main continental mass.

The tectonic evolution of the Great Basin during the early Paleozoic and latest Precambrian may involve elements of both the stable continental margin model and the marginal sea model. An appealing interpretation is that the Great Basin was a stable continental margin during most of latest Precambrian and early Paleozoic time and that a subduction zone and marginal sea did not develop until late in the early Paleozoic, perhaps in the Devonian. According to this interpretation, Ordovician and Silurian rocks in the Klamath Mountains (Irwin, 1966; Churkin, 1974; Hopson, 1973) and Silurian and related rocks in the northern Sierra Nevada (McMath, 1966) are largely oceanic deposits unrelated to an island arc system, whereas younger rocks, such as the Devonian of the Klamath Mountains, which include thousands of meters of volcanic flows and pyroclastics, are island arc deposits.

The development of the Cordilleran miogeocline, which rests on Precambrian crystalline rocks and Precambrian supracrustal detrital rocks, marks a distinct change in the tectonic setting of western North America (Stewart, 1972). The pregeosynclinal supracrustal detrital rocks occur in platform facies and in major west- and northwest-trending aulacogens within which 3000–6000 m (10,000–20,000 ft) of strata accumulated. The miogeocline cuts across these older structures. If aulacogens are offshoots of geosynclines, as generally believed (Salop and Scheinmann, 1969, p. 586; Hoffman, 1971), then a pre-Cordilleran geosyncline may have been present off North America, but we know of no evidence that would indicate the trend or position of such a geosyncline, if it existed.

A rifting event appears to be the most likely way to form the Cordilleran geosyncline. This rifting could involve a major continental separation (Stewart, 1972) or the reshaping of the margin by rifting and migration of a continental fragment westward. In either case, volcanic rocks in uppermost Precambrian and Lower Cambrian strata in the eastern Great Basin, which are unique for Paleozoic and early Mesozoic rocks of the eastern Great Basin, are believed to mark zones of extension in the continental basement that tapped sources of basalt in the mantle below. These volcanic rocks may be related to extensional events similar to those that formed grabens and led to the eruption of basaltic rocks in the Triassic Newark Group in the eastern United States. The Triassic grabens, presumed to be related to the breakup of the supercontinent of Pangaea and opening of the Atlantic basin (Dietz and Holden, 1970), lie about 320 km (200 mi) inland of the continental slope, a distance comparable to that of upper

Precambrian and Lower Cambrian volcanic rocks of the Great Basin east from the inferred Cordilleran continental margin. According to this interpretation, the Wasatch line marks the eastern edge of continental extension and thinning during late Precambrian rifting.

The outer shelf of the Cordilleran miogeocline is characterized by carbonate shoals and reefs, thinning of sedimentary units, relatively abrupt changes in facies, and unconformities. Shoal or reeflike deposits included bioherms near the shelf edge in the Ordovician (Ross and Cornwall, 1961) and Silurian (Winterer and Murphy, 1960) and perhaps during most periods. The Precambrian Reed Dolomite (Nelson, 1962), for example, in Inyo County, California, changes facies from carbonate and quartzite into dominantly carbonate within a relatively short distance; this may indicate a nearness to a shoal or reef. Most of the rock systems are thinner along the miogeoclinal shelf edge (figs. 8–12), and some individual units, such as the Eureka Quartzite (Ketner, 1968) are missing or thin. Facies changes and unconformities, for example, those in Ordovician rocks in central Nevada (Webb, 1958; Lowell, 1960; Merriam, 1963; McKee and Ross, 1969), are also characteristics of the shelf edge. These shelf edge features may be related to a basement-ridge barrier similar to that on the eastern margin of the North American continent and elsewhere (Drake and others, 1959; Emery and others, 1970; Hedberg, 1970; Burk, 1968).

REFERENCES CITED

ABBOTT, W. O., 1951, Cambrian diabase flow in central Utah: Compass, v. 29, p. 5–10.

ALBERS, J. P., 1967, Belt of sigmoidal bending and right-lateral faulting in the western Great Basin: Geol. Soc. America Bull., v. 78, p. 143–156.

———, AND STEWART, J. H., 1973, Geology and mineral resources of Esmeralda County, Nevada: Nevada Bur. Mines and Geology Bull. 78, 80 p.

ANDERSON, R. E., LONGWELL, C. R., ARMSTRONG, R. E., AND MARVIN, R. F., 1972, Significance of K-Ar ages of Tertiary rocks from Lake Mead region, Nevada-Arizona: Geol. Soc. America Bull., v. 83, p. 273–288.

ARMSTRONG, F. C., 1969, Geologic map of the Soda Springs quadrangle, southeastern Idaho: U.S. Geol. Survey Misc. Geol. Inv. Map I-557.

ARMSTRONG, R. L., 1968a, The Cordilleran miogeosyncline in Nevada and Utah: Utah Geol. and Mineralog. Survey Bull. 78, 58 p.

———, 1968b, Sevier orogenic belt in Nevada and Utah: Geol. Soc. America Bull., v. 79, p. 429–458.

———, AND HILLS, F. A., 1967, Rubidium-strontium and potassium-argon geochronologic studies of mantled gneiss domes, Albion Range, southern Idaho, U.S.A.: Earth and Planetary Sci. Letters, v. 3, p. 114–124.

BAKER, A. A., 1947, Stratigraphy of the Wasatch Mountains in the vicinity of Provo, Utah: U.S. Geol. Survey Oil and Gas Inv. Prelim. Chart no. 30.

BARNES, HARLEY, AND POOLE, F. G., 1968, Regional thrust-fault system in Nevada Test Site and vicinity, *in* Eckel, E. B. (ed.), Nevada Test Site: Geol. Soc. America Mem. 110, p. 233–238.

BAYLEY, R. W., AND MUEHLBERGER, W. R., 1968, Basement rock map of the United States, exclusive of Alaska and Hawaii: U.S. Geol. Survey.

BERRY, W. B. N., AND BOUCOT, A. J., 1970, Correlation of the North American Silurian rocks *with contributions by* Berdan, J. M., Klappes, Gilbert, Martinsson, Anders, Rexroad, C. B., Sheldon, R. P., Ziegler, A. M.: Geol. Soc. America Special Paper 102, 289 p.

BICK, K. F., 1966, Geology of the Deep Creek Mountains, Tooele and Juab Counties, Utah: Utah Geol. and Mineralog. Survey Bull. 77, 120 p.

BRADY, M. J., 1965, Thrusting in the Southern Wasatch Mountains, Utah: Brigham Young Univ. Geology Studies, v. 12, p. 3–53.

BURCHFIEL, B. C., AND DAVIS, G. A., 1972, Structural framework and evolution of the southern part of the Cordilleran orogen, western United States: Am. Jour. Sci., v. 272, p. 97–118.

BURK, C. A., 1968, Buried ridges within continental margins: New York Acad. Sci. Trans., ser. 2, v. 30, p. 397–409.

BUSHNELL, KENT, 1967, Geology of the Rowland quadrangle, Elko County, Nevada: Nevada Bur. Mines Bull. 67, 38 p.

CARLISLE, DONALD, MURPHY, M. A., NELSON, C. A., AND WINTERER, E. L., 1957, Devonian stratigraphy of Sulphur Springs and Pinyon ranges, Nevada: Am. Assoc. Petroleum Geologists Bull., v. 41, p. 2175–2191.

CHURKIN, MICHAEL, JR., 1974, Paleozoic marginal ocean basin volcanic arc systems in the Cordilleran foldbelt, *in* Dott, R. H., Jr., and Shauer, R. H. (eds.), Modern and Ancient Geosynclinal Sedimentation: Soc. Econ. Paleontologists and Mineralogists Special Pub. 19, p. 174–192.

———, AND KAY, MARSHALL, 1967, Graptolite-bearing Ordovician siliceous and volcanic rocks, northern Independence Range, Nevada: Geol. Soc. America Bull., v. 78, p. 651–668.

COHENOUR, R. E., 1959, Sheeprock Mountains, Tooele and Juab Counties: Utah Geol. Mineralog. Survey Bull. 63, 201 p.

COMPTON, R. R., 1972, Geologic map of the Yost quadrangle, Box Elder County, Utah, and Cassia County, Idaho: U.S. Geol. Survey Misc. Geol. Inv. Map I-672.

CORNWALL, H. R., AND KLEINHAMPL, F. J., 1964, Geology of Bullfrog quadrangle and ore deposits related to Bullfrog Hills caldera, Nye County, Nevada, and Inyo County, California: *ibid.,* Prof. Paper 454-J, p. J1–J25.

Crittenden, M. D., Jr., 1961, Magnitude of thrust faulting in northern Utah: *ibid.*, 424-D, p. D128–D131.
———, McKee, E. H., and Peterman, Zell, 1971, 1.5-billion-year-old rocks in the Willard thrust sheet, Utah: Geol. Soc. America Abstracts with Programs, v. 3, p. 105–106.
———, Schaeffer, F. E., Trimble, D. E., and Woodward, L. A., 1971, Nomenclature and correlation of some upper Precambrian and basal Cambrian sequences in western Utah and southeastern Idaho: *ibid.*, v. 82, p. 581–602.
———, Sharp, B. J., and Calkins, F. C., 1952, Geology of the Wasatch Mountains east of Salt Lake City, Parleys Canyon to Transverse Range, *in* Marsell, R. E. (ed.), Geology of the central Wasatch Mountains, Utah: Utah Geol. Soc. Guidebook no. 8, p. 1–37.
———, Stewart, J. H., and Wallace, C. A., 1972, Regional correlation of upper Precambrian strata in western North America: 24th Internat. Geol. Cong. Rept., sec. 1, p. 334–341.
Decker, R. W., 1962, Geology of the Bull Run quadrangle, Elko County, Nevada: Nevada Bur. Mines Bull. 60, 65 p.
Dietz, R. S., 1972, Geosynclines, mountains, and continent building: Scientific American, v. 226, p. 30–38.
———, and Holden, J. C., 1966, Miogeoclines (miogeosynclines) in space and time: Jour. Geology, v. 74, p. 566–583.
———, 1970, The breakup of Pangaea: Scientific American, v. 223, p. 30–41.
Drake, C. L., Ewing, Maurice, and Sutton, G. H., 1959, Continental margins and geosynclines—the east coast of North America north of Cape Hatteras, *in* Ahrens, L. H., Press, Frank, Rankama, Kalervo, and Runcorn, S. K. (eds.), Physics and chemistry of the earth: London, Pergamon Press, v. 3, p. 110–198.
Eardley, A. J., 1947, Paleozoic Cordilleran geosyncline and related orogeny: Jour. Geology, v. 55, p. 309–342.
Ekren, E. B., Anderson, R. E., Rogers, C. L., and Noble, D. C., 1971, Geology of northern Nellis Air Force Base Bombing and Gunnery Range, Nye County, Nevada: U.S. Geol. Survey Prof. Paper 651, 91 p.
———, Rogers, C. L., Anderson, R. E., and Orkild, P. P., 1968, Age of basin and range normal faults in Nevada Test Site and Nellis Air Force Range, Nevada, *in* Eckel, E. B. (ed.), Nevada Test Site: Geol. Soc. America Mem. 110, p. 247–250.
Emery, K. O., Uchupi, Eleazar, Phillips, J. D., Bowin, C. O., Bunce, E. T., and Knott, S. T., 1970, Continental rise off eastern North America: Am. Assoc. Petroleum Geologists Bull., v. 54, p. 44–108.
Evans, J. G., 1972, Preliminary geologic map of the Welches Canyon quadrangle, Nevada: U.S. Geol. Survey Misc. Field Studies Map MF-326.
———, and Cress, L. D., 1972, Preliminary geologic Nevada map of the Schroeder Mountain quadrangle, Nevada: *ibid.*, MF-324.
Ferguson, H. G., 1924, Geology and ore deposits of the Manhattan district, Nevada: *ibid.*, Bull. 723, 163 p.
———, 1933, Geology of the Tybo district, Nevada: Nevada Univ. Bull., v. 27, 61 p.
Fleck, R. J., 1967, The magnitude, sequence, and style of deformation in southern Nevada and eastern California (Ph.D. thesis) : Univ. California, Berkeley, 92 p.
———, 1970a, Age and possible origin of the Las Vegas Valley shear zone, Clark and Nye Counties, Nevada: Geol. Soc. America Abstracts with Programs, v. 2, p. 333.
———, 1970b, Tectonic style, magnitude, and age of deformation in the Sevier orogenic belt in southern Nevada and eastern California: *ibid.*, Bull., v. 81, p. 1705–1720.
Flint, R. F., Sanders, J. F., and Rogers, John, 1960, Diamictite, a substitute term for symmictite: *ibid.*, v. 71, p. 1809.
Ford, T. D., and Breed, W. J., 1973, Late Precambrian Chuar Group, Grand Canyon, Arizona: *ibid.*, v. 84, p. 1242–1243.
Gardner, D. H., and Peterson, B. L., 1968, Paleozoic thrust faults exposed in the northern Snake Mountains, Elko County, northeastern Nevada (abs.) : *ibid.*, Special Paper 121, p. 506.
Gilluly, James, 1967, Geologic map of the Winnemucca quadrangle, Pershing and Humboldt Counties, Nevada: U.S. Geol. Survey Geologic Quadrangle Map GQ-656.
———, and Gates, Olcott, 1965, Tectonic and igneous geology of the northern Shoshone Range, Nevada *with sections on* Gravity in Crescent Valley *by* Donald Plouff *and* Economic geology *by* K. B. Ketner: *ibid.*, Prof. Paper 465, 53 p.
———, and Masursky, Harold, 1965, Geology of the Cortez quadrangle, Nevada *with a section on* Gravity and aeromagnetic surveys by D. R. Mabey: *ibid.*, Bull. 1175, 117 p.
Hamilton, Warren, 1969, Mesozoic California and the underflow of Pacific mantle: Geol. Soc. America Bull., v. 80, p. 2409–2429.
Hansen, W. R., 1965, Geology of the Flaming Gorge area, Utah-Colorado-Wyoming: U.S. Geol. Survey Prof. Paper 490, 196 p.
Hedberg, H. D., 1970, Continental margins from viewpoint of the petroleum geologist: Am. Assoc. Petroleum Geologists Bull., v. 54, p. 3–43.
Hewett, D. F., 1956, Geology and mineral resources of the Ivanpah quadrangle, California and Nevada: U.S. Geol. Survey Prof. Paper 275, 172 p.
Hintze, L. F., 1954, Mid-Ordovician erosion in Utah (abs.): Geol. Soc. America Bull., v. 65, p. 1343.
———, 1959, Ordovician regional relationships in north-central Utah and adjacent areas *in* Williams, N. C. (ed.), Guidebook to the geology of the Wasatch and Uinta Mountains transition area: Intermountain Assoc. Petroleum Geologists, 10th Ann. Field Conf. Guidebook, p. 46–53.
———, 1963, Summary of Ordovician stratigraphy of Utah *in* Crawford, A. L. (ed.), Surface, structure, and stratigraphy of Utah: Utah Geol. and Mineralog. Survey Bull. 54, p. 51–61.
Hoffmann, Paul, 1971, Aulacogenes and orthogeosynclines in the lower Proterozoic of the northwestern Canadian Shield: Geol. Soc. America Abstracts with Programs, v. 3, p. 601.

Hope, R. A., 1970, Preliminary geologic map of Elko County, Nevada: U.S. Geol. Survey open-file map.
Hopson, C. A., 1973, Ordovician and Late Jurassic ophiolitic assemblages in the Pacific Northwest: Geol. Soc. America Abstracts with Programs, v. 5, p. 57.
Hose, R. K., and Blake, M. C., Jr., 1970, Geologic map of White Pine County, Nevada: U.S. Geol. Survey open-file map.
Hotz, P. E., and Willden, Ronald, 1964, Geology and mineral deposits of the Osgood Mountains quadrangle, Humboldt County, Nevada: *ibid.,* Prof. Paper 431, 128 p.
Howard, K. A., 1971, Paleozoic metasediments in the northern Ruby Mountains, Nevada: Geol. Soc. America Bull., v. 82, p. 259–264.
Humphrey, F. L., 1960, Geology of the White Pine mining district, White Pine County, Nevada: Nevada Bur. Mines Bull. 57, 119 p.
Irwin, W. P., 1966, Geology of the Klamath Mountains province, *in* Bailey, E. H. (ed.), Geology of northern California: California Div. Mines and Geology Bull. 190, p. 19–38.
Jaffe, H. W., Gottfried, David, Waring, C. L., and Worthing, H. W., 1959, Lead-alpha age determinations of accessory minerals of igneous rocks (1953–1957): U.S. Geol. Survey Bull. 1097-B, p. 65–148.
James, H. L., 1972, Subdivision of Precambrian: an interim scheme to be used by the U.S. Geological Survey: Am. Assoc. Petroleum Geologists Bull., v. 56, p. 1128–1133.
Johnson, J. G., 1965, Lower Devonian stratigraphy and correlation, northern Simpson Park Range, Nevada: Canadian Petroleum Geology Bull., v. 13, p. 365–381.
———, 1971, Timing and coordination of orogenic, epeirogenic, and eustatic events: Geol. Soc. America Bull., v. 82, p. 3263–3298.
Kay, Marshall, 1951, North American geosynclines: Geol. Soc. America Mem. 48, 143 p.
———, 1960, Paleozoic continental margin in central Nevada, western United States: 21st Internat. Geol. Cong. Rept., pt. 12, p. 94–103.
Kay, Marshall, and Crawford, J. P., 1964, Paleozoic facies from the miogeosynclinal to the eugeosynclinal belt in thrust slices, central Nevada: Geol. Soc. America Bull., v. 75, p. 425–454.
Kellogg, H. E., 1963, Paleozoic stratigraphy of the southern Egan Range, Nevada: *ibid.,* v. 74, p. 685–708.
Kepper, J. C., 1972, Paleoenvironmental patterns in middle to lower Upper Cambrian interval in eastern Great Basin: Am. Assoc. Petroleum Geologists Bull., v. 56, p. 503–527.
Kerr, J. W., 1962, Paleozoic sequences and thrust slices of the Seetoya Mountains, Independence Range, Elko County, Nevada: Geol. Soc. America Bull., v. 73, p. 439–460.
Ketner, K. B., 1966, Comparison of Ordovician eugeosynclinal and miogeosynclinal quartzites of the Cordilleran geosyncline: U.S. Geol. Survey Prof. Paper 550-C, p. C54–C60.
———, 1968, Origin of Ordovician quartzite in the Cordilleran miogeosyncline: *ibid.,* 600-B, p. B169–B177.
———, 1969, Ordovician bedded chert, argillite, and shale of the Cordilleran eugeosyncline in Nevada and Idaho: *ibid.,* 650-B, p. B32–B34.
King, P. B., 1969, The tectonics of North America—A discussion to accompany the Tectonic Map of North America: *ibid.,* 628, 95 p.
Klein, G. de V., 1972, Determination of paleotidal range in clastic sedimentary rocks: 24th Internat. Geol. Cong. Rept., sec. 6, p. 397–405.
Kleinhampl, F. J., and Ziony, J. I., 1967, Preliminary geologic map of northern Nye County, Nevada: U.S. Geol. Survey open-file map.
Langenheim, R. L., Jr., Hill, J. D., and Waines, R. H., 1960, Devonian stratigraphy of the Ely area *in* Boettcher, J. W. and Sloan, W. W., Jr. (eds), Geology of east-central Nevada: Intermountain Assoc. Petroleum Geologists 11th Ann. Field Conf. Guidebook, p. 63–71.
Lanphere, M. A., Wasserburg, G. J., and Albee, A. L., 1963, Redistribution of strontium and rubidium isotopes during metamorphism, World Beater Complex, Panamint Range, California, *in* Craig, Harmon (ed.), Isotopic and cosmic chemistry: North-Holland Publishing Co., Amsterdam, p. 269–320.
Lovejoy, D. W., 1959, Overthrust Ordovician and the Nannie's Peak intrusive, Lone Mountain, Elko County, Nevada: Geol. Soc. America Bull., v. 70, p. 539–563.
Lowell, J. D., 1958, Lower and Middle Ordovician stratigraphy in eastern and central Nevada (Ph.D. thesis): Columbia Univ., New York.
———, 1960, Ordovician miogeosynclinal margin in central Nevada: 21st Internat. Geol. Cong. Rept., pt. 7, p. 7–17.
McKee, E. H., 1968, Geology of the Magruder Mountain area, Nevada-California: U.S. Geol. Survey Bull. 1251-H, p. H1–H40.
———, and Ross, R. J., Jr., 1969, Stratigraphy of eastern assemblage rocks in a window in Roberts Mountains thrust, northern Toquima Range, central Nevada: Am. Assoc. Petroleum Geologists Bull., v. 53, p. 421–429.
McMath, V. E., 1966, Geology of the Taylorsville area, northern Sierra Nevada, *in* Bailey, E. H. (ed.), Geology of northern California: California Div. Mines and Geology Bull. 190, p. 173–183.
McNair, A. H., 1951, Paleozoic stratigraphy of part of northwestern Arizona: Am. Assoc. Petroleum Geologists Bull., v. 35, p. 503–541.
———, 1952, Summary of the pre-Coconino stratigraphy of southwestern Utah, northwestern Arizona, and southeastern Nevada, *in* Thune, H. W. (ed.), Guidebook to the geology of Utah, no. 7, Cedar City, Utah to Las Vegas, Nevada: Intermountain Assoc. Petroleum Geologists, 3rd Ann. Field Conf. Guidebook, p. 45–51.
Maxson, J. H., 1961, Geologic map of the Bright Angel quadrangle, Grand Canyon National Park, Arizona: Grand Canyon Natural History Assoc.
Means, W. D., 1962, Structure and stratigraphy in the central Toiyabe Range, Nevada: Univ. California Pub. Geol. Sci., v. 42, p. 71–110.
Merriam, C. W., 1940, Devonian stratigraphy and paleontology of the Roberts Mountains region, Nevada: Geol. Soc. America Special Paper 25, 114 p.

——, 1963, Paleozoic rocks of Antelope Valley, Eureka and Nye Counties, Nevada: U.S. Geol. Survey Prof. Paper 423, 67 p.
——, 1973, Middle Devonian rugose corals of the central Great Basin: *ibid.*, 799 (in press).
——, and Anderson, C. A., 1942, Reconnaissance survey of the Roberts Mountains, Nevada: Geol. Soc. America Bull., v. 53, p. 1675–1727.
Misch, Peter, and Hazzard, J. C., 1962, Stratigraphy and metamorphism of Late Precambrian rocks in central northeastern Nevada and adjacent Utah: Am. Assoc. Petroleum Geologists Bull., v. 46, p. 289–343.
Moores, E. M., 1970, Ultramafics and orogeny, with models of the U.S. Cordillera and the Tethys: Nature, v. 228, p. 837–842.
Morris, H. T., and Lovering, T. S., 1961, Stratigraphy of the East Tintic Mountains, Utah: U.S. Geol. Survey Prof. Paper 361, 145 p.
Nelson, C. A., 1962, Lower Cambrian-Precambrian succession, White-Inyo Mountains, California: Geol. Soc. America Bull., v. 73, p. 139–144.
Nolan, T. B., Merriam, C. W., and Williams, J. S., 1956, The stratigraphic section in the vicinity of Eureka, Nevada: U.S. Geol. Survey Prof. Paper 276, 77 p.
Oriel, S. S., 1965, Preliminary geologic map of the SW¼ of the Bancroft quadrangle, Bannock and Caribou Counties, Idaho: *ibid.*, Mineral Inv. Field Studies Map MF-299.
——, 1969, Geology of the Fort Hill quadrangle, Lincoln County, Wyoming: *ibid.*, Prof. Paper 594-M, 40 p.
——, and Armstrong, F. C., 1971, Uppermost Precambrian and lowest Cambrian rocks in southeastern Idaho: *ibid.*, 394, 52 p.
Osmond, J. C., Jr., 1954, Dolomites in Silurian and Devonian of east-central Nevada: Am. Assoc. Petroleum Geologists Bull., v. 38, p. 1911–1956.
——, 1962, Stratigraphy of Devonian Sevy Dolomite in Utah and Nevada: Am. Assoc. Petroleum Geologists Bull., v. 46, p. 2033–2056.
Oversby, Brian, 1972, Thrust sequences in the Windermere Hills, northeastern Elko County, Nevada: Geol. Soc. America Bull., v. 83, p. 2677–2688.
Palmer, A. R., 1971, The Cambrian of the Great Basin and adjoining areas, western United States, *in* Holland, C. H. (ed.), Cambrian of the new world: Wiley-Interscience, New York, p. 1–78.
Poole, F. G., Baars, D. L., Drewes, Harald, Hayes, P. T., Ketner, K. B., McKee, E. D., Teichert, Curt, and Williams, J. S., 1967, Devonian of the southwestern United States, *in* Oswald, D. H. (ed.), International Symposium on the Devonian System, v. 1: Alberta Soc. Petroleum Geologists, Calgary, p. 879–912.
——, Brobst, D. A., and Shawe, D. R., 1968, Sedimentary origin of bedded barite in central Nevada: Geol. Soc. America Special Paper 121, p. 242.
——, and Desborough, G. A., 1973, Alpine-type serpentinites in Nevada and their tectonic significance: *ibid.*, Abstracts with Programs, v. 5, p. 90–91.
Proffett, J. M., Jr., 1971, Late Cenozoic structure in the Yerington district, Nevada and the origin of the Great Basin: *ibid.*, v. 3, p. 181.
Rigo, R. J., 1968, Middle and Upper Cambrian stratigraphy in the autochthon and allochthon of northern Utah: Brigham Young Univ. Geology Studies, v. 15, p. 31–66.
Rinehart, C. D., and Ross, D. C., 1964, Geology and mineral deposits of the Mount Morrison quadrangle, Sierra Nevada, California: U.S. Geol. Survey Prof. Paper 385, 106 p.
Riva, John, Thrusted Paleozoic rocks in the northern and central HD Range, northeastern Nevada: Geol. Soc. America Bull., v. 81, p. 2689–2716.
Roberts, R. J., 1964, Stratigraphy and structure of the Antler Peak quadrangle, Humboldt and Lander Counties, Nevada: U.S. Geol. Survey Prof. Paper 459-A, 93 p.
——, 1968a, Tectonic framework of the Great Basin, *in* V. H. McNutt-Geology Dept. Colloquium Ser. 1: Univ. Missouri, Rolla, Jour., no. 1, p. 101–119.
——, 1968b, Distribution of facies within the Cordilleran geosyncline, Silurian and Devonian time (abs.): Geol. Soc. America Special Paper 121, p. 551.
——, 1972, Evolution of the Cordilleran foldbelt: *ibid.*, v. 83, p. 1989–2004.
——, Crittenden, M. D., Jr., Tooker, E. W., Morris, H. T., Hose, R. K., and Cheney, T. M., 1965, Pennsylvanian and Permian basins in northwestern Utah, northeastern Nevada and south-central Idaho: Am. Assoc. Petroleum Geologists Bull., v. 49, p. 1926–1956.
——, Hotz, P. E., Gilluly, James, and Ferguson, H. G., 1958, Paleozoic rocks of north-central Nevada: *ibid.*, v. 42, p. 2813–2857.
Robison, R. A., 1960, Lower and Middle Cambrian stratigraphy of the eastern Great Basin *in* Boettcher, J. W. and Sloan, W. W., Jr. (eds.), Guidebook to the geology of east-central Nevada: Intermountain Assoc. Petroleum Geologists, 11th Ann. Field Conf. Guidebook, p. 43–52.
——, 1964, Upper Middle Cambrian stratigraphy of western Utah: Geol. Soc. America Bull., v. 75, p. 995–1010.
Ross, D. C., 1965, Geology of the Independence quadrangle, Inyo County, California: U.S. Geol. Survey Bull. 1181-0, 64 p.
——, 1966, Stratigraphy of some Paleozoic formations in the Independence quadrangle, Inyo County, California: *ibid.*, Prof. Paper 396, 64 p.
Ross, R. J., Jr., 1964a, Middle and Lower Ordovician formations in southernmost Nevada and adjacent California, *with a section on* Paleotectonic significance of Ordovician sections south of the Las Vegas shear zone by R. J. Ross, Jr., and C. R. Longwell: *ibid.*, Bull. 1180-C, p. C1–C101.
——, 1964b, Relations of Middle Ordovician time and rock units in basin ranges, western United States: Am. Assoc. Petroleum Geologists Bull., v. 48, p. 1526–1554.

———, 1970, Ordovician brachiopods, trilobites, and stratigraphy in eastern and central Nevada: U.S. Geol. Survey Prof. Paper 639, 103 p.
———, and Berry, W. B. N., 1963, Ordovician graptolites of the basin ranges in California, Nevada, Utah, and Idaho: *ibid.,* Bull. 1134, 177 p.
———, and Cornwall, H. R., 1961, Bioherms in the upper part of the Pogonip in southern Nevada: *ibid.,* Paper 424-B, p. B231–B233.
Rubey, W. W., 1958, Geology of the Bedford quadrangle, Wyoming: *ibid.,* Geol. Quadrangle Map GQ-109.
———, and Hubbert, M. K., 1959, Role of fluid pressure in mechanics of overthrust faulting, [Pt. 2] Overthrust belt in geosynclinal area of western Wyoming in light of fluid-pressure hypothesis: Geol. Soc. America Bull., v. 70, p. 167–206.
Salop, L. I., and Scheinmann, Yu. M., 1969, Tectonic history and structures of platforms and shields: Tectonophysics, v. 7, p. 565–597.
Seeland, D. A., 1968, Paleocurrents of the late Precambrian to Early Ordovician (basal Sauk) transgressive clastics of the western and northern United States (Ph.D. thesis): Univ. Utah, Salt Lake City, 276 p.
———, 1969, Marine current directions in upper Precambrian and Cambrian rocks of the southwestern United States *in* Baars, D. L. (ed.), Geology and natural history of the Grand Canyon region: Four Corners Geol. Soc., Fifth Field Conf. Guidebook, p. 123–126.
Shride, A. F., 1967, Younger Precambrian geology in southern Arizona: U.S. Geol. Survey Prof. Paper 566, 89 p.
Smith, J. D., and Hopkins, T. S., 1971, Sediment transport on the continental shelf off of Washington and Oregon in light of recent current measurements: Geol. Soc. America Abstracts with Programs, v. 3, p. 710–711.
Smith, J. F., Jr., and Ketner, K. B., 1968, Devonian and Mississippian rocks and the date of the Roberts Mountains thrust in the Carlin-Piñon Range area, Nevada: U.S. Geol. Survey Bull. 1251-I, p. I1–I18.
Staatz, M. H., and Carr, W. J., 1964, Geology and mineral deposits of the Thomas and Dugway Ranges, Juab and Tooele Counties, Utah: *ibid.,* Prof. Paper 415, 188 p.
Stevens, C. H., and Olson, R. C., 1972, Nature and significance of the Inyo thrust fault, eastern California: Geol. Soc. America Bull., v. 83, p. 3761–3768.
Stewart, J. H., 1967, Possible large right-lateral displacement along fault and shear zones in Death Valley-Las Vegas area, California and Nevada: *ibid.,* v. 78, p. 131–142.
———, 1970, Upper Precambrian and Lower Cambrian strata in the southern Great Basin, California and Nevada: U.S. Geol. Survey Prof. Paper 620, 206 p.
———, 1971, Basin and range structure: a system of horsts and grabens produced by deep-seated extension: Geol. Soc. America Bull., v. 82, p. 1019–1044.
———, 1972, Initial deposits in the Cordilleran geosyncline: evidence of a late Precambrian (<850 m.y.) continental separation: *ibid.,* v. 83, p. 1345–1360.
———, Albers, J. P., and Poole, F. G., 1968, Summary of regional evidence for right-lateral displacement in the western Great Basin: *ibid.,* v. 79, p. 1407–1413.
———, and Palmer, A. R., 1967, Callaghan window—A newly discovered part of the Roberts thrust, Toiyabe Range, Lander County, Nevada: U.S. Geol. Survey Prof. Paper 575-D, p. D56–D63.
———, Ross, D. C., Nelson, C. A., and Burchfiel, B. C., 1966, Last Chance thrust—A major fault in the eastern part of Inyo County, California: *ibid.,* 550-D, p. D23–D34.
Theodore, T. G., and Roberts, R. J., 1971, Geochemistry and geology of deep drill holes at Iron Canyon, Lander County, Nevada *with a section on* Geophysical logs of drill hole DDH-2, by C. J. Jablocki: *ibid.,* Bull. 1318, 32 p.
Thorman, C. H., 1970, Metamorphosed and nonmetamorphosed Paleozoic rocks in the Wood Hills and Pequop Mountains, northeastern Nevada: Geol. Soc. America Bull., v. 81, p. 2417–2448.
Trimble, D. E., and Carr, W. J., 1962, Paleozoic rocks measured southwest of Pocatello, Idaho: U.S. Geol. Survey open-file rept. no. 651.
Troxel, B. W., 1967, Sedimentary rocks of late Precambrian and Cambrian age in the southern Salt Spring Hills, southeastern Death Valley, California: California Div. Mines and Geology Spec. Rept. 92, p. 33–41.
———, and Wright, L. A., 1968, Precambrian stratigraphy of the Funeral Mountains, Death Valley, California (abs.): Geol. Soc. America Special Paper 121, p. 574–575.
Walcott, C. D., 1894, Precambrian igneous rocks of the Unkar terrane, Grand Canyon of the Colorado, Arizona: U.S. Geol. Survey 14th Ann. Rept., p. 503–519.
Wallace, C. A., 1972, A basin analysis of the upper Precambrian Uinta Mountain Group, Utah (Ph.D. thesis): Univ. California, Santa Barbara, 412 p.
———, and Crittenden, M. D., Jr., 1969, The stratigraphy, depositional environment and correlation of the Precambrian Uinta Mountain Group, western Uinta Mountains, Utah, *in* Lindsay, J. B. (ed.), Geologic guidebook of the Uinta Mountains, Utah's maverick range: Intermountain Assoc. Petroleum Geologists, 16th Ann. Field Conf. Guidebook, p. 127–141.
Washburn, R. H., 1970, Paleozoic stratigraphy of Toiyabe Range, southern Lander County, Nevada: Am. Assoc. Petroleum Geologists Bull., v. 54, p. 275–284.
Wasserburg, G. J., Wetherill, G. W., and Wright, L. A., 1959, Ages in the Precambrian terrane of Death Valley, California: Jour. Geology, v. 67, p. 702–708.
Webb, G. W., 1958, Middle Ordovician stratigraphy in eastern Nevada and western Utah: Am. Assoc. Petroleum Geologists Bull., v. 42, p. 2335–2377.
Winterer, E. L., and Murphy, M. A., 1960, Silurian reef complex and associated facies, central Nevada: Jour. Geology, v. 68, p. 117–139.
Woodward, L. A., 1962, Structure and stratigraphy of the central northern Egan Range, White Pine County, Nevada (Ph.D. thesis): Univ. Washington, Seattle, 145 p.

———, 1963, Late Precambrian metasedimentary rocks of Egan Range, Nevada: Am. Assoc. Petroleum Geologists Bull., v. 47, p. 814–822.

———, 1965, Late Precambrian stratigraphy of northern Deep Creek Range, Utah: *ibid.,* v. 49, p. 310–316.

———, 1967, Stratigraphy and correlation of late Precambrian rocks of Pilot Range, Elko County, Nevada, and Box Elder County, Utah: *ibid.,* v. 51, p. 235–243.

———, 1968, Lower Cambrian and upper Precambrian strata of Beaver Mountains, Utah: *ibid.,* v. 52, p. 1279–1290.

Wright, L. A., and Troxel, B. W., 1966, Strata of late Precambrian-Cambrian age, Death Valley region, California-Nevada: *ibid.,* v. 50, p. 846–857.

———, 1967, Limitations on right-lateral strike-slip displacement, Death Valley and Furnace Creek fault zones, California: Geol. Soc. America Bull., v. 78, p. 933–950.

———, 1970, Summary of regional evidence for right-lateral displacement in the western Great Basin: discussion: *ibid.,* v. 81, p. 2167–2174.

Young, J. C., 1960, Structure and stratigraphy in north-central Schell Creek Range, *in* Boettcher, J. W., and Sloan, W. W., Jr. (eds.), Guidebook to the geology of east-central Nevada: Intermountain Assoc. Petroleum Geologists, 11th Annual Field Conf. Guidebook, p. 158–172.

FLYSCH DEPOSITS OF ANTLER FORELAND BASIN, WESTERN UNITED STATES

F. G. POOLE
U.S. Geological Survey, Denver, Colorado

ABSTRACT

In Late Devonian and Mississippian times, as much as 4500 m (15,000 ft) of well-bedded neritic and bathyal marine, flyschlike mudstone, siltstone, sandstone, conglomerate, and subordinate impure limestone was deposited in a subsiding, elongate, foreland basin (exogeosynclinal trough) on the continental shelf (Cordilleran miogeocline) east of the Antler orogenic belt and west of the cratonic platform. The term Antler flysch is applied to these Upper Devonian and Mississippian, and related Lower Pennsylvanian, homotaxial deposits that filled the Antler foreland basin in the western United States. The major source of siliceous flysch sediments in the foreland trough was terrigenous detritus derived from a rising cordillera to the west composed of strongly deformed Devonian and older oceanic rocks that during the Antler Orogeny was deformed and subsequently obducted eastward onto the outer carbonate shelf as the Roberts Mountains Allochthon. Significant amounts of westerly derived detritus in Upper Devonian deposits reflect early Antler orogenic activity along the continental margin. Recurring uplift of the cordillera followed Antler obduction, as indicated by the presence of chert and quartzite detritus derived from the allochthon in Mississippian and Pennsylvanian deposits of the foreland basin.

Continued orogenic compressive stress that was directed continentward during the Mississippian and Pennsylvanian resulted in a general eastward shift in sites of thick sedimentation. Near the end of Antler flysch deposition in Late Mississippian time, clastic sediments filled the foreland trough and spread eastward across the carbonate shelf onto the craton. Retarded subsidence of the foreland basin and significant decrease in volume of detritus from the reduced cordillera in latest Mississippian time are evidenced by widespread carbonate deposition in the Pennsylvanian.

Most of the flysch sediment within the foreland basin was deposited in a relatively deep-water trough by sediment gravity flows originating in relatively shallow water. Proximal and distal turbidites, debris-flow deposits, and hemipelagic deposits are recognized in the Antler flysch; these facies and their associations in the flysch trough indicate a complex system of submarine slope-fan-basin floor environments.

INTRODUCTION

This paper is concerned with one of the most significant flysch sequences in the western United States. The site of flysch deposition was a linear structural depression that developed in the Upper Devonian and Mississippian continental shelf owing to cratonward compression during Antler orogenic deformation along the western margin of North America. The linear depression is referred to as the Antler foreland basin or exogeosynclinal trough and is synonymous with exogeosyncline of Kay (1947, 1951). It was filled with as much as 4500 m (15,000 ft) of flysch sediment, and was flanked on the west by an orogenic highland and on the east by a carbonate shelf. Owing to the importance of Antler flysch in the interpretation of Upper Devonian and Mississippian sedimentation, paleogeography, and paleotectonics, some generalized results of field studies are presented in this paper in advance of a more detailed report.

The term "flysch" as used herein follows the definition proposed by Hsü (1970) and designates interbedded coarse and fine sediments deposited in a marine geosynclinal environment. Siliceous sediments predominate but impure lime sediments or calcareous flysch locally form a significant part. Although sedimentary structures are important in the recognition of flysch, they are not included in its definition. Turbidite and other sedimentary structures reflect modes of sediment transport and deposition but are not necessarily restricted to flysch basins. Tectonic framework has a direct effect on flysch sediments. In order to form a thick flysch sequence, a rapidly rising source area and a subsiding complementary basin or trough seem essential.

The term Antler flysch is used herein for Upper Devonian, Mississippian, and related Lower Pennsylvanian homotaxial flysch deposits that filled the Antler foreland basin or exogeosynclinal trough on the continental shelf or Cordilleran miogeocline cratonward of the Antler orogenic belt or highland in the western United States. Flysch deposits of the Antler foreland basin or exogeosynclinal trough in the western United States occur in a generally north-trending sinuous belt 160 to 320 km (100 to 200 mi) wide for a distance of about 1600 km (1000 mi), from southern California to northeastern Washington (fig. 1).

Available geologic information on the Cordil-

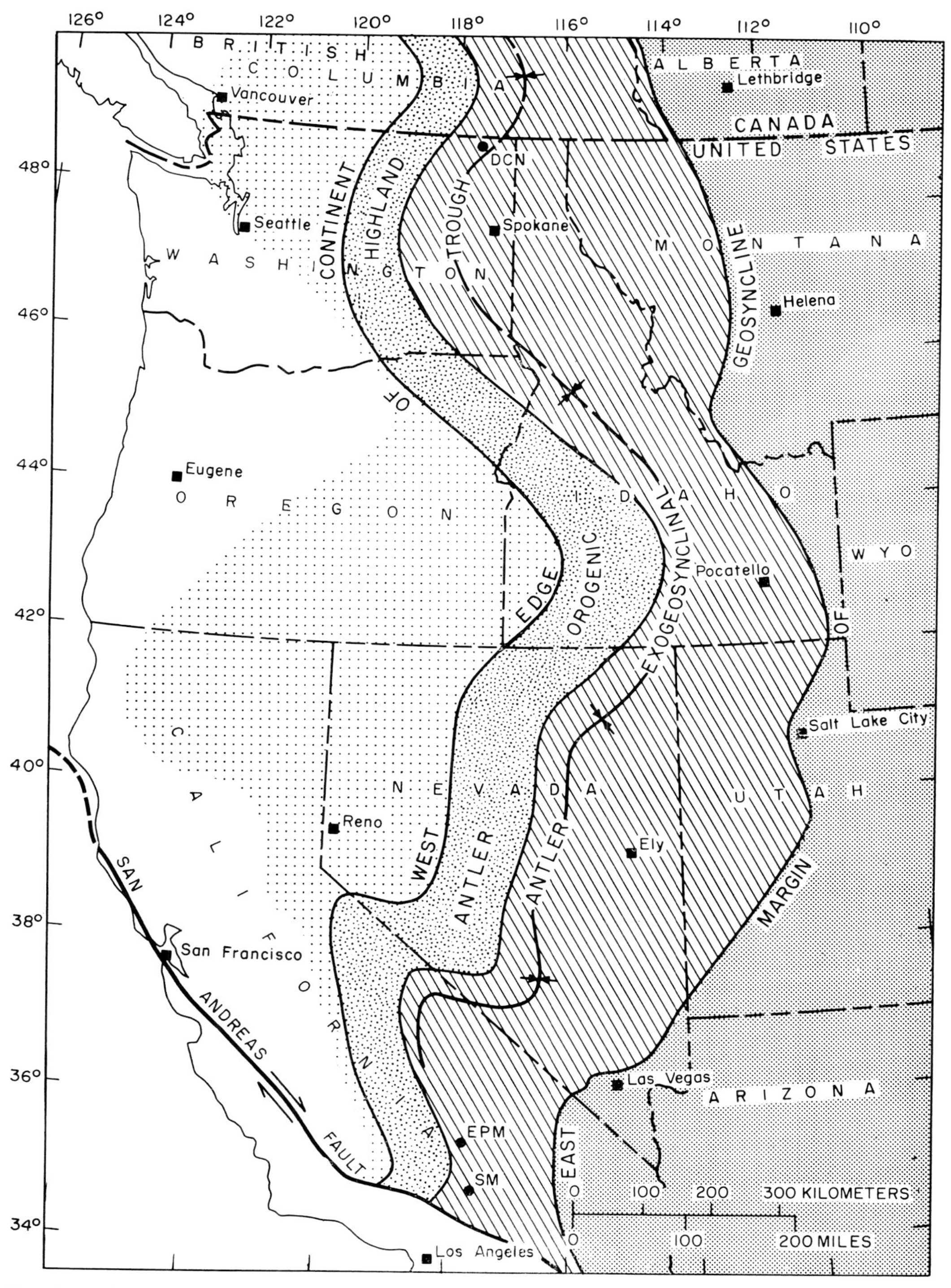

FIG. 1.—Index map of western United States showing Upper Devonian and Mississippian structural and depositional framework (partly restored). Cratonic platform in compact stippling; Devonian continental shelf in hatching and random stippling; oceanic area in diffuse stippling. Antler orogenic highland developed on outer continental shelf in latest Devonian time. SM, Shadow Mountains; EPM, El Paso Mountains; DCN, Deep Creek—Northport area.

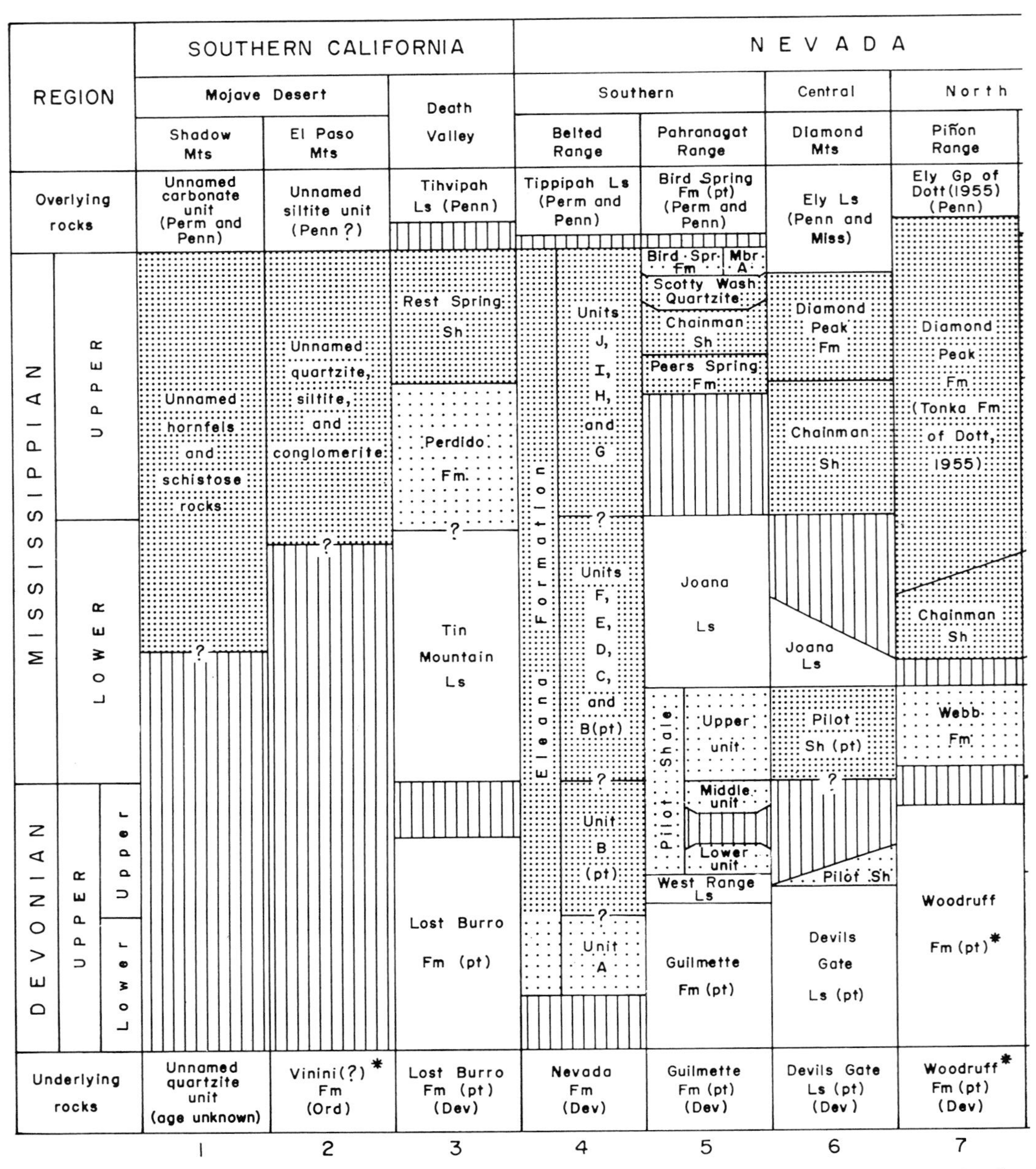

FIG. 2.—Generalized chart showing stratigraphic position and some local names applied to Antler flysch (stippled) and associated rocks in the western United States. Hatching indicates strata absent because of nondeposition or erosion. Compact stippling indicates that strata consist mostly of orogenic sediments; diffuse stippling indicates that strata consist of either minor orogenic sediments or mostly of reworked orogenic sediments. Asterisk (*) indicates allochthonous eugeosynclinal- and transitional-facies rocks in Roberts Mountains Allochthon. Sources of data: 1. Troxel and Gundersen (1970); F. G. Poole and J. H. Stewart (unpub. data); 2. Poole (unpub. data); 3. McAllister (1952); Mackenzie Gordon, Jr., and Poole (unpub. data); 4. Poole, Houser, and Orkild (1961); Poole, Orkild, Gordon, and Duncan (1965); Gordon and Poole (1968); Ekren, Anderson, Rogers, and Noble (1971); Poole and Gordon (unpub. data); 5. Reso (1963); Sandberg

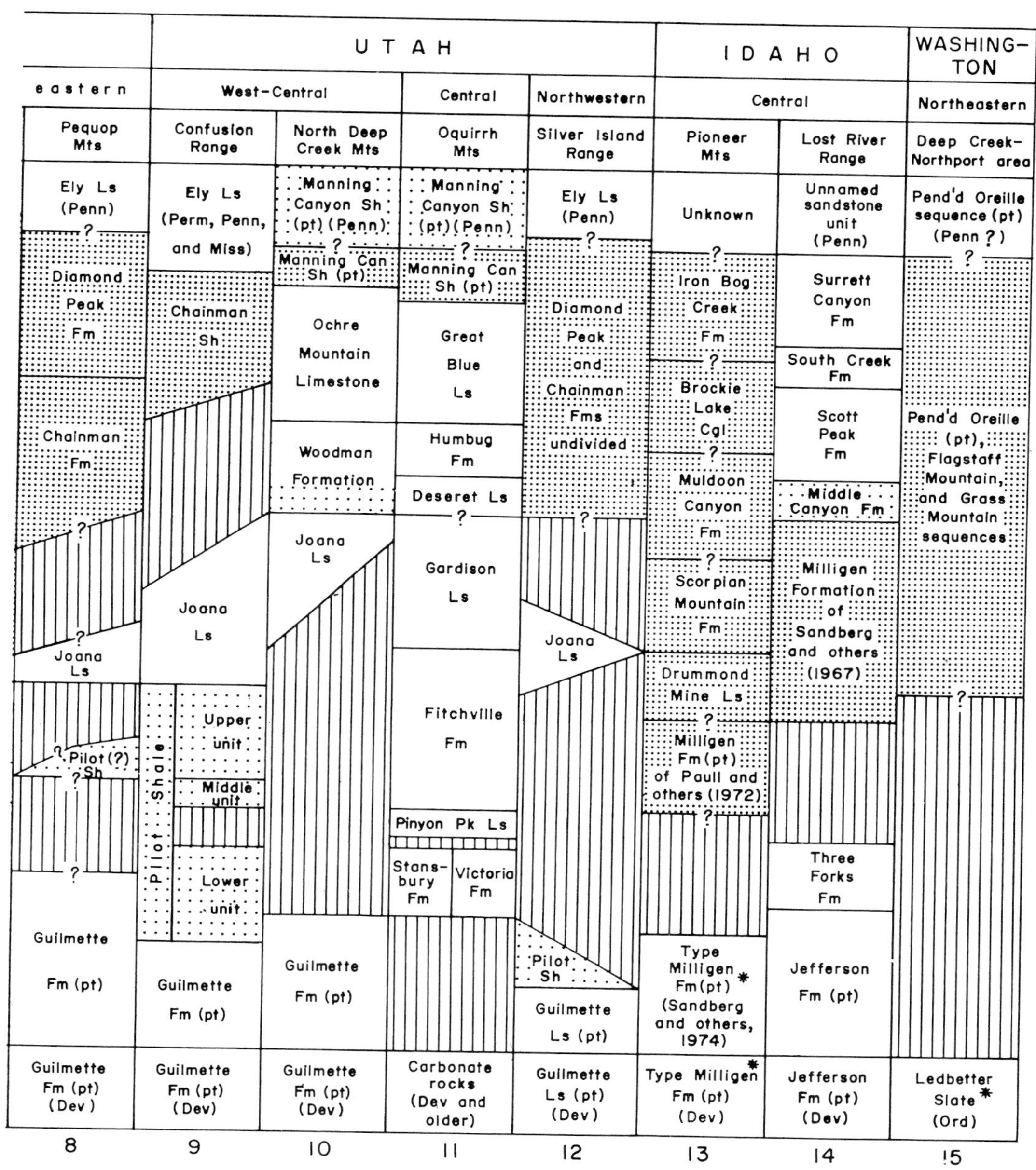

and Poole (1970); Poole (unpub. data); 6. Nolan, Merriam, and Brew (1971); Brew (1971); Poole (unpub. data); 7. Dott (1955); Smith and Ketner (1968, 1972, and unpub. data, 1974); Poole (unpub. data); 8. Thorman (1970); Poole (unpub. data); 9. Hose (1966); Poole and C. A. Sandberg (unpub. data); 10. Nolan (1935); Poole (unpub. data); 11. Gilluly (1932); Rigby (1959); Moyle (1959); Morris and Lovering (1961); C. A. Sandberg and Poole (unpub. data); 12. Schaeffer (1960); Poole and C. A. Sandberg (unpub. data); 13. Paull and others (1972); C. A. Sandberg, W. E. Hall, J. N. Batchelder, and Claus Axelsen (unpub. data, 1974); R. A. Paull, Betty Skipp, C. A. Sandberg, and Poole (unpub. data); 14. Sandberg and others (1967); Mamet and others (1971); 15. Yates (1964, 1971).

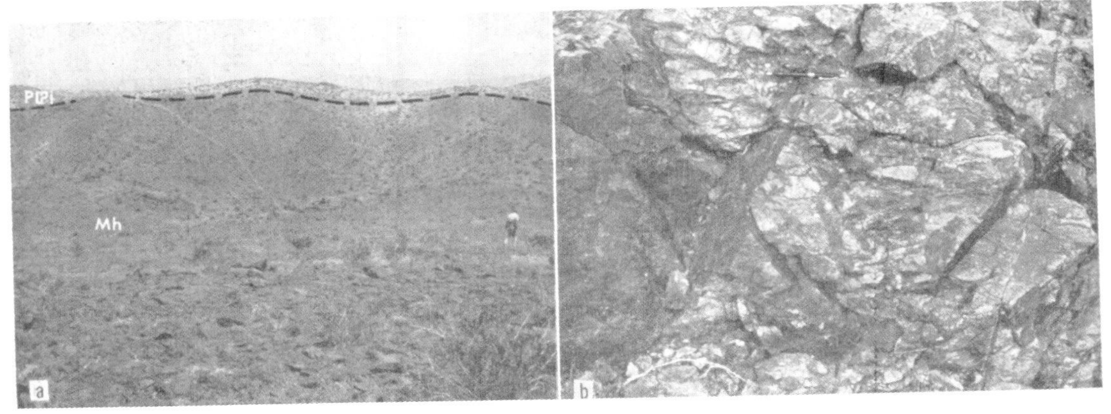

FIG. 3.—Unnamed flysch deposits in Shadow Mountains, southwestern Mojave Desert, southern California. *a*, Metamorphosed dark-colored flysch (mostly hornfels of Mississippian age, Mh) overlain by light-colored limestone (mostly calcite marble of Pennsylvanian and Permian age, PIPl). *b*, Tectonically stretched quartzite pebbles in a dark-colored conglomeratic quartzite bed within the "schistose rocks" map unit of Troxel and Gundersen (1970).

leran region of western Canada indicates that the Antler flysch belt continues from the Kootenay Arc area of southern British Columbia northwestward through British Columbia, and possibly into Yukon territory. In the southwestern United States the Antler flysch belt is truncated by the Cenozoic San Andreas fault in southern California (fig. 1). The flysch sequence has not been recognized in the Salinian structural block west of the San Andreas fault, and its absence indicates an offset of at least 650 km (400 mi) northwestward beyond the present continental margin.

Upper Devonian and Mississippian Antler flysch exposed in the western United States is known under many different formal and informal stratigraphic names (fig. 2, stippled units). Lithologic descriptions of several sedimentary units assigned to the Antler flysch may be found in the publications listed in sources of data for figure 2. Flysch deposits recently recognized in southern California and inferred in eastern Washington will be discussed briefly here because they are not described elsewhere in the literature.

Southernmost exposures of the Antler flysch occur in the western Mojave Desert of southern California. In the Shadow Mountains (fig. 1) northwest of Victorville, brown fine-grained and minor coarse-grained detrital sedimentary rocks herein assigned to the Antler flysch have been metamorphosed to hornfels, schist, and thin marbles (fig. 2, col. 1). This unnamed metamorphosed flysch sequence (fig. 3a), which is estimated to be a few thousand feet thick, was mapped as "hornfels" and "schistose rocks" by Troxel and Gundersen (1970). Tectonically stretched quartzite-pebble conglomerates (fig. 3b) occur in some thin units of coarse-grained flysch. Overlying the flysch sequence is a thick succession of metamorphosed carbonate rocks (fig. 3a) that are similar in lithology to the Bird Spring Formation (Permian and Pennsylvanian) of the southern Great Basin. In the northwestern Mojave Desert, a fine- to coarse-grained flysch sequence as much as 1,500 feet thick occurs within the Garlock Formation (Dibblee, 1967; formerly the Garlock Series of Dibblee, 1952) in the El Paso Mountains (fig. 1). This flysch consists of brown interbedded quartzite and siltite and subordinate argillite and conglomerite (fig. 4) that is correlative with the Antler flysch (fig. 2, col. 2). The flysch sequence overlies strongly deformed Ordovician oceanic rocks provisionally assigned to the Roberts Mountains Allochthon.

Northernmost exposures of the Antler flysch in the western United States occur in the Deep Creek-Northport area of northeastern Washington (fig. 1). In this area, dark-gray to black argillite and subordinate siltite, quartzite, impure limestone, and conglomerate were grouped into major rock units and referred to informally as Grass Mountain, Flagstaff Mountain, and Pend 'd Oreille sequences by Yates (1964, 1971); they were provisionally assigned to the Carboniferous (fig. 2, col. 15). The Grass Mountain, Flagstaff Mountain, and part of the Pend 'd Oreille sequences seem to be flyschlike in character and herein are provisionally considered to represent Antler flysch.

FIG. 4.—Unnamed flysch deposits in El Paso Mountains, northwestern Mojave Desert, southern California. *a,* Cyclically interbedded quartzite, siltite, and argillite unit (qsa) overlain by chert-pebble conglomerite (cg). Ledge-forming conglomerite unit is about 7.5 m (25 ft) thick. *b,* Closeup of cyclical unit showing thin turbidite beds of quartzite-siltite (hammer on one of thicker beds) separated by interturbidite beds of laminated argillite-siltite.

GEOLOGIC SETTING

Figure 1 shows the structural and depositional framework for the Upper Devonian and Mississippian Antler flysch in the western United States, based on both present and restored facies distributions. The Devonian miogeosyncline or continental shelf extended from the outer edge of the craton or stable platform (compact stippling) westward to the outer edge of the continent. The synclinal axis depicted on the continental shelf, in front of the Antler orogenic highland, represents the axial trace of the Antler foreland basin or exogeosyncline.

The snakelike bending of the Antler orogenic highland and exogeosynclinal trough is believed to be mainly the result of Mesozoic and Cenozoic oroflexural bending. The prominent bending along the California-Nevada state line is clearly oroflexural bending related to the Walker Lane dextral strike-slip structural zone. Prominent eastward bending of Antler structural and depositional features in northeastern Nevada and southwestern and central Idaho is believed to be due mainly to large-scale oroflexural bending and overthrusting (see Stewart and Poole, this volume). Some of the curvature of the highland may reflect the configuration of the Paleozoic continental edge.

The Antler orogenic belt and exogeosynclinal trough, where they trend northwesterly between central Idaho and central Washington, coincide approximately with two regional lineaments known as the Trans-Idaho Discontinuity (Yates, 1968) and the Olympic-Wallowa Lineament (Raisz, 1945; Wise, 1963; and Skehan, 1965). The Trans-Idaho Discontinuity has been interpreted by Yates (1968) and Jones, Irwin, and Ovenshine (1972) as a northwest-trending sinistral transcurrent fault zone. The subparallel Olympic-Wallowa Lineament, which is located 75–150 km (50–100 mi) farther southwest and which trends northwest from west-central Idaho through northeastern Oregon into west-central Washington, also has been interpreted by many geologists as a sinistral strike-slip fault and by Skehan (1965) as the surface expression of a deep crustal boundary reflecting the change from continental crust on the northeast to oceanic crust on the southwest. The Olympic-Wallowa Lineament nearly coincides with a segment of the inferred western edge of the middle Paleozoic continent shown in figure 1.

FLYSCH SEDIMENTATION

The axial portion of the Antler foreland basin contains a flysch sequence whose thickness is generally 1500 to 3000 m (5000 to 10,000 ft) but locally is in excess of 4500 m (15,000 ft). Lithologic sequences and thicknesses vary areally, indicating a complex system of submarine fan and basin floor environments. Sedimentary features within the flysch sequence provide insight into provenance, mechanisms of sediment transport, and environments of deposition.

Lithology and Provenance

The Antler flysch is composed principally of (1) hemipelagic mudstones, siltstones, and minor limestones; (2) turbidite sandstones, siltstones, and minor conglomerates and limestones; (3) massive, ungraded to poorly graded (disorganized) conglomerates and pebbly sandstones; and (4) crudely stratified, graded (organized) conglomerates and pebbly sandstones. Generally mudstone and siltstone are more abundant than sandstone, but at certain intervals sandstone and conglomerate or limestone beds dominate.

Shaly mudstone and fine-grained siltstone units, randomly distributed in the flysch succession and intercalated cyclically between turbidite beds, probably represent fine sediment that was introduced into the flysch basin by density currents and then deposited from dilute suspension. These mudstones and siltstones generally contain horizontal meandering trails that belong to Seilacher's (1964) *Nereites*-facies. Many black mudstones, siltstones, and certain impure limestones contain as much as a few percent organic carbon and disseminated pyrite or marcasite that suggest an environment of restricted circulation in these parts of the flysch basin. These hemipelagic and low-density turbidity current deposits are similar to facies G of Walker and Mutti (1973).

Most thin-bedded sandstones, conglomerates, coarse-grained siltstones, and detrital limestones in the flysch succession contain sedimentary structures characteristic of turbidites described by Bouma (1962), Walker (1967), and other workers. An outcrop of siliceous fine-grained flysch near the axis of the Antler foreland basin seen in figure 5a is a sequence of cyclical sandstone and siltstone turbidites and mudstone interturbidites in the Chainman Shale (Upper Mississippian) in central Nevada. Generally, the silt and sand grains in the flysch are composed of quartz, chert, and quartzite. Sparse sand-sized grains of feldspar have been seen in some thin sections of siliceous flysch rocks. Many graded conglomerate, sandstone and limestone beds contain displaced fossil fragments of shallow-water organisms. An outcrop of calcareous fine-grained flysch near the axis of the Antler foreland basin seen in figure 5b is a sequence of cyclical impure limestone turbidites and spicular limestone interturbidites in the Drummond Mine Limestone (Lower Mississippian) of Paull and others (1972) in central Idaho.

Gravels in conglomerate, conglomeratic sandstone, siltstone, and mudstone generally are composed of subrounded to rounded chert, quartzite, and sparse limestone or dolomite, and of angular to subrounded intraformational "rip-up" clasts of argillite and siltite. Most gravels are of pebble to small cobble size, but locally sparse boulders are conspicuous. Outcrops of siliceous coarse-grained flysch deposits along the west margin of the Antler foreland basin are shown in figure 6. These coarse-grained flysch deposits occur within a southeast-trending submarine channel about 300 m (1000 ft) wide and 150 m (500 ft) deep at the base of the Diamond Peak Formation in north-central Nevada. Fig-

FIG. 5.—Lithology and bedding features of turbidite sequences in fine-grained flysch. *a,* Interbedded sandstone to coarse-grained siltstone turbidites (ledges) and mudstone interturbidites (recesses) of siliceous basinal flysch in the lower part of the Chainman Shale, Diamond Mountains, central Nevada; hammer is on a sandstone turbidite bed; stratigraphic top is to the right (overturned section). *b,* Interbedded impure limestone turbidites (dark-colored interval above pencil) and light-colored spicular limestone interturbidites of calcareous flysch in the lower part of the Drummond Mine Limestone, Pioneer Mountains, central Idaho.

FIG. 6.—Lithology and bedding features of coarse-grained flysch debris-flow deposits in a submarine channel in the lower part of the Diamond Peak Formation, Piñon Range, north-central Nevada. *a,* Massive gravelflow conglomerate composed of quartzite and chert clasts. *b,* Bedded sandstone and pebbly sandstone units (sps) overlain by scouring massive conglomeratic sandstone (cgs).

ure 6a shows a massive disorganized gravelflow conglomerate in the lower medial part of the channelfill and figure 6b shows a bedded sandstone and pebbly sandstone unit overlain by a scouring massive disorganized conglomeratic sandstone in the upper medial part of the channelfill. An isolated rounded quartzite boulder 1 m (3 ft) across was seen in the coarse conglomerate unit pictured in figure 6a. Some pebbly mudstone-siltstone units also occur locally in the Antler flysch.

The major source of siliceous flysch sediments in the Antler foreland basin was terrigenous detritus derived from a rising cordillera to the west. Chert and quartzite detritus in the flysch deposits is similar to Devonian and older eugeosynclinal cherts and quartzites in the Roberts Mountains Allochthon exposed along the Antler orogenic belt. A general easterly decrease of grain size and coarse/fine sediment ratio in the siliceous flysch sequence also indicates that detritus was shed toward the craton from lands to the west.

The major source of calcareous flysch sediments in Lower Mississippian (Kinderhookian) limestone turbidite sequences—such as the Camp Creek sequence of Ketner (1970) and the correlative Tripon Pass Limestone of Oversby (1973) in northeastern Nevada, and the Drummond Mine Limestone of Paull and others (1972) in central Idaho—was limestone detritus probably derived mainly from the west edge of the shallow-water carbonate shelf or bank near the eastern margin of the Antler foreland basin. Current sole marks suggest a southwest direction of transport for the Camp Creek limestone turbidites (Ketner, 1970). Paleocurrent directions have not been determined from the turbidite sequences in the Tripon Pass and Drummond Mine Limestones.

Sedimentary Structures and Sediment Transport

"Bouma sequences" of internal sedimentary structures in thin-bedded sandstones, conglomerates, siltstones, and limestones, displaced shallow-water organisms in graded beds, and sole markings indicate that turbidity currents were an important process for relatively deep-water sedimentation in the Antler foreland basin. Bouma (1962) defined a complete turbidite as one which contains a graded division (A), lower division of parallel lamination (B), division of current ripple lamination and/or convolute lamination (C), upper division of parallel lamination (D), and pelitic division (E). Some turbidites of the Antler flysch contain divisions A-D of the ideal Bouma sequence (fig. 7a); however, most turbidites examined contain an incomplete Bouma sequence beginning with division B or C (fig. 8). Interturbidite pelitic division E usually occurs between major turbidite sequences. Figure 7 shows examples of complete Bouma sequences of sedimentary structures in sandstone and limestone turbidites; figure 7a represents a sandstone turbidite (A-D) and 7b represents a limestone turbidite (A-D). The specimen pictured in 7b is from the lower unit of the Pilot Shale, which is believed to contain a significant amount of early flysch. Figure 8 shows examples of incomplete Bouma sequences of sedimentary structures in sandstone and impure limestone turbidites; figure 8a repre-

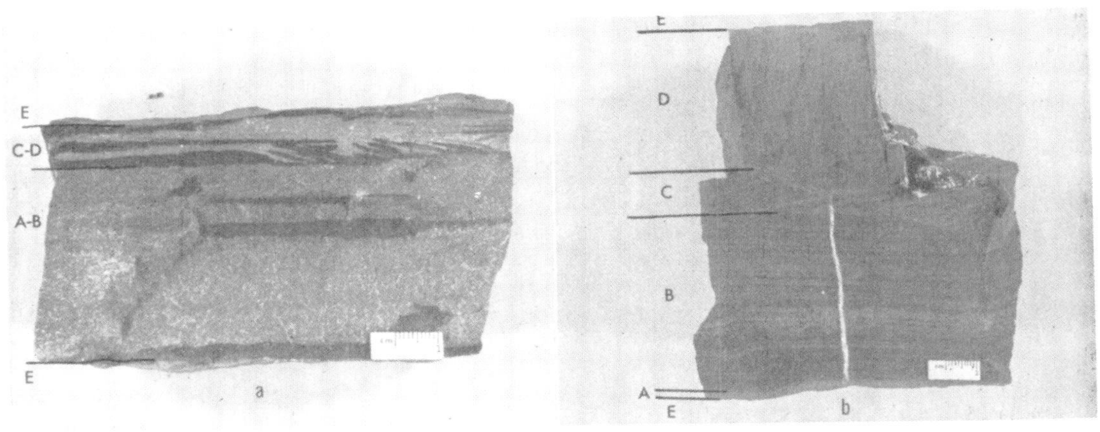

FIG. 7.—Internal structure of proximal turbidite beds showing complete Bouma sequences. *a,* Specimen of coarse to fine sandstone turbidite sequence showing graded and horizontal laminae of divisions A and B, current ripple laminae of division C and parallel laminae of division D, and interturbidite pelitic division E with meandering trails of *Nereites* community; dark-gray areas are iron oxide stains; specimen from Chainman Formation, Pequop Mountains, northeastern Nevada. *b,* Specimen of impure limestone turbidite sequence showing graded thin sandy layer of division A, horizontal laminae of division B, current ripple laminae and minor convolute laminae of division C, and upper parallel laminae of division D; interturbidite shaly calcsiltite of division E (not attached to specimen) contains styliolinids and tentaculitids; specimen from lower unit of Pilot Shale, Leppy Range, Nevada, near Utah State line.

sents a sandstone turbidite (B-D) and 8b represents two amalgamated calcarenitic sandstone-siltstone turbidites (B-D underlain by C). Interturbidite mudstones and siltstones of Bouma division E shown in figure 7a contain meandering trails of the *Nereites* community and those shown in figure 7b, styliolinids and tentaculitids.

Many sandstone turbidites contain sole markings including scour marks (mainly flute casts) and tool marks (mainly groove casts). Figure 9 illustrates two current structures commonly found as sole marks on sandstone beds in unit C of the Eleana Formation in southern Nevada. These flute and groove casts occur on the lower parallel laminated division B of incomplete Bouma sequences.

Many chert-pebble and quartzite-pebble conglomerates and pebbly sandstones are stratified and show grading with clast orientation and imbrication. These organized conglomerates and pebbly sandstones are similar to facies A2 and A4 of Walker and Mutti (1973).

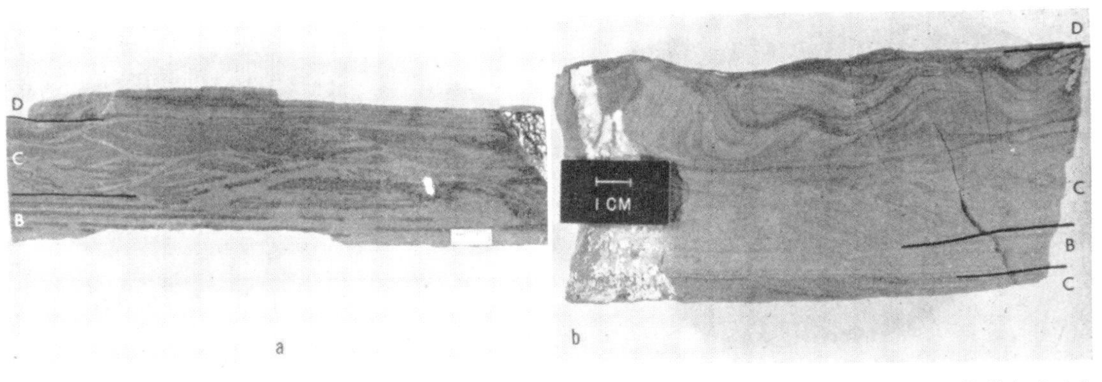

FIG. 8.—Internal structure of distal turbidite beds showing incomplete Bouma sequences. *a,* Polished slab of sandstone turbidite showing horizontal parallel laminae of division B, current ripple laminae of division C, and upper parallel laminae of division D; specimen from unit C of Eleana Formation, Belted Range, south-central Nevada. *b,* Specimen of amalgamated limy sandstone-siltstone turbidite sequences showing current ripple laminae of division C directly overlain by horizontal parallel laminae of division B, current ripple laminae, wavy and convoluted laminae of division C, and thin upper parallel laminae of division D; specimen from base of Scorpion Mountain Formation, Pioneer Mountains, central Idaho.

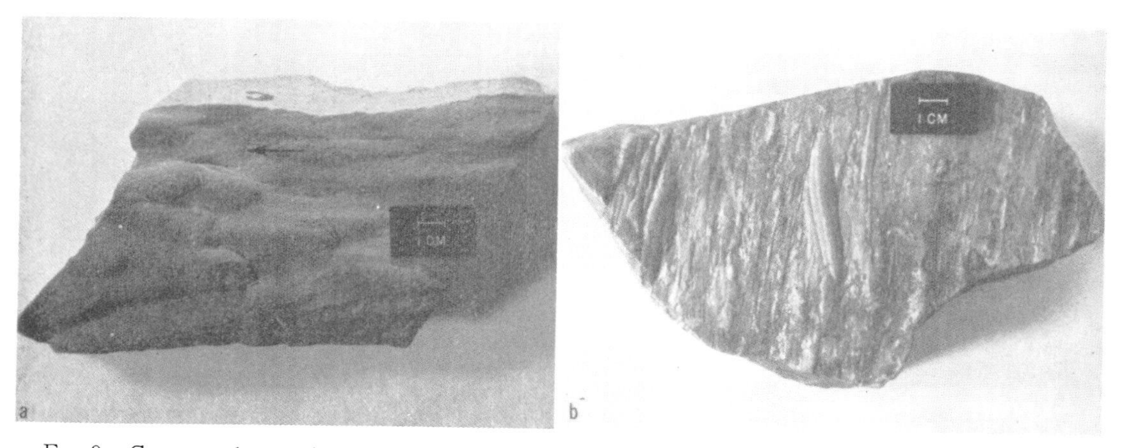

FIG. 9.—Current sole markings on lower parallel laminated distal turbidite sandstone of incomplete Bouma sequence; specimens from unit C of Eleana Formation, Belted Range, south-central Nevada. *a,* Flute casts. Arrow indicates current direction. *b,* Striation and groove casts.

Thick sandstones and conglomerates show sedimentary structures different from those within the thin-bedded sandstones of the flysch succession where it is dominated by mudstone and siltstone. The Bouma sequence is not applicable to disorganized and many organized conglomerates and pebbly coarse-grained sandstones. Many thick conglomerates (fig. 6a) and pebbly sandstones (fig. 6b) along the west margin of the Antler foreland basin are massive to irregularly bedded, commonly scoured, and ungraded to poorly graded. These disorganized conglomerates and pebbly sandstones are similar to facies A1 and A3 of Walker and Mutti (1973).

Several transport mechanisms can be inferred for the coarse- and fine-grained flysch sediments filling the Antler foreland basin. Most of the sediments in the flysch trough are believed to have been transported by sediment gravity flows (Middleton and Hampton, 1973), mainly debris flows and density currents, that moved sediment from shallow-water environments flanking the foreland basin to deep-water environments forming a foreland trough in the axial portion.

Internal sedimentary structures and sole markings of many thin beds of sandstone, coarse-grained siltstone, and detrital limestone indicate deposition by turbidity currents. Bouma (1962) and Walker (1967) have pointed out that turbidites which begin with division B or higher divisions represent more distal deposition than those which begin with division A. Walker (1967) studied the variability of turbidite sequences in several formations and related them to the flow regimes of depositing currents. He described several differences between proximal and distal turbidity-current flow. Bouma and Hollister (1973) defined proximal turbidites as turbidites that initially result from high flow regime conditions, whereas distal turbidites are totally deposited under low flow regime conditions. They emphasized the fact that varying conditions may force one current to travel far into a basin before significant deposition begins, while another current may lack the properties to travel very far. Consequently, some distal turbidites may be deposited near the source, whereas some larger flows may deposit a proximal turbidite far into the basin; however, it seems logical to expect more proximal turbidites nearer the source.

The thick-bedded pebbly sandstones and conglomerates may have been deposited by a mechanism differing from a turbidity current. Many beds are massive and contain sedimentary structures different from those in the thin-bedded sandstones. Some of these massive sandstones and conglomerates exhibit features comparable to debris flows in modern submarine channels described by Normark (1970), Haner (1971), Nelson and Kulm (1973), and other workers. Thick, poorly graded gravels, sands, and pebbly sands are common deposits in submarine canyons, and in channels incised into the upper part of submarine fans (Haner, 1971). Hampton (1972) has suggested that turbidity currents may originate from submarine landslides and debris flows by incorporation of water and sufficient agitation.

Unsorted pebbly and cobbly mudstones and siltstones occur locally in the flysch succession in the Antler foreland basin. Such rocks may have originated from rapid deposition of sand and gravel on top of water-saturated mud (Crowell, 1957) that subsequently moved downslope forming a chaotic mixture of mud, sand, and gravel; they belong to facies F of Walker and Mutti (1973). These deposits, which reflect unstable slope conditions, are common locally in the Antler flysch along the western margin of the foreland basin.

Fossils and Paleoecology

Fossil animals and plants in the flysch succession provide important clues regarding depositional environments and bathymetry in the Antler foreland basin. The most common fossils in the fine-grained siliceous and calcareous Antler flysch are meandering indigenous grazing trails and displaced fragments of shallow-marine invertebrates and vascular land plants. Most of the meandering trails (fig. 10) belong to the *Nereites*-facies or community described by Seilacher (1964, 1967) and, according to him, indicate a deep-water marine environment. Meandering trails of the *Nereites* community are confined to nonturbidite mudstones, siltstones, and impure limestones; they are especially common in pelitic units between major turbidite sequences. Sparse horizontal *Scalarituba* trails (fig. 10f) and *Zoophycos* burrows occur in some flysch units. Various land plant fragments (fig. 11) occur sparingly to commonly in both fine and coarse detrital sediments in the flysch basin. These plants are abraded and commonly show a preferred orientation; they obviously were washed into the marine basin from a land area, presumably the Antler mountainous terrane to the west. Local concentrations of plant debris have produced some thin layers of carbonized wood and sparse thin seams of coaly material.

Although meandering trails of the *Nereites* community, siliceous sponge spicules, and radiolarian tests suggest bathyal marine environments, neritic marine environments in other parts of the flysch basin are evidenced by the local abundance of shallow-water benthonic fauna containing pelmatozoans, bryozoans, brachiopods, pelecypods, trilobites, and corals. Shallow-water marine invertebrates and land plants occur sparingly in fine- to coarse-grained siliceous and calcareous flysch sediments, but commonly it can be demonstrated sedimentologically that they were transported into deeper water by wave action and density currents. The occurrence of invertebrate fossils in sandstone and limestone turbidites intercalated with mudstones and siltstones containing indigenous meandering trails of *Nereites* community indicates that the invertebrates were transported by density currents from their shallow-water living sites. Most land plants found in the Antler flysch are believed to have floated basinward, where they became waterlogged, sank, and became incorporated in the flysch succession. Many plants in the flysch basin may have been redistributed by sediment gravity flows.

Summary of Depositional Environments

Much of the flysch sediment within the Antler foreland basin was deposited in deep water by turbid flows originating in relatively shallow water. Proximal and distal turbidites, debris-flow deposits, and hemipelagic deposits are recognized in the flysch sequence. Alluvial and deltaic environments must have occurred along the eastern margin of the Antler highland; however, deposits characteristic of these environments have not been positively identified.

Many shaly mudstone- and siltstone-dominated sections in the medial portion of the Antler foreland basin contain both distal and proximal turbidite sequences comparable to modern middle and outer submarine fan deposits and basin plain deposits.

Sedimentary features of thick sandstones, pebbly sandstones, and ungraded to poorly graded conglomerates along the western margin of the Antler foreland basin indicate deposition as debris flows resembling those found in modern submarine canyons and in channels of modern submarine inner fans and channels incised into middle fans. Some thin beds of coarse-grained sandstone, conglomeratic sandstone, and organized conglomerate associated with fine-grained flysch facies in the medial part of the foreland basin may represent deposits in submarine inner fan and middle fan systems.

Many shaly mudstone, fine-grained siltstone, and micritic limestone units in the foreland basin are inferred to represent very fine-grained sediment deposited from dilute suspension. Most of the mud may have entered suspension through turbulence associated with turbidity current movement and then was deposited by normal hemipelagic sedimentation. These fine-grained flysch units commonly contain horizontal meandering trails assignable to the *Nereites*-facies indicative of a quiet, bathyal environment.

Antler flysch turbidite facies and facies associations indicate that debris flows and turbidity currents originating in coastal waters near the west margin of the Antler foreland basin adjacent to the highland flowed eastward down

FIG. 10.—Indigenous meandering trails of *Nereites* community in hemipelagic mudstone and siltstone (division E of Bouma sequence) between sandstone turbidites. *a,* Meandering trails in Muldoon Canyon Formation, Pioneer Mountains, central Idaho. *b,* Meandering trails and small fragments of macerated plant debris in Milligen Formation of Sandberg and others (1967), Lost River Range, central Idaho. *c,* Meandering trails in Chainman Formation, Pequop Mountains, northeastern Nevada. *d,* Meandering trails in unit D of Eleana Formation, Belted Range, south-central Nevada. *e,* Narrow and wide meandering trails in Diamond Peak Formation, Diamond Mountains, central Nevada. *f,* Meandering trails and segmented trail (*Scalarituba?*) in basal beds of Scorpion Mountain Formation, Pioneer Mountains, central Idaho.

FIG. 11.—Displaced land plants in marine siltstone of Milligen Formation of Sandberg and others (1967), Lost River Range, central Idaho. *a*, Bedding surface showing relatively large stem with some branch attachments and small fragments of macerated plant debris. *b*, Bedding surface showing abraded stems with preferred orientation and macerated plant debris.

the basin slope into bathyal waters of the basin floor. Also, some turbidity currents originating near the east margin of the Antler foreland basin at the outer (west) edge of the shallow-water carbonate shelf or bank flowed westward down the basin slope into deep water of the basin floor. In both cases, the turbidity currents probably were diverted on the basin floor and flowed parallel to the basin axis, a possibility suggested by a few directional measurements on current structures in distal turbidites.

Near the end of Antler flysch deposition, clastic sediments filled the foreland trough and spread eastward across the carbonate shelf onto the cratonic platform. During this final infilling stage, shallow neritic environments above storm-wave base dominated the foreland basin and carbonate shelf areas resulting in the reworking of younger flysch and carbonate deposits. Finally, due to retarded subsidence of the foreland basin and significant decrease in volume of detritus shed from the reduced highland, widespread shallow-water carbonate deposition prevailed in the area of the former foreland trough and expanded westward onto the subdued highland.

REGIONAL DEPOSITIONAL PATTERNS

Generalized isopach and facies maps of the lower Upper Devonian, upper Upper Devonian, Lower Mississippian, and Upper Mississippian reveal important features in the evolutionary development and character of the flysch basin.

Lower Upper Devonian

The isopach and facies map of the lower Upper Devonian (fig. 12) delineates areas of thinning and thickening associated with ancestral ridges and furrows on the continental shelf that were mostly inherited from older trends. The Uinta uplift in northern Utah was a major positive area at the end of early Late Devonian time. Most of the quartz sand in the limestone and sandstone units on the inner shelf was derived from older quartzites exposed on the craton to the east, whereas the shale and siltstone deposits near the continental edge (fig. 12, open hatching) may represent fine detritus from areas of uplift along the continental margin in addition to normal continental rise sediments transported seaward from cratonic uplifts. In the southeastern part of the map the open-hatched area represents sandstone, siltstone, and shale deposits in an intracratonic basin.

Figure 13 shows an outcrop of cliff-forming Devils Gate Limestone in central Nevada that represents typical shallow-water limestone deposits on the continental shelf near the west edge of the limestone and sandy limestone area (diffuse stippling) shown on the map.

Figure 14 shows an outcrop of slope-forming Pilot Shale in northeastern Nevada that represents limy siltstone deposits in an incipient flysch basin on the shelf shown by the compact-stippled area west of the Uinta uplift (fig. 12). The basal part of the Pilot Shale in this shelf basin is laterally equivalent to the upper part of the type Devils Gate Limestone of figure 13. The lower unit of the Pilot in this shelf basin may represent extra-basinal detritus derived from both early Antler orogenic uplifts along the continental margin and local uplifts along the cratonic margin.

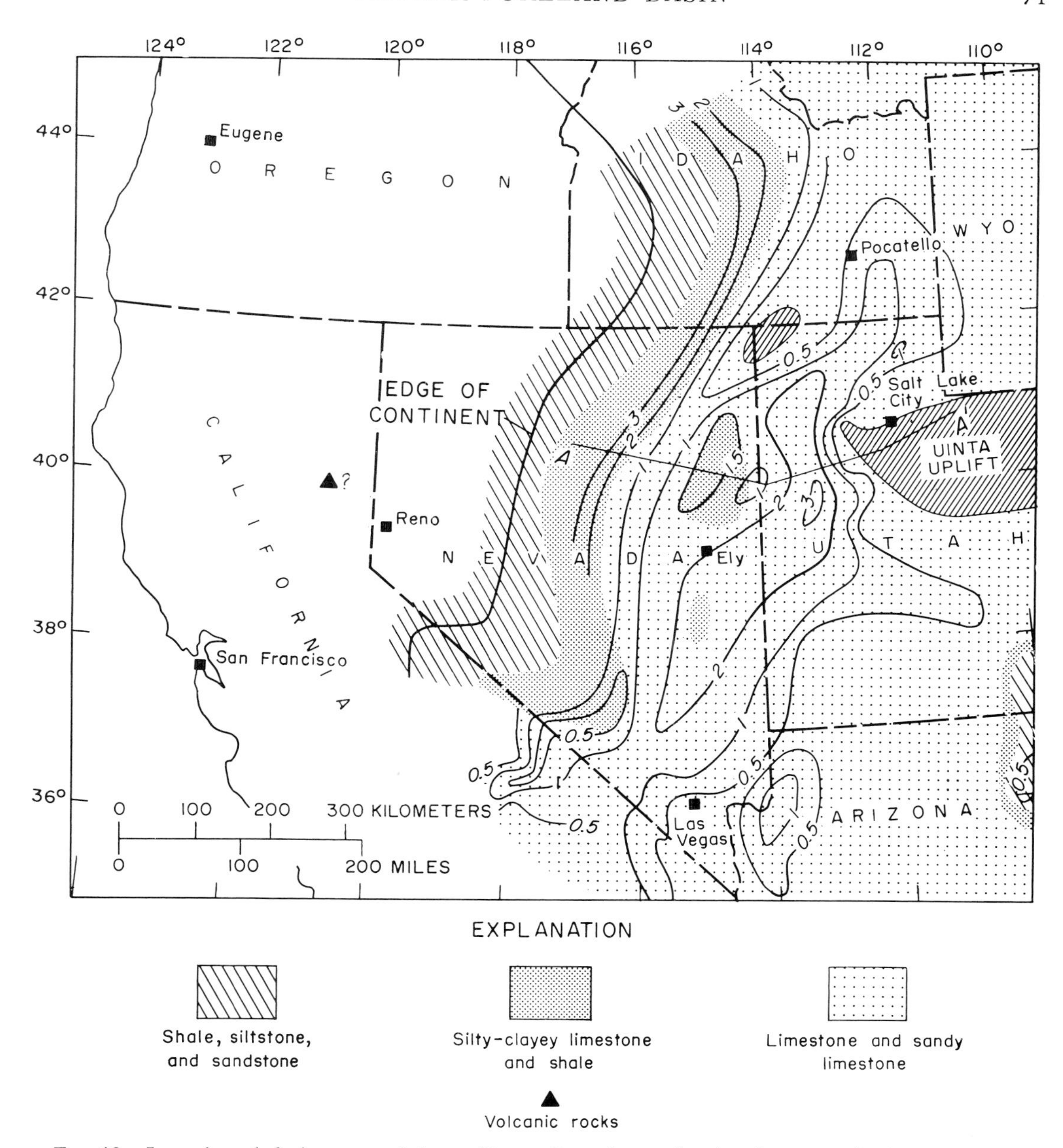

FIG. 12.—Isopach and facies map of lower Upper Devonian rocks (partly restored). Emergent areas shown by compact hatching. Isopachs in thousands of feet. Thickness and facies west of continental edge not shown owing to insufficient stratigraphic data. Cross section AA′ shown by figure 25a.

Upper Upper Devonian

The map of the upper Upper Devonian (fig. 15) reveals broad regional uplift of the shelf and platform with many emergent areas. Near the beginning of this time interval, the local but prominent Stansbury uplift (fig. 15, small area of compact hatching southwest of Salt Lake City) and its complementary subsiding basin formed at the west edge of the craton in northern Utah. The Stansbury basin adjacent to the uplift received over 450 m (1500 ft) of carbonate- and quartzite-clast conglomerate and quartzose sandstone derived from erosion of lower Paleozoic rocks in the core of the uplift

FIG. 13.—Cliff-forming thin-bedded Devils Gate Limestone at Devils Gate, central Nevada, represents shallow-water limestone deposits typical of outer carbonate shelf.

(Rigby, 1958; Stokes and Arnold, 1958). Figure 16 shows an outcrop of one of the carbonate-clast conglomerate units in the Stansbury Mountains. The Stansbury uplift and adjacent high area in northwestern Utah divided the previously continuous shelf basin into two parts, outlined by 500-foot isopach lines in figure 15 (compare figs. 12 and 15).

Figure 17 shows an outcrop of Pilot Shale in the southern of the two basins near the Nevada-Utah State line. The ledge shown in the center of the picture is a sandy limestone turbidite unit that is conglomeratic and contains flute casts, displaced shallow-marine fossils, and slump folds. The limy siltstones above and below the ledge are late Late Devonian in age, and like similar limy siltstones in the early Late Devonian part of the Pilot may represent extra-basinal detritus derived from uplifted areas both east and west of the shelf basins. Thin sandstone and limestone turbidite beds are common in the lower unit of the Pilot Shale within the shelf basins.

A major hiatus has been recognized between the lower and middle units of the Pilot Shale in the eastern Great Basin (Sandberg and Poole, 1970). This widespread hiatus records an interruption in deposition on the upper Upper Devonian continental shelf in the western United States (fig. 2) that is believed to reflect regional warping concomitant with emplacement of the Roberts Mountains Allochthon. Emplacement of the allochthon onto the outer continental shelf greatly restricted the shelf seas east of the orogenic highland, and created widespread euxinic and hypersaline conditions that resulted in some euxinic and evaporitic deposits in latest Devonian time.

Figure 18 shows an outcrop of strongly deformed upper Upper Devonian radiolarian oceanic chert and argillite of the Slaven Chert that occurs in the Roberts Mountains Allochthon in central Nevada. The major source of chert, argillite, and quartzite detritus deposited in the Antler foreland basin included chert and argillite of the Devonian Slaven Chert and type Milligen Formation, chert, argillite, and quartzite of the Ordovician Valmy, Vinini, Phi Kappa, and related formations, and several Cambrian formations exposed in the Antler orogenic highland.

In southwestern Nevada and adjacent California the open-hatched area on figure 15 is a fine-grained flysch sequence probably derived from uplifts along the continental edge. In the northeastern part of the map the open-hatched area represents evaporitic deposits.

Lower Mississippian

The Lower Mississippian map (fig. 19) shows a well-developed flysch trough directly east of the Antler orogenic highland or Roberts Mountains Allochthon. East of the exogeosynclinal flysch trough or foreland trough carbonate sand, silt, and mud deposits, all of the shallow-water type, covered the shelf and platform including many former emergent areas. The eastern margin of the allochthon was overlapped by Lower Mississippian detrital sediments. Sedimentological data on Lower Mississippian deposits west of the Antler highland indicate a marginal

FIG. 14.—Slope-forming lower unit of Pilot Shale (Dpl) in Red Hills, northeastern Nevada, represents shaly limy siltstone deposits in an incipient basin on the carbonate shelf. Guilmette Limestone (Dg) ledges in lower foreground; Joana Limestone (Mj) cliff in upper right.

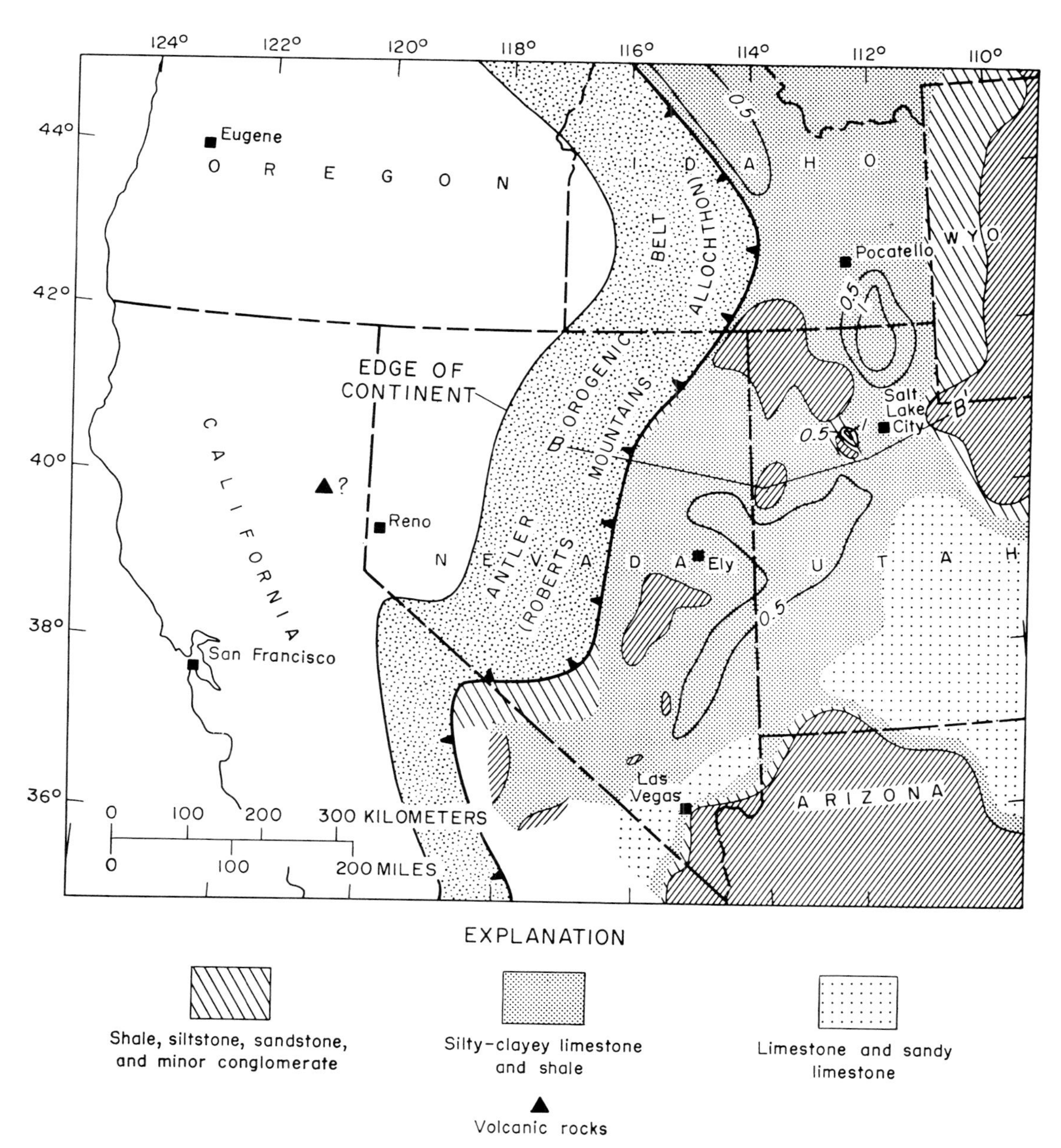

FIG. 15.—Isopach and facies map of upper Upper Devonian rocks (partly restored). Emergent areas shown by compact hatching. Isopachs in thousands of feet. Thickness and facies west of continental edge not shown owing to insufficient stratigraphic data. Roberts Mountains Allochthon emplaced near end of late Late Devonian time. Cross section BB′ shown by figure 25b.

FIG. 16.—Upper Upper Devonian carbonate-clast conglomerate in Stansbury Formation, Stansbury Mountains, northwestern Utah, represents coarse detritus from local Stansbury uplift at outer edge of cratonic platform.

ocean basin environment that included some volcanism. Note the proto-Oquirrh basin outlined by the 1,000-foot isopach in north-central Utah, south of Salt Lake City (fig. 19).

Figure 20 shows an outcrop of slope-forming Lower Mississippian Milligen Formation of Sandberg and others (1967) in central Idaho that represents fine-grained flysch deposits east of the trough axis.

Figure 21 shows a typical cliffy outcrop of Lower Mississippian Joana Limestone in westernmost Utah that represents relatively pure limestone sediments deposited on the carbonate shelf between the subsiding foreland trough on the west and the stable cratonic platform on the east. The Lower Mississippian upper unit of the Pilot Shale consists of thin-bedded sandy and silty limestone and limy siltstone, which in many areas represent a depositional setting similar to that of the Upper Devonian lower unit of the Pilot. The upper unit of the Pilot is believed to contain a significant amount of detritus derived from the Antler orogenic highland.

Upper Mississippian

The Upper Mississippian map (fig. 22) shows the full development of the flysch basin east of the Antler orogenic highland. Locally very thick deposits of chert and quartzite gravels in the foreland basin indicate major uplift of the Antler mountainous terrane to the west in Late Mississippian time. The compact-stippled narrow band on the shelf (fig. 22) represents the change from dominantly limestone deposits on the east to dominantly fine-grained detrital sediments of the siliceous flysch on the west. By very late Mississippian time, the westerly derived clastic sediments filled the foreland trough and spread eastward across the limestone shelf onto the cratonic platform.

Several volcanic-rich Upper Mississippian deposits have been recognized in north-central Nevada and in northern California (fig. 22). Possibly, Mississippian formations in Nevada that contain submarine lava flows are allochthonous and were thrust eastward across the Antler orogenic belt in post-Mississippian time.

Figure 23 pictures an outcrop in north-central

FIG. 17.—Sandy limestone turbidite (ledge 1 m thick in center of picture) that is conglomeratic and contains flute casts, displaced shallow-marine fossils, and slump folds within lower unit of the Pilot Shale in Confusion Range, Utah, near Nevada State line. Note large slump fold behind tape.

FIG. 18.—Strongly deformed upper Upper Devonian radiolarian oceanic chert and mudstone of Slaven Chert in Roberts Mountains Allochthon, Toquima Range, central Nevada.

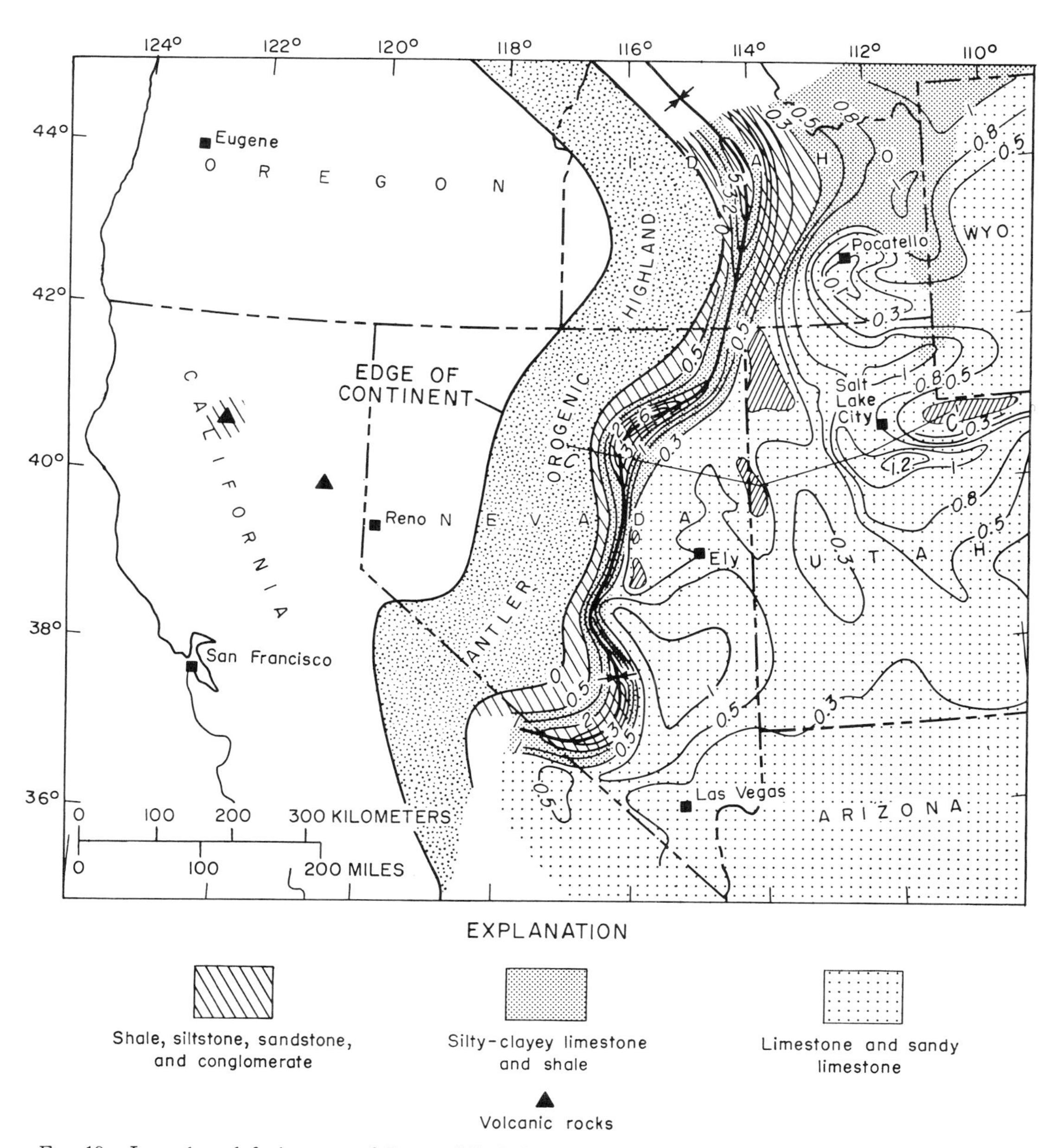

FIG. 19.—Isopach and facies map of Lower Mississippian rocks (partly restored). Emergent areas shown by compact hatching. Isopachs in thousands of feet. Thickness and facies west of continental edge not shown owing to insufficient stratigraphic data. Cross section CC′ shown by figure 25c.

FIG. 20.—Slope-forming argillite, siltite, and minor quartzite of Lower Mississippian Milligen Formation (Mm) of Sandberg and others (1967) in Lost River Range, central Idaho, represents fine-grained flysch in eastern part of foreland trough. Milligen underlain by shaly limestone of upper Upper Devonian Three Forks Formation (Dt) which directly overlies the prominent limestone cliffs of lower Upper Devonian part of the Jefferson Formation (Dj) of carbonate shelf. Milligen overlain by Upper Mississippian slope-forming and cliffy limestone units in upper right (Middle Canyon Formation, Mmc, and Scott Peak Formation, Msp). Photograph by C. A. Sandberg.

Nevada showing nearly vertical conglomerate beds of a part of the Upper Mississippian Diamond Peak Formation that are overlain with angular unconformity by steeply dipping limestones of the Pennsylvanian and Permian Strathearn Formation of Dott (1955). This Upper Mississippian conglomerate sequence was deposited along the west margin of the Antler foreland trough adjacent to the Antler highland, and is believed to represent mostly submarine upper fan deposits. The angular unconformity seen in figure 23 records medial Pennsylvanian uplift and erosion along the east margin of the Antler highland.

Figure 24 shows an outcrop of slope-forming Upper Mississippian Indian Springs Formation of Webster and Lane (1967) in southeastern Nevada that may contain a significant amount of reworked flysch sediments transported eastward across the carbonate shelf onto the platform margin which, in this picture, is represented by the underlying Upper Mississippian Battleship Wash Formation of Langenheim and Langenheim (1965). The upper contact of the Indian Springs Formation shown in figure 24 is placed at the base of a limestone-pebble conglomerate bed about 14 m (46 ft) above the upper contact selected by Webster and Lane (1967). The conglomerate bed apparently marks a regional unconformity correlative with the major unconformity at the top of the formation described by Gordon and Poole (1968) in southwestern Nevada. No turbidites have been recognized in the Indian Springs Formation in southern Nevada. Numerous *in situ* rhizomorph compressions of *Stigmaria* (Pfefferkorn, 1972) penetrate the top surface of the Battleship Wash Formation at Arrow Canyon. The occurrence of *Stigmaria,* which presumably is a rhizomorph of *Lepidodendron* and similar lycopods, indicates estuarine environmental conditions in this area in late Late Mississippian time.

ORIGIN OF ANTLER FORELAND BASIN

The four cross sections in figure 25 were constructed on an east-west line extending from north-central Nevada to north-central Utah across geosynclinal isopach and facies trends (figs. 12, 15, 19, 22). This upward chronological sequence of sections reveals several significant regional tectonic features. Major east-directed horizontal forces in the upper crust apparently caused eastward translation of axes of ridges and basins on the continental shelf during Mississippian time as depicted by the slanted dashed arrows in figure 25c and d. Figures 19, 22, and 25c-d indicate eastward expansion and translation of the Antler foreland basin, which received thick sediment accumulation, and its east-bordering complementary arch that received

FIG. 21.—Cliff-forming Lower Mississippian Joana Limestone (Mj) in Burbank Hills, westernmost Utah, represents relatively pure shallow-water dominantly bioclastic limestone deposits of carbonate shelf. Slope-forming unit between base of Joana cliff and the prominent dark sandstone ledge in middle of slope is Lower Mississippian upper unit of the Pilot Shale (Mpu). Sandstone ledge represents most of middle unit of Pilot (Dpm) and underlying slope-forming siltstones and limestones represent lower unit of Pilot (Dpl).

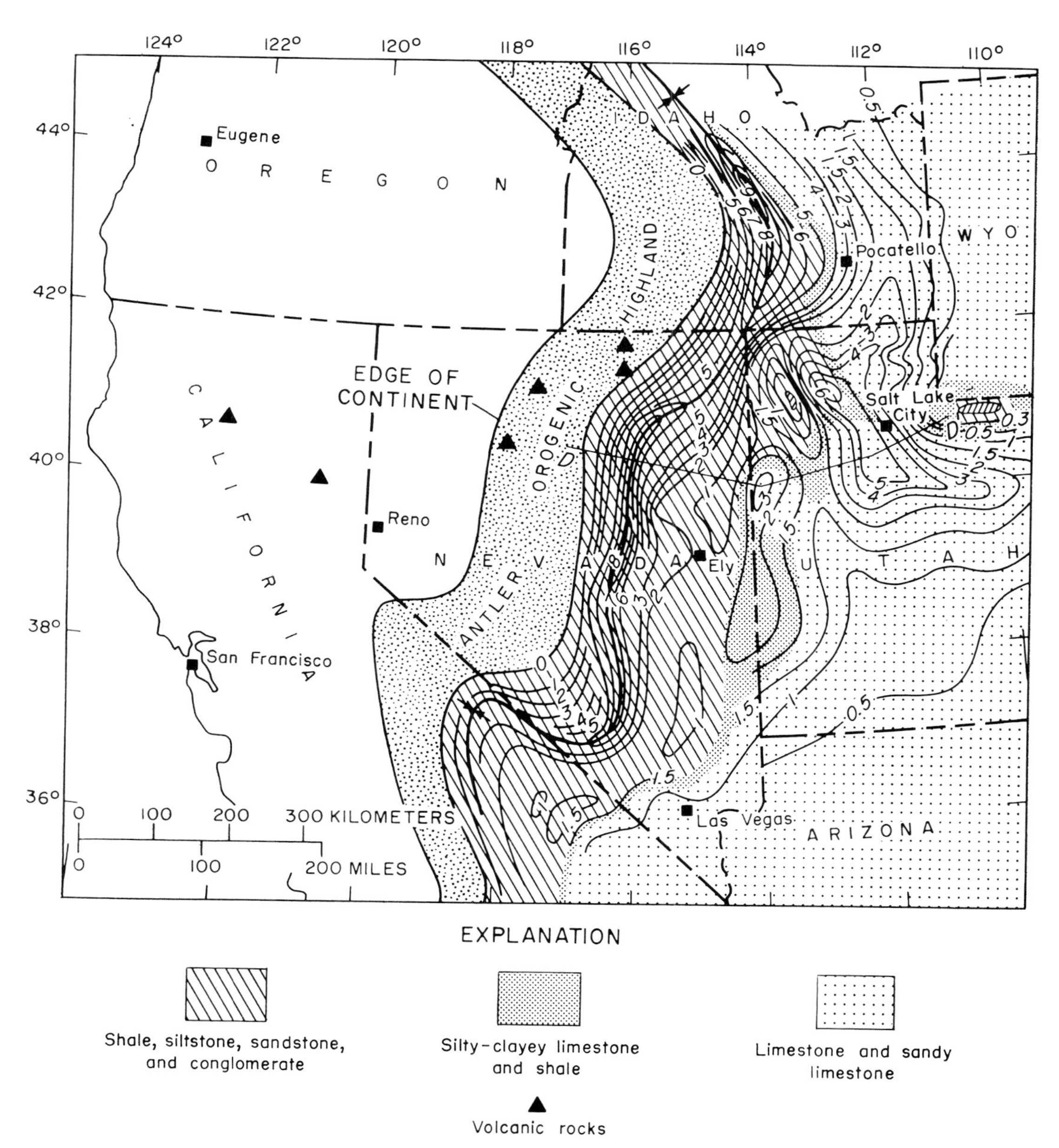

FIG. 22.—Isopach and facies map of Upper Mississippian rocks (partly restored). Emergent areas shown by compact hatching. Isopachs in thousands of feet. Thickness and facies west of continental edge not shown owing to insufficient stratigraphic data. Cross section DD′ shown by figure 25d.

FIG. 23.—Nearly vertical conglomerate beds of Upper Mississippian part of Diamond Peak Formation (Mdp) overlain with angular unconformity by steeply dipping limestones of Pennsylvanian and Permian Strathearn Formation (PIPs) in west Carlin Canyon, north-central Nevada—Upper Mississippian conglomerate represents coarse-grained flysch deposited along west margin of Antler foreland trough adjacent to the Antler highland.

thin sediment accumulation. Most likely, continental margin deformation during the Antler Orogeny warped the continental crust under the shelf and initiated sites destined for subsequent major subsidence and uplift; continued orogenic compressive stress that was directed continentward resulted in a general eastward shift in sites of thick sedimentation (fig. 25c-d) during the Carboniferous.

Plate-Tectonic Model

The geological evolution of western United States can be explained by plate-tectonic models even though many features remain obscure. Paleozoic paleogeography of the United States Cordillera seems compatible with the suggestion of Burchfiel and Davis (1972) that an offshore island arc complex above an east-dipping subduction zone was separated from the continental edge by a small ocean basin. Figure 26 shows a hypothetical and generalized plate-tectonic model which may help explain the origin of the Antler foreland basin and its flysch deposits. Thickness trends and facies patterns indicate a relatively stable Atlantic-type continental margin from late Precambrian to Early Devonian time (fig. 26a; also see Stewart and Poole, this volume).

Figure 26b shows Antler orogenic deformation and subsequent overthrust or obduction of Devonian and older oceanic sedimentary and mafic volcanic rocks from the inner arc or marginal ocean basin onto the outer continental shelf. Major deformation of the eugeosynclinal rocks occurred along the continental margin during early Antler orogenic activity, and this strong deformation is believed to predate emplacement of the Roberts Mountains Allochthon. Small slices of serpentinized ultramafic rocks are tectonically interleaved with oceanic crustal and sedimentary rocks of the allochthon (Poole and Desborough, 1973). Poole and Desborough (1973) considered these serpentinites to be fragments of early Paleozoic or late Precambrian upper mantle that were detached at the eastern Pacific margin along active subduction or obduction zones during middle Paleozoic lithospheric plate convergence. The east-directed overthrusting may have been related to compressive stress transmitted continentward from the underthrusting oceanic plate which resulted in partial closure of the marginal ocean basin (Burchfiel and Davis, 1972). This compressive stress may have warped the continental crust under the shelf and initiated development of the structural exogeosynclinal trough or foreland basin directly east of the Antler orogenic belt.

Although the model of figure 26 gives a good first-order fit with the geologic data available in the United States Cordillera, it may be somewhat oversimplified. Nevertheless, it seems evident from regional patterns of paleogeography and paleotectonics that the origin of the Upper Devonian and Mississippian foreland trough and flysch deposits in the western United States is related to the interaction of oceanic and continental plates along the Paleozoic continental margin.

FIG. 24.—Slope-forming Upper Mississippian Indian Springs Formation (Mis) at Arrow Canyon, southeastern Nevada, represents flysch sediments transported eastward across carbonate shelf onto the platform margin which is represented by underlying limestone cliff of the Upper Mississippian Battleship Wash Formation (Mbw) behind truck. Limestone cliff above Indian Springs Formation is basal part of Bird Spring Formation of Pennsylvanian and Permian age (PIPbs).

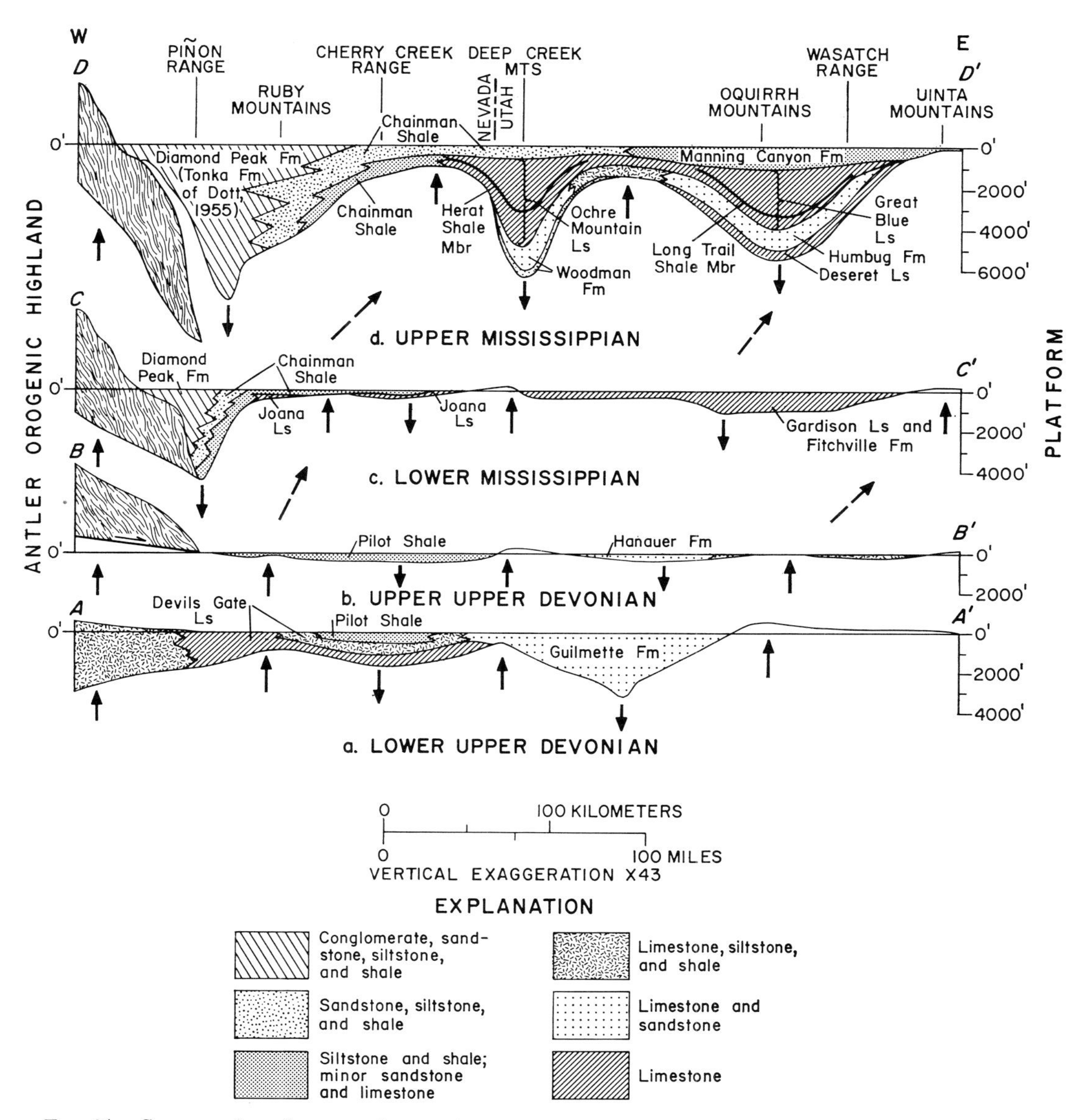

FIG. 25.—Cross sections from north-central Nevada to north-central Utah showing development of structural and depositional features on the continental shelf during the Antler Orogeny. Roberts Mountains Allochthon on left side of sections B, C, and D is depicted by compact wavy lines.

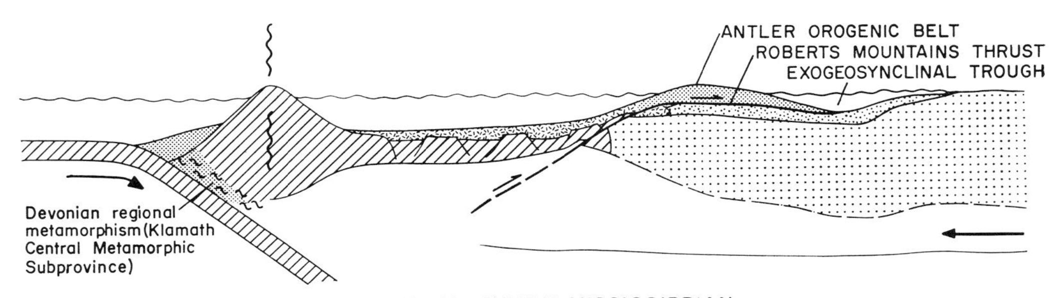

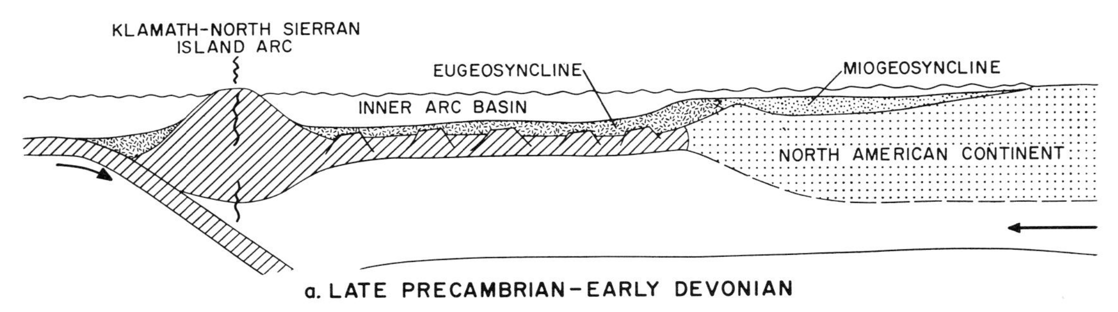

FIG. 26.—Hypothetical and generalized sketch showing relationship between early Paleozoic island arc and the North American continent based on the hypothesis of an east-dipping subduction zone (modified from Burchfiel and Davis, 1972). Continental crust in diffuse stippling; oceanic crust in hatching; upper mantle below crust is unpatterned; strongly deformed transitional and eugeosynclinal rocks in Antler orogenic belt shown by compact stippling; metamorphism in subduction zone shown as short squiggly lines. *a,* Relatively stable continental margin. *b,* Antler orogenic deformation and emplacement of Roberts Mountains Allochthon from the inner arc basin.

REFERENCES CITED

BOUMA, A. H., 1962, Sedimentology of some flysch deposits: Elsevier, Amsterdam, Netherlands, 168 p.
———, AND HOLLISTER, C. D., 1973, Deep ocean basin sedimentation, *in* Middleton, G. V., and Bouma, A. H. (chm.), Turbidites and deep water sedimentation: Pacific Sec. Soc. Econ. Paleontologists and Mineralogists, Los Angeles, p. 79–118.
BREW, D. A., 1971, Mississippian stratigraphy of the Diamond Peak area, Eureka County, Nevada: U.S. Geol. Survey Prof. Paper 661, 84 p.
BURCHFIEL, B. C., AND DAVIS, G. A., 1972, Structural framework and evolution of the southern part of the Cordilleran orogen, western United States: Am. Jour. Sci., v. 272, p. 97–118.
CROWELL, J. C., 1957, Origin of pebbly mudstones: Geol. Soc. America Bull., v. 68, p. 993–1009.
DIBBLEE, T. W., JR., 1952, Geology of the Saltdale quadrangle, California: California Div. Mines and Geology Bull. 160, 66 p.
———, 1967, Areal geology of the western Mojave Desert, California: U.S. Geol. Survey Prof. Paper 522, 153 p.
DOTT, R. H., JR., 1955, Pennsylvanian stratigraphy of Elko and northern Diamond Ranges, northeastern Nevada: Am. Assoc. Petroleum Geologists Bull., v. 39, p. 2211–2305.
EKREN, E. B., ANDERSON, R. E., ROGERS, C. L., AND NOBLE, D. C., 1971, Geology of northern Nellis Air Force Base Bombing and Gunnery Range, Nye County, Nevada: U.S. Geol. Survey Prof. Paper 651, 91 p.
GILLULY, JAMES, 1932, Geology and ore deposits of the Stockton and Fairfield quadrangles, Utah: *ibid.* 173, 171 p.
GORDON, MACKENZIE, JR., AND POOLE, F. G., 1968, Mississippian-Pennsylvanian boundary in southwestern Nevada and southeastern California, *in* Eckel, E. B. (ed.), Nevada Test Site: Geol. Soc. America Memoir 110, p. 157–168.
HAMPTON, M. A., 1972, The role of subaqueous debris flow in generating turbidity currents: Jour. Sediment. Petrology, v. 42, p. 775–793.
HANER, B. E., 1971, Morphology and sediments of Redondo submarine fan, southern California: Geol. Soc. America Bull., v. 82, p. 2413–2432.
HOSE, R. K., 1966, Devonian stratigraphy of the Confusion Range, west-central Utah: U.S. Geol. Survey Prof. Paper 550-B, p. B36–B41.

Hsü, K. J., 1970, The meaning of the word flysch—a short historical search, *in* Lajoie, J. (ed.), Flysch sedimentology in North America: Geol. Assoc. Canada Special Paper 7, p. 1–11.
Jones, D. L., Irwin, W. P., and Ovenshine, A. T., 1972, Southeastern Alaska—a displaced continental fragment?: U.S. Geol. Survey Prof. Paper 800-B, p. B211–B217.
Kay, Marshall, 1947, Geosynclinal nomenclature and the craton: Am. Assoc. Petroleum Geologists Bull., v. 31, p. 1289–1293.
——— 1951, North American geosynclines: Geol. Soc. America Memoir 48, 143 p.
Ketner, K. B., 1970, Limestone turbidite of Kinderhook age and its tectonic significance, Elko County, Nevada: U.S. Geol. Survey Prof. Paper 700-D, p. D18–D22.
Langenheim, V. A. M., and Langenheim, R. L., Jr., 1965, The Bird Spring Group, Chesterian through Wolfcampian, at Arrow Canyon, Arrow Canyon Range, Clark County, Nevada: Illinois State Acad. Sci. Trans., v. 58, p. 225–240.
Mamet, B. L., Skipp, Betty, Sando, W. J., and Mapel, W. J., 1971, Biostratigraphy of Upper Mississippian and associated Carboniferous rocks in south-central Idaho: Am. Assoc. Petroleum Geologists Bull., v. 55, p. 20–33.
McAllister, J. F., 1952, Rocks and structure of the Quartz Spring area, northern Panamint Range, California: California Div. Mines and Geology Special Rept. 25, 38 p.
Middleton, G. V., and Hampton, M. A., 1973, Sediment gravity flows: mechanics of flow and deposition, *in* Middleton, G. V., and Bouma, A. H. (chm.), Turbidites and deep water sedimentation: Pacific Sec. Soc. Econ. Paleontologists and Mineralogists, Los Angeles, p. 1–38.
Morris, H. T., and Lovering, T. S., 1961, Stratigraphy of the East Tintic Mountains, Utah: U.S. Geol. Survey Prof. Paper 361, 145 p.
Moyle, R. W., 1959, Mississippian and Pennsylvanian rocks, Manning Canyon Shale: Utah Geol. Soc. Guidebook to the Geology of Utah No. 14, p. 59–92.
Nelson, C. H., and Kulm, L. D., 1973, Submarine fans and deep-sea channels, *in* Middleton, G. V., and Bouma, A. H. (chm.), Turbidites and deep water sedimentation: Pacific Sec. Soc. Econ. Paleontologists and Mineralogists, Los Angeles, p. 39–78.
Nolan, T. B., 1935, The Gold Hill mining district, Utah: U.S. Geol. Survey Prof. Paper 177, 172 p.
———, Merriam, C. W., and Brew, D. A., 1971, Geologic map of the Eureka quadrangle, Eureka and White Pine Counties, Nevada: *ibid.*, Misc. Geol. Inv. Map I-612.
Normark, W. R., 1970, Growth patterns of deep-sea fans: Am. Assoc. Petroleum Geologists Bull., v. 54, p. 2170–2195.
Oversby, Brian, 1973, New Mississippian formation in northeastern Nevada and its possible significance: *ibid.*, v. 57, p. 1779–1783.
Paull, R. A., Wolbrink, M. A., Volkmann, R. G., and Grover, R. L., 1972, Stratigraphy of Copper Basin Group, Pioneer Mountains, south-central Idaho: *ibid.*, v. 56, p. 1370–1401.
Pfefferkorn, H. W., 1972, Distribution of *Stigmaria wedingtonensis* (Lycopsida) in the Chesterian (Upper Mississippian) of North America: Am. Midland Naturalist, v. 88, p. 225–231.
Poole, F. G., and Desborough, G. A., 1973, Alpine-type serpentinites in Nevada and their tectonic significance: Geol. Soc. America Abstracts with Programs, v. 5, p. 90–91.
———, Houser, F. N., and Orkild, P. P., 1961, Eleana Formation of Nevada Test Site and vicinity, Nevada: U.S. Geol. Survey Prof. Paper 424-D, p. D104–D111.
———, Orkild, P. P., Gordon, Mackenzie, Jr., and Duncan, Helen, 1965, Age of the Eleana Formation (Devonian and Mississippian) in the Nevada Test Site, *in* Cohee, G. V., and West, W. S. (eds.), Changes in stratigraphic nomenclature by the U.S. Geological Survey, 1964; *ibid.* Bull. 1224-A, p. A51–A53.
Raisz, Erwin, 1945, The Olympic-Wallowa lineament: Am. Jour. Sci., v. 243-A (Daly Volume), p. 479–485.
Reso, Anthony, 1963, Composite columnar section of exposed Paleozoic and Cenozoic rocks in the Pahranagat Range, Lincoln County, Nevada: Geol. Soc. America Bull., v. 74, p. 901–918.
Rigby, J. K., 1958, Geology of the Stansbury Mountains, eastern Tooele County, Utah: Utah Geol. Soc., Guidebook to the Geology of Utah No. 13, p. 1–134.
———, 1959, Stratigraphy of the southern Oquirrh Mountains, lower Paleozoic succession: *ibid.*, Guidebook to the Geology of Utah No. 14, p. 9–36.
Sandberg, C. A., Mapel, W. J., and Huddle, J. W., 1967, Age and regional significance of basal part of Milligen Formation, Lost River Range, Idaho: U.S. Geol. Survey Prof. Paper 575-C, p. C127–C131.
———, and Poole, F. G., 1970, Conodont biostratigraphy and age of West Range Limestone and Pilot Shale at Bactrian Mountain, Pahranagat Range, Nevada: Geol. Soc. America Abstracts with Programs, v. 2, p. 139.
Seilacher, Adolf, 1964, Biogenic sedimentary structures, *in* Imbrie, John, and Newell, Norman (eds.), Approaches to Paleoecology: Wiley, N.Y., p. 296–316.
———, 1967, Bathymetry of trace fossils, *in* Hallam, A. (ed.), Depth indicators in marine sedimentary environments: Marine Geology, v. 5, p. 413–428.
Schaeffer, F. E., 1960, Stratigraphy of the Silver Island Mountains: Utah Geol. Soc. Guidebook to the Geology of Utah No. 15, p. 15–113.
Skehan, J. W., 1965, A continental-oceanic crustal boundary in the Pacific northwest: Air Force Cambridge Research Laboratories Scientific Rept. 3, 52 p.
Smith, J. F., Jr., and Ketner, K. B., 1968, Devonian and Mississippian rocks and the date of the Roberts Mountains thrust in the Carlin-Piñon Range area, Nevada: U.S. Geol. Survey Bull. 1251-I, 18 p.
Smith, J. F., Jr., and Ketner, K. B., 1972, Generalized geologic map of the Carlin, Dixie Flats, Pine Valley, and Robinson Mountains quadrangles, Elko and Eureka Counties, Nevada: *ibid.* Misc. Field Studies Map MF-481.

STEWART, J. H., AND POOLE, F. G., 1974, Lower Paleozoic and uppermost Precambrian Cordilleran miogeocline, Great Basin, western United States: this volume.
STOKES, W. L., AND ARNOLD, D. E., 1958, Northern Stansbury Range and the Stansbury Formation: Utah Geol. Soc. Guidebook to the Geology of Utah No. 13, p. 135–149.
THORMAN, C. H., 1970, Metamorphosed and nonmetamorphosed Paleozoic rocks in the Wood Hills and Pequop Mountains, northeast Nevada: Geol. Soc. America Bull., v. 81, p. 2417–2447.
TROXEL, B. W., AND GUNDERSEN, J. N., 1970, Geology of the Shadow Mountains and northern part of the Shadow Mountains southeast quadrangles, western San Bernardino County, California: California Div. Mines and Geology Prelim. Rept. 12.
WALKER, R. G., 1967, Turbidite sedimentary structures and their relationship to proximal and distal depositional environments: Jour. Sediment. Petrology, v. 37, p. 25–43.
———, AND MUTTI, EMILIANO, 1973, Turbidite facies and facies associations, *in* Middleton, G. V., and Bouma, A. H. (chm.), Turbidites and deep water sedimentation: Pacific Sec. Soc. Econ. Paleontologists and Mineralogists, Los Angeles, p. 119–157.
WEBSTER, G. D., AND LANE, N. G., 1967, Mississippian-Pennsylvanian boundary in southern Nevada: Kansas Univ. Dept. Geology Special Pub. 2, p. 503–522.
WISE, D. U., 1963, An outrageous hypothesis for the tectonic pattern of the North American cordillera: Geol. Soc. America Bull., v. 74, p. 357–362.
YATES, R. G., 1964, Geologic map and sections of the Deep Creek area, Stevens and Pend Oreille Counties, Washington: U.S. Geol. Survey Misc. Geol. Inv. Map I-412.
———, 1968, The trans-Idaho discontinuity: 23d Internat. Geol. Cong. Rept., v. 1, p. 117–123.
———, 1971, Geologic map of the Northport quadrangle, Washington: U.S. Geol. Survey Misc. Geol. Inv. Map I-603.

TECTONIC CONTROL OF LATE PALEOZOIC AND EARLY MESOZOIC SEDIMENTATION NEAR THE HINGE LINE OF THE CORDILLERAN MIOGEOSYNCLINAL BELT[1]

H. J. BISSELL
Brigham Young University, Provo, Utah

ABSTRACT

Mississippian, Pennsylvanian, Permian and Lower Triassic strata in the eastern Great Basin aggregate almost 12,250 m (40,000 ft) of dominantly marine clastics and carbonates that accumulated in the eastern part of the miogeosynclinal belt of the Cordilleran geosyncline. Subsidence and hypersubsidence created depocenters in the Oquirrh, Sublett, Arcturus, Park City, Bird Spring, and other sedimentary basins within this region. Late Paleozoic (Antler and Sonoma) and early Mesozoic (ancestral Sevier) tectonism within and west of but adjacent to the miogeosynclinal belt controlled contrasting realms of clastic and carbonate sedimentation within and near these basins. Various highlands in western and northwestern Utah, in eastern and northeastern Nevada, in southern Nevada, and in orogenic belts (Antler and Sonoma) lying still farther west were stripped, in some instances into their Precambrian cores, to provide some sediment to the adjacent mobile depocenters. The Emery, Uncompahgre, and Kaibab uplifts southeast of the miogeosynclinal belt also were source areas at times. However, the continental craton to the east and northeast provided most of the sediment which makes up the abnormal thickness of clastic strata in the basins described. Sediment derived from Precambrian and younger rocks in Wyoming, Montana, and Alberta was transported southerly, then westerly, and dumped into the various depocenters.

Although the Sonoma orogeny was a major tectonic event farther west during Late Permian and Early Triassic times, the transition from Permian to Early Triassic was not marked by major deformation in the eastern part of the miogeosynclinal belt. Only a paraconformity marks the Permian-Triassic boundary at many places in the miogeosynclinal belt, although significant disconformities are present locally. Up to 1225 m (4000 ft) of Lower Triassic sediments accumulated during the waning stages of deposition, but by mid-Triassic time a major tectonic reversal occurred. The region that had been a negative mobile belt since late Precambrian time was then uplifted, whereas the region east of the Las Vegas-Wasatch hinge line became negative, and the hinge line was the fulcrum. The Cordilleran geosyncline was destroyed before the end of the Triassic, and throughout much of the Late Triassic, Jurassic, and Cretaceous some of its sediments were stripped away, to be recycled and deposited in the Rocky Mountain geosyncline to the east. The tectonic behavior of the Las Vegas-Wasatch hinge line thus controlled sedimentation over a large region throughout Paleozoic and Mesozoic time.

INTRODUCTION

The Cordilleran miogeosynclinal belt was a major downwarp of the earth's crust in what is now the eastern Great Basin of the western conterminus United States and consisted of a variety of depocenters and intrageosynclinal positive features. This dominantly negative repository of clastics and carbonates originated by at least late Precambrian time, was initially a miogeocline open to the west (Stewart and Poole, this volume), responded to various episodes of tectonism in the middle and late Paleozoic, and was phased out in about medial Triassic time. Walcott (1893, p. 346, 357) defined and illustrated the "Cordilleran sea" for Paleozoic times as one that covered an area of about 1×10^6 km^2 (4×10^5 mi^2) in the western United States. He indicated that a stratigraphic section in excess of 9000 m (30,000 ft) accumulated in this subsiding area. Willis (1909) published paleogeographic maps depicting the distribution of landmasses and epeiric seaways in this area without, however, using the name Cordilleran. Seemingly it was Schuchert (1923) who first used the name Cordilleran geosyncline; his paleogeographic maps identify the various depositional basins in which Paleozoic sediments accumulated. In more recent years Eardley (1947, 1951, p. 43) also used the term Cordilleran geosyncline for depocenters in which Paleozoic sediments accumulated to great thicknesses; he stated that this geosyncline consisted of two main troughs, a western or Pacific trough and an eastern or Rocky Mountain trough. These correspond in a general manner to what Kay (1947, p. 1291) called the Frazer

[1] The writer acknowledges contributions of many colleages, too numerous to list, who have given of their time and interpretations of the geology of this vast area. Some of the concepts presented here are the outgrowth of field and laboratory investigations made over many years of study in the eastern Great Basin. Acknowledgments are accorded the National Science Foundation for financial assistance for some field seasons. The Research Division of Brigham Young University also provided funds for certain field and laboratory expenses.

Belt and the Millard Belt, respectively. Bissell (1959, 1960, 1962a, 1964), Dott (1955), and others have recognized also that the Cordilleran geosyncline consists of a western part, the eugeosynclinal belt and an eastern part, the miogeosynclinal belt. A complex group of late Paleozoic and early Mesozoic depocenters that accumulated mainly shallow-marine sediments in the eastern part of the miogeosynclinal belt will be discussed in this paper. Accordingly, attention will be focused on the tectonic control of late Paleozoic and early Mesozoic sedimentation near the hinge line along the eastern side of the Cordilleran geosyncline.

MISSISSIPPIAN TECTONICS AND SEDIMENTATION

Although the Antler orogenic belt characterized part of central Nevada and western Idaho (Roberts and others, 1958; Roberts, 1964, 1968), this tectonic belt exercised some degree of control of patterns of sedimentation as far east as western Utah. For example, eastward-thinning tongues of chert-pebble conglomerate, grit, and sandstone of the Diamond Peak Formation are present near Wendover, Utah-Nevada, various clastics occur in the Mississippian succession near Gold Hill, Utah, and still farther south along the Utah-Nevada border there are clastic sedimentary rocks that seemingly had a western source (see Poole, this volume).

An isopach map for the total Mississippian System in eastern Nevada and western Utah reveals an interesting pattern of depocenters (fig. 1). Noteworthy was the presence of three basins with thick sediments immediately east of and parallel to the eastern flank of the Antler orogenic belt; the depoaxes of these basins parallel the trend of the orogenic belt. Brew (1971) has studied the Mississippian sediments in the southern depocenter near Eureka, Nevada, stating (p. 1): "Synorogenic clastic rocks of Mississippian age deposited in an elongate, rapidly subsiding trough east of the Antler orogenic belt in east-central Nevada consist of about 7000 ft [2100 m] of Chainman and Diamond Peak Formations." Roberts and Thomasson (1964) inferred that the Antler orogenic belt extends into southwestern Idaho. The subovate northern depocenter of Mississippian sedimentation in central Idaho accumulated a sequence of clastic and carbonate rocks in response to varying degrees of tectonism in the orogenic belt to the west. Recently, the thick succession of Lower Mississippian to Middle(?) Pennsylvanian strata, dominantly clastics, in the Pioneer Mountains of south-central Idaho has been named the Copper Basin Group (Paul and others, 1972). This thick pile (about 5500 m or 18,000 ft) of sediments probably also accumulated in response to tectonism in the Antler orogenic belt to the west. The elongate Mississippian depocenter that lies in northeastern Nevada is smaller than the two depocenters to the south and north, but nonetheless Mississippian sedimentation in it was controlled also by tectonism in the Antler orogenic belt to the west.

Madison-Brazer basin.—The depocenter that dominated the eastern part of the Cordilleran miogeosynclinal belt during Mississippian time was an asymmetric subovate basin in northern Utah and southeastern Idaho (fig. 1) in which at least 2100 m (7000 ft) of carbonates and clastic sediments accumulated. In essence, this is a composite of the Madison basin and Brazer basin (Eardley, 1947, fig. 2, p. 311), and is bounded on the east by the Las Vegas-Wasatch Line (fig. 1), a tectonic hinge that separated the discrete sedimentary basins of the Cordilleran miogeosynclinal belt from the shelf and craton to the east. The Madison-Brazer basin of northern Utah and southeastern Idaho was the forerunner of the Oquirrh basin of Pennsylvanian and Permian times. Patterns of tectonism and concomitant sediment that were established in Mississippian time were perpetuated, with various modifications, along somewhat similar lines in later Paleozoic time, even into the Early Triassic. Armstrong (1968a) presented an isopach map for the Mississippian in the Cordilleran miogeosynclinal belt suggesting also that the Oquirrh basin had originated by Mississippian time, but that it was somewhat triangular, being longer in an east-west direction than shown in figure 1.

Significant downwarp of the crust occurred as early as late Precambrian time in the area directly west of the Las Vegas-Wasatch Line in northern Utah and southeastern Idaho, but it was not until about Mississippian time that a true sedimenitary basin with an identifiable depocenter came into existence. Its depoaxis lay generally north-south parallel to the Las Vegas-Wasatch Line. Variations in tectonism along the Las Vegas-Wasatch Line to the east and in the Wendover uplift west of the Madison-Brazer basin are reflected in the thickness and lithology of the sediments in the basin. Quartz and calcite sands, shales, and a wide spectrum of carbonate types comprise the succession of strata. Similar lithologies typify the Mississippian sedimentary rocks of the area near Mackay, Idaho (Skipp, 1961). The Las Vegas-Wasatch Line was an important lineament and significant tectonic hinge line that separated contrasting realms of tectonism and sedimentation during the late Paleozoic, and into early

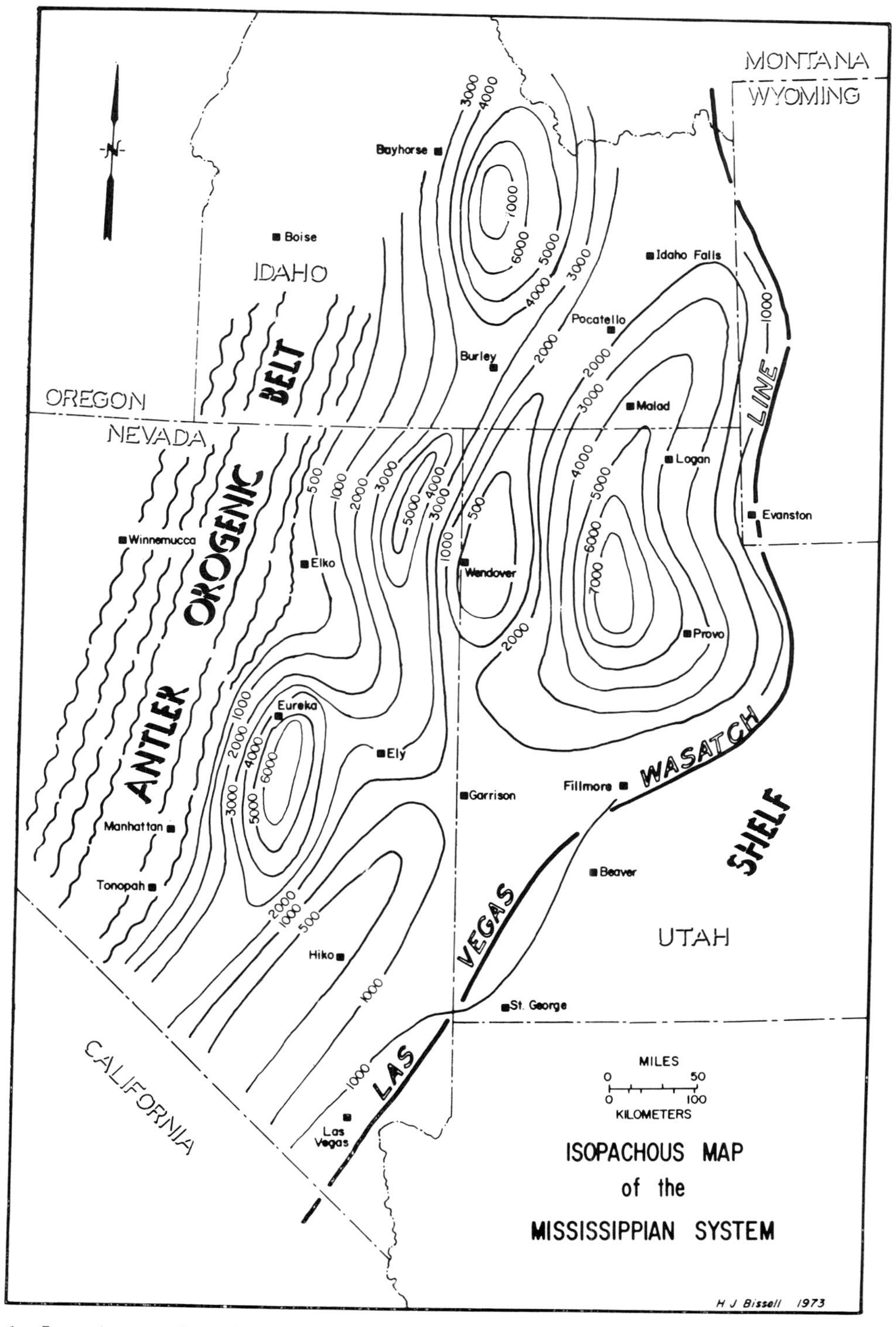

FIG. 1.—Isopach map of the Mississippian System in western Utah, eastern Nevada and adjacent Idaho. See text for discussion of depocenters.

Mesozoic times. Its significance will be stressed also for the Pennsylvanian, Permian and Early Triassic from the standpoint of tectonic control of realms of sedimentation.

PENNSYLVANIAN TECTONICS AND SEDIMENTATION

Diastrophic activity apparently continued along the Antler orogenic belt in varying degrees of intensity throughout Pennsylvanian time, but not in the form of the earlier thrust-faulting in which upper plates were translated easterly. However, clastic sediments were transported into parts of the Cordilleran miogeosynclinal belt that lay farther east. For the most part, two main depocenters accumulated most of this sediment (fig. 2). The northern one was the ovate Elko basin centered north of Elko, Nevada. The southern one was the Bird Spring basin, elongated north-south in southern Nevada. Clastics and carbonates of the Ely Group and the Strathearn Formation formed in the Elko basin; whereas shales, carbonates of many varieties, sandstones, and orthoquartzites accumulated in the Bird Spring basin. The depoaxis of the Elko basin was closer to the Antler orogenic belt than that of the Bird Spring basin. Accordingly, tectonic activity along the Antler orogenic belt to the west exerted a strong influence on the dispersal and types of Pennsylvanian sediments in the Elko basin; by contrast, the Bird Spring basin was so close to the Las Vegas-Wasatch Line that this tectonic hinge line to the east seemingly controlled sedimentation also in some measure at least.

Oquirrh basin.—The principal sedimentary basin in central and northern Utah directly adjacent to the Las Vegas-Wasatch Line was the Oquirrh basin (fig. 2). Eardley (1947, 1951) illustrated this depocenter as a hook-shaped basin that occupied central and western Utah, northeastern Nevada, and much of western Idaho for Pennsylvanian time. Stokes and Heylmun (1958) indicated that this basin was aligned northwesterly, related in some manner to an extension of the Uncompahgre uplift. Bissell (1962c) presented isopach maps for the Oquirrh basin and believed that it was confined to central and northwestern Utah (Bissell, 1970), was connected to the Ely basin of east-central Nevada across a broad sill, and connected also to the Bird Spring basin of southern Nevada. Subsequently, revisions were made in the map pattern (Bissell, 1970), and still further refinements are presented herein (fig. 2). Roberts and others (1965) illustrated the Oquirrh basin as an east-west elongated trough that extended from about the present Wasatch mountain front westerly into northeastern Nevada. Armstrong (1968a) indicated, on an isopach map for the Pennsylvanian, that the Oquirrh basin was confined largely to central and northern Utah but extended into southern Idaho. The present writer does not subscribe to the interpretation that the depoaxis of the Oquirrh basin was aligned east-west, but by contrast believes that the data obtained in measured sections favors the map pattern shown on figure 2 (see also Bissell, 1964, 1970).

The isopach map indicates that as much as 5000 m (16,000 ft) of Pennsylvanian sedimentary rocks are present in the Oquirrh basin. The depoaxis is aligned northwesterly from the Las Vegas-Wasatch Line near Provo, but veers northerly into southeastern Idaho. For the most part, the Oquirrh Formation (Morrowan through Virgilian, inclusive, and Wolfcampian) designates a thick pile of orthoquartzite, unique to this sedimentary basin, together with various types of associated limestone. Petrologic and petrographic studies of the sediments that accumulated in the Oquirrh basin refute the concept that much of the material was derived from the Antler orogenic belt. On the contrary, the quartz sand, calcite sands, and some other sediment were derived from northeasterly, easterly, and southeasterly source areas. Both the Emery and Unrompahgre uplifts (fig. 2) as well as the Weber shelf farther north, provided significant amounts of clastic and chemical sediment. There is a strong possibility that much material was derived from Montana and Wyoming to the northeast, swept southward by currents in the neritic zone and then dumped westward into the Oquirrh basin. Currents and prograding deltas that moved northwesterly from the Emery and Uncompahgre uplifts aided in diverting the southerly dispersal to induce a systematic filling of the Oquirrh basin. Stewart (1972) made a similar interpretation for the source and mode of sediment supply to the Cordilleran miogeocline during late Precambrian and Early Cambrian times. In addition, a substantial amount of the sediment that accumulated in the Oquirrh basin was derived from intrageosynclinal highs (Bissell, 1962a, 1964). Positive areas in west-central Utah, east-central Nevada, and northeastern Nevada to northwestern Utah evidently were source areas at times. The principal tectonic control for sedimentation in the Oquirrh basin, however, is believed to have been the Las Vegas-Wasatch Line (fig. 2). A restored cross section (fig. 3) that relies partly on this inference has been constructed across the Cordilleran miogeosynclinal belt from the Antler orogenic belt on the

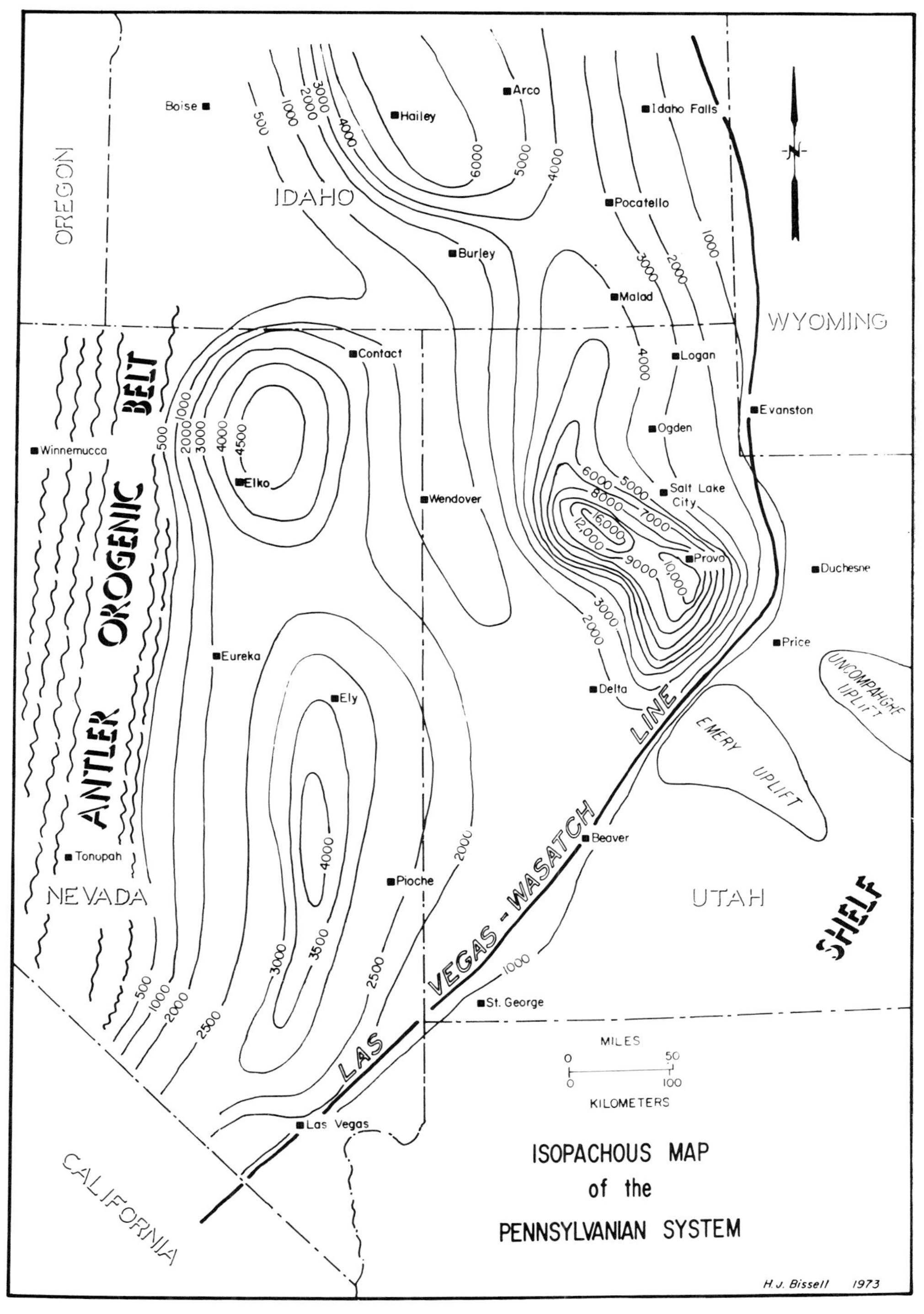

FIG. 2.—Isopach map of the Pennsylvanian System in western Utah, eastern Nevada, and adjacent Idaho. See text for discussion of depocenters, and refer to Bissell (1964, 1970) for location and thickness of the measured sections of late Paleozoic and early Mesozoic strata.

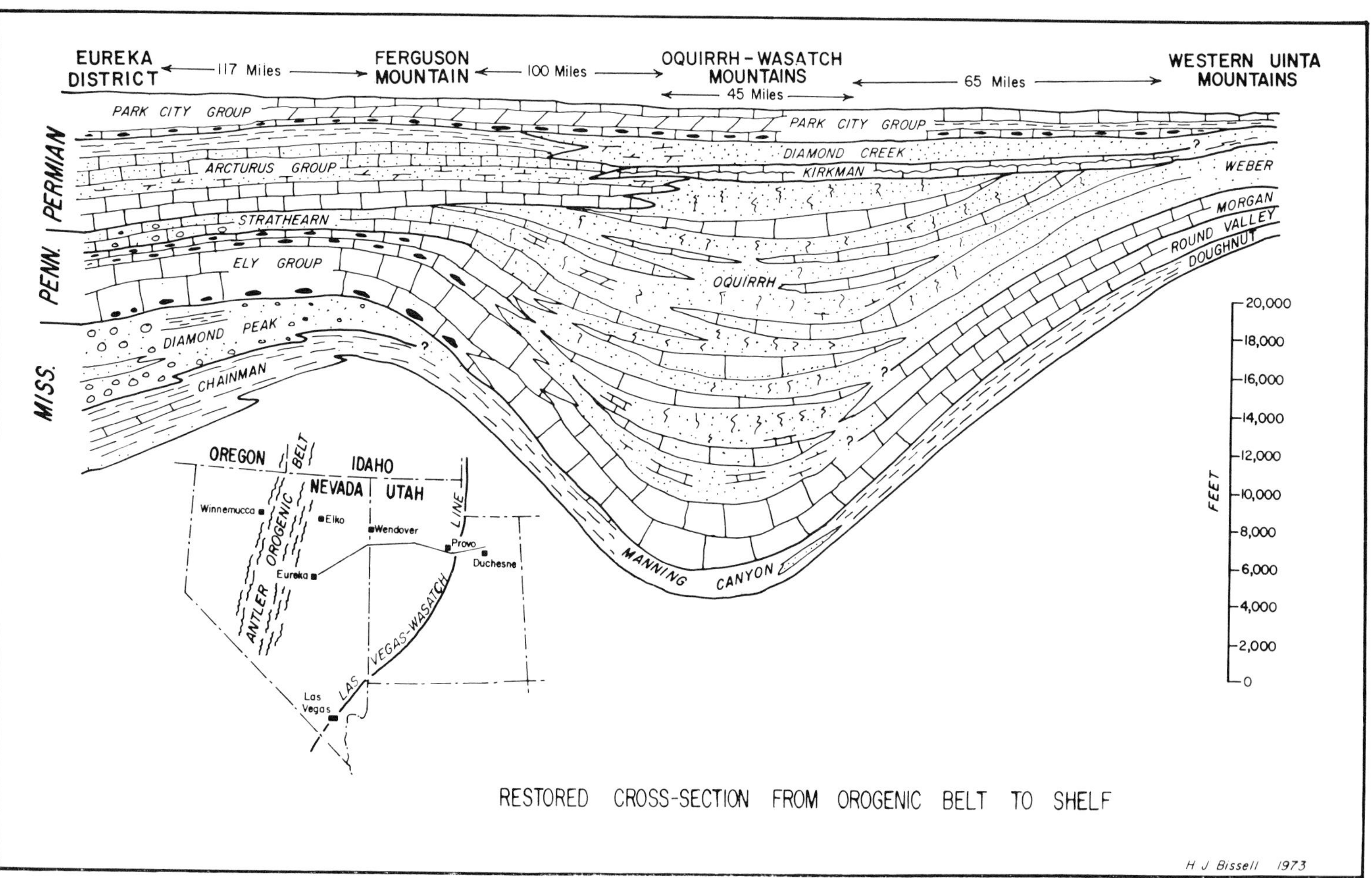

FIG. 3.—Restored cross-section of late Mississippian, Pennsylvanian, and Permian sedimentary rocks across the Cordilleran miogeosynclinal belt from the Antler orogenic belt on the west to the shelf and craton on the east.

west, across the Oquirrh basin and the hinge line to the shelf on the craton to the east.

PERMIAN TECTONICS AND SEDIMENTATION

Details of Permian tectonism and sedimentation in the Cordilleran miogeosynclinal belt are presented elsewhere (Bissell, 1970, 1972) and are not repeated here. In the light of some new information however, a new isopach map of the Permian System has been constructed outlining depocenters for eastern Nevada and western Utah, and also extending into adjacent Idaho (fig. 4). This map differs in many respects from that presented by Armstrong (1968a) for essentially the same area. The Sonoma orogenic belt, Permian successor to the Antler orogenic belt, was the dominant tectonic feature along the western continental margin.

Arcturus basin.—The Arcturus basin was the dominant depocenter in east-central Nevada, and adjacent western Utah, and northeastern Nevada. When considered from the standpoint of separate Wolfcampian, Leonardian, and Guadalupian depocenters, such names as Pequop-Diamond Creek basin, Montello-Sublett basin, and others can be applied (Bissell, 1970). For the sake of simplicity, however, the total Permian section in the composite Arcturus basin is treated as a whole here. The Sonoma orogenic belt exerted the most important tectonic control on sedimentation in the Arcturus and related basins in eastern and northeastern Nevada; seemingly, this tectonic belt extended into western Idaho where it influenced the patterns of sediment dispersal and accumulation there. Up to 3000 m (10,000 ft) of orthoquartzite, limestone, dolostone, sandstone, shale, and evaporites accumulated in these sedimentary basins. The names that identify the Permian formations are too formidable to list here, but are given elsewhere (Bissell, 1962a, 1962b, 1962c, 1964, 1970; McKee, and others, 1967).

Oquirrh-Sublett basin.—The Oquirrh-Sublett basin was a major elongate depocenter near the Las Vegas-Wasatch Line in northeastern Utah and southeastern Idaho. Roberts and others (1965) envisioned that the Oquirrh basin was elongated east-west, extending into central Utah during Permian time. This interpretation has been challenged (Bissell, 1967), and a more realistic isopach map of the Permian sedimentary rocks was constructed (Bissell, 1970). Recently, Cramer (1971) proposed the new name "Sublett basin" to identify the depocenter in south-central Idaho in which more than 6750 m (22,000 ft) of Permian sedimentary rocks accumulated. Accordingly, the isopach map for the entire Permian System presented here includes the Oquirrh basin of central and northwestern Utah and its continuation into southern Idaho as the Sublett basin. The tectonic control of sedimentation in this depocenter elongated north-south was not the Sonoma orogenic belt far to the west, but rather the nearby Las Vegas-Wasatch Line bounding the shelf and craton to the east. In addition, certain intrageosynclinal positive features reflected by the presence of Permian highlands in western Utah, eastern Nevada, northeastern Nevada, and northwestern Utah (Bissell, 1960, 1962a, 1970; Steele, 1960; and others) were the sources of much clastic and perhaps some chemical sediment to the western part of the Oquirrh-Sublett basin, and perhaps also to the eastern margin of the Arcturus basin. Yochelson and Fraser (1973) have pointed out that a major structural high was present within and adjacent to the present Pequop Mountains of northeastern Nevada, and emphasized that this high affected the stratigraphy of Pennsylvanian and lower Permian beds and had a lesser influence on later Permian sedimentation. Maps constructed by Brill (1963) demonstrate that a southwest-trending arch in southwestern Utah and adjacent Nevada affected sedimentation in that region. In emphasizing this latter fact, Armstrong (1968a) pointed out that the inferred arch coincides with the belt of later Mesozoic thrusting. He stated (1968a, p. 37): "There is a lack of data pertaining to later times, but the arch may be a precursor of later deformation in the area. It would be an ancestral Sevier arch." Accordingly, Armstrong placed this arch, with query, on his isopach map for the Permian (1968a, fig. 10, p. 36).

The Oquirrh-Sublett basin was slightly more than 500 km (300 mi) long and measured about 150 km (100 mi) in width at the widest point (fig. 4). Its southern lobe in the present Wasatch Mountains near Provo, Utah received 4000 m (13,000 ft) of orthoquartzite, limestone, sandstone, and minor shale and phosphorite. Similar lithic types characterized the Sublett portion of the trough, although orthoquartzites dominate the sequence there. The relationships of the formations in the southeastern area to those of northwestern Utah have been presented elsewhere (Bissell, 1970, 1972) and need not be repeated here. Some source areas of sediment for the Oquirrh-Sublett basin lay along the Las Vegas-Wasatch Line and the shelf and craton east of it, as well as in the nearby Emery and Uncompahgre uplifts southeast of the southern part of the basin. However, the Weber shelf and the Montana-Wyoming shelf farther north were also important source

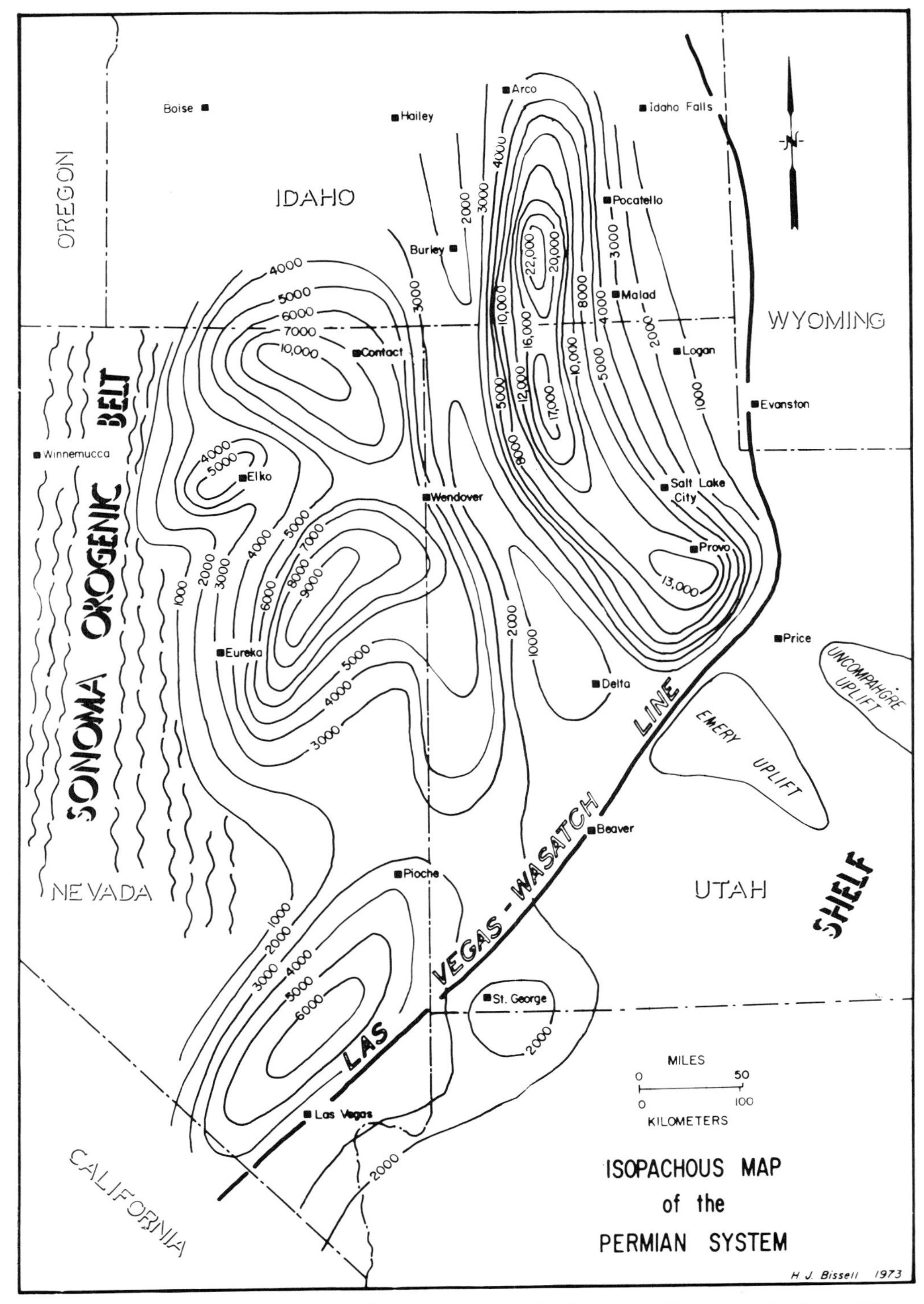

FIG. 4.—Isopach map of the Permian System in western Utah, eastern Nevada, and adjacent Idaho. See text for discussion of depocenters.

areas; sediment derived from positive features beyond them was swept southerly along the neritic zone, then transported westerly and dumped into the subsiding basin.

The Oquirrh-Sublett basin was the site of the deepest downwarp of crust within the Cordilleran miogeosynclinal belt. It is a classic example of *hypersubsidence,* a concept introduced by Kamen-Kaye (1967) for such negative repositories of sediment. His definition is repeated here (1967, p. 1838): "*Hypersubsidence:* A phase of subsidence which is attained when the algebraic sum of terms in the epeirogenic cycle is equivalent to a final basement depression of 15,000 ft [4500 m], an amount which may be definitive in the quantitative classification of major epeirogenic sedimentary basins." If the data presented here for the isopach maps of the Mississippian, Pennsylvanian, and Permian are summed for the portions of the Madison-Brazer and Oquirrh-Sublett basins (figs. 1, 2, 4 inclusive) in central and northwestern Utah just west of the Las Vegas-Wasatch Line (a tectonic hinge), the result is 12,200 m (40,000 ft). Separate totals are 2100 m (7000 ft) for the Mississippian, 4900 m (16,000 ft) for the Pennsylvanian, and 5300 m (17,000 ft) for the Permian. No one complete section of Mississippian-Pennsylvanian-Permian strata amounts to the full thickness at any one locality, but there are areas where more than 4500 m (15,000 ft) of sedimentary rocks are present. In addition, it now appears that the Oquirrh-Sublett basin contains one of the finest suites of orthoquartzites and calcarenaceous orthoquartzites of which we have record. In fact, this lithologic assemblage is characteristic of this depocenter.

Field studies are underway by various geologists in an effort to determine the paleobathymetry during late Paleozoic time for the Oquirrh-Sublett basin. Ichnology, the overall study of traces made by organisms, seemingly is proving of value in appraising fossil traces (ichnofossils) in the late Paleozoic sedimentary rocks. If interpreted correctly, these fossil traces may enable the geologist to demonstrate that at times and at places the Oquirrh-Sublett basin, and perhaps some other sedimentary basins of the late Paleozoic Cordilleran miogeosynclinal belt, had water depths in excess of 100 fathoms (Chamberlain and Clark, 1973). Sediment that was swept from the craton across the shelf east of the Las Vegas-Wasatch Line may have poured into relatively deep water in the depocenters farther west. This is an oversimplification, however, for sedimentary structures and fossil assemblages suggest that many units of Mississippian, Pennsylvanian, and Permian age in these basins evidently accumulated in peritidal environments. Additional discussions pertaining to this problem, and germane to an appraisal of tectonism and sedimentation of Permian sedimentary rocks in the Cordilleran miogeosynclinal belt, are provided by Hose and Repenning (1959), Stevens (1965, 1966, 1967), and Zabriskie (1970).

Bird Spring basin.—During Permian time the Bird Spring basin was the dominant depocenter in southern Nevada, and its depoaxis was aligned northeasterly parallel to the Las Vegas-Wasatch Line. Various types of limestone and dolostone, sandstone, orthoquartzite, and minor shale accumulated to a thickness in excess of 1800 m (6000 ft). This sedimentary basin was "sandwiched" between the southern part of the Sonoma orogenic belt on the west, and the Las Vegas-Wasatch tectonic hinge line on the east. Both these tectonic elements controlled sedimentation in the Bird Spring basin. It has been pointed out that in this part of the Cordilleran miogeosynclinal belt it is possible to reconstruct the shelf-to-basin transition in Pennsylvanian, Permian, and Lower Triassic sedimentary rocks (Bissell and Chilingar, 1968; Bissell, 1969). When traced from the shelf or platform area westerly across the hinge line into the Bird Spring basin, marine units expand at the expense of some non-marine units, and in the basin most formations are of marine origin. Figure 5 is a restored cross-section which depicts the nature of this shelf-to-basin transition for the late Paleozoic and early Mesozoic sedimentary rocks in southern Nevada. The craton and shelf areas were the source areas for much of the sediment that was dumped into the Bird Spring basin, although the southern part of the Sonoma orogenic belt on the west must have also contributed copious amounts of sediment to this sedimentary basin.

The Keystone thrust shown on Figure 5 separates some contrasting sedimentary types, mostly in the Leonardian-age Permian; it has been suggested that the upper plate, consisting of Paleozoic and Lower Triassic basinal facies, has been moved about 40 km (25 mi) from a westerly source to its present position (Longwell and others, 1965). After studying Figure 5, one might gain a different interpretation, namely, that the telescoping of basinal facies easterly upon transitional facies could have involved a shorter distance. Regardless, it demonstrates that a westerly thickening of upper Paleozoic and Lower Triassic sedimentary rocks occurred.

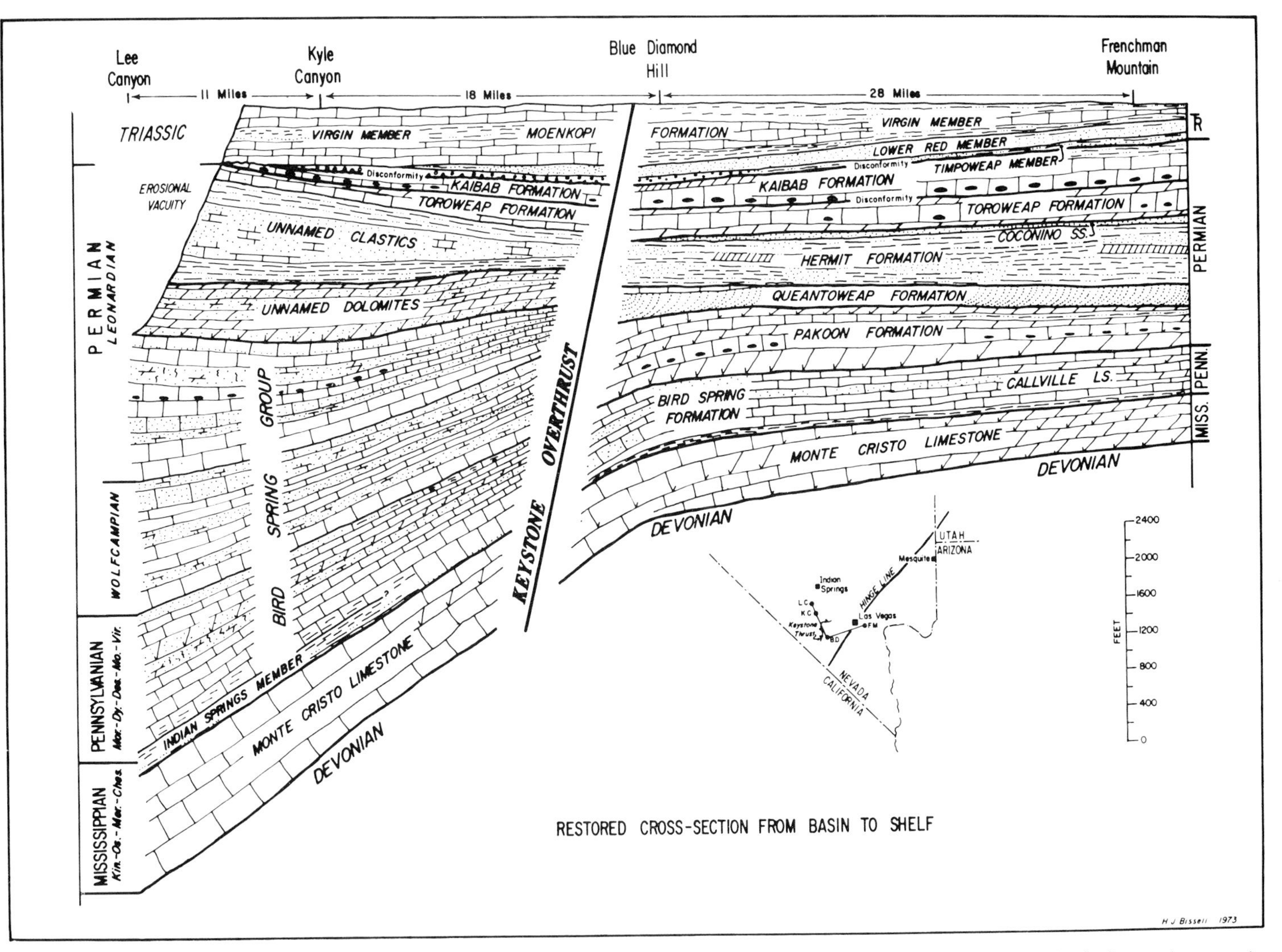

FIG. 5.—Restored cross-section of late Paleozoic and early Mesozoic sedimentary rocks from the shelf on the east to the basin on the west in the southern part of the Cordilleran miogeosynclinal belt.

EARLY TRIASSIC TECTONICS AND SEDIMENTATION

The nature of the Permian-Triassic boundary in the region under discussion has been treated elsewhere (Bissell, 1972); it is marked by paraconformity generally, but by disconformities locally. An isopach map (fig. 6) constructed for Lower Triassic sedimentary rocks in the Cordilleran miogeosynclinal belt shows the Sonoma orogenic belt as a tectonic welt along the western margin of North America; the Koipato Formation accumulated there during Triassic time (Silberling, 1971). Tectonism in the southern part of the Sonoma orogenic belt controlled patterns of sediment dispersal and accumulation along the western part of the basin in southern Nevada, and the tectonic hinge line and shelf to the east also profoundly affected the types and thicknesses of Lower Triassic sediments in the eastern part. Carbonates of the Virgin Member of the Moenkopi Formation have been studied in detail to further document the interpretation of shelf and basinal sedimentation (Bissell, 1970). Uplift of the eastern part of the Sonoma orogenic belt occurred during Early Triassic time, as shown by two lines of evidence: (1) chert-pebble and cobble conglomerates are present in the Virgin Member of the Moenkopi in the Spring Mountains; and (2) cobble and boulder conglomerates together with local erosional relief are commonly present along the Permian-Triassic contact. In places in the Spring Mountains, channels at least 30 m (100 ft) deep were eroded into and through the Permian Kaibab and Toroweap Formations, and even locally into the underlying unnamed redbeds of Permian age (Bissell, 1969). In commenting on this relationship, Armstrong (1968a, p. 37) stated: "At least 1400 ft [425 m] of Permian strata are missing below the disconformity in westernmost exposures of the Spring Mountains perhaps as a result of erosion on the ancestral Sevier arch." As the Sonoma orogeny died out, the ancestral Sevier arch was apparently initiated, and epeirogenic uplift moved progressively eastward across the Cordilleran miogeosynclinal belt.

During Early Triassic time the Thaynes and other formations accumulated in a broad ovate depocenter in northeastern Nevada, northeastern Nevada, northwestern Utah, and southeastern Idaho. The axis of the depocenter was aligned generally north-south, and maximum subsidence accounted for about 1225 m (4000 ft) of interbedded carbonates and clastics. This depocenter, along with the one farther south in southern Nevada, represented the culmination of marine (and some nonmarine) sedimentation in the Cordilleran miogeosynclinal belt.

The Sonoma orogenic belt was the source of much clastic sediment in the Cordilleran miogeosynclinal belt during Early Triassic time. Some geologists have not adequately distinguished the effects of the Sonoma orogenic belt from those of the Antler orogenic belt; for example, in discussing the Mesozoic history for western Nevada, Silberling and Roberts (1962, p. 37) stated: "Throughout Middle and Late Triassic time and probably during the Early Jurassic as well, the Antler orogenic belt was intermittently a local source of clastic debris for the flanking areas. Nevertheless, the major portion of the fine-grained terrigenous sediment deposited in the early Mesozoic seas of western Nevada must have been transported across the beveled orogenic belt from the continental area farther east and southeast." An examination of the isopach map for the Lower Triassic sedimentary rocks suggests that the major source of fine-grained terrigenous sediment at that time also was the continental area farther east. Again, the Montana-Wyoming shelf and craton east of the geosyncline was an important provenance, and the area now occupied by the Colorado Plateau was another important highland that contributed detritus to the miogeosynclinal belt.

MID-TRIASSIC TECTONIC TRANSITION

Epeirogenic upwarping after culmination of Early Triassic sedimentation effected withdrawal of marine waters from the miogeosynclinal belt, although sedimentation in the eugeosynclinal belt of western Nevada and adjacent areas continued throughout the remainder of the Triassic and into the Jurassic. Stanley, Jordan, and Dott (1971) presented a modified interpretation of the paleogeography of the western United States in Late Triassic and Early Jurassic time. They envision a shallow sea across the Cordilleran miogeosynclinal belt from the eugeosynclinal belt west of it. The writer favors the viewpoint of Armstrong (1968a, p. 38) who said: "By Middle Jurassic time, the area [miogeosyclinal belt] was undergoing orogenic deformation and supplying clastic sediments to depositional basins on the Colorado Plateau . . . Therefore, it is unlikely that the Cordilleran geosyncline [miogeosynclinal belt] was an area of sediment accumulation for any great length of time after the end of the Triassic." Armstrong (1968a, p. 38) added this note: "The youngest deposits of the Cordilleran geosyncline are of Triassic and possibly Early Jurassic age. About the middle of the Triassic Period, marine waters withdrew from the Nevada portion of the eastern Great Basin for

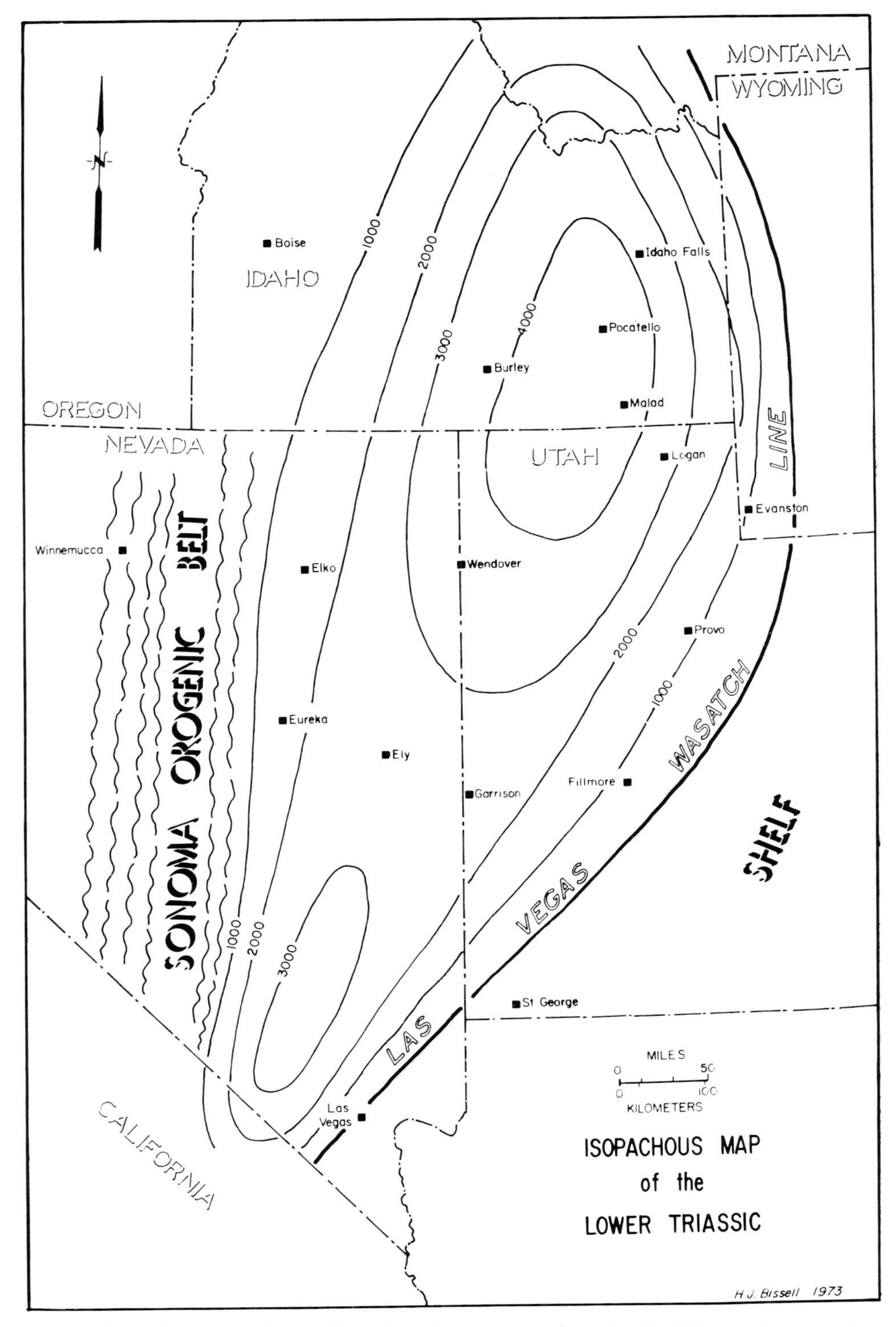

FIG. 6.—Isopach map for Lower Triassic sedimentary rocks of the Cordilleran miogeosynclinal belt in parts of Utah, Nevada, and Idaho.

the last time. This marked the beginning of the orogenic chapter in the history of this region [the miogeosynclinal belt]."

It thus appears from stratigraphic data that the Cordilleran geosyncline "closed its doors" at about mid-Triassic time when the areally extensive blanket of the Shinarump Conglomerate (basal member of the Chinle Formation) was spread discontinuously throughout parts of the eastern Great Basin as well as in the Colorado Plateau region and other areas of the shelf and craton. Orogenic deformation referred to by Armstrong (1968a, 1968b) effectively destroyed the Cordilleran miogeosynclinal belt as a repository of sediment accumulation, uplifted the area above ocean waters, and instituted an easterly drainage system towards the newly forming Rocky Mountain geosyncline. The Cordilleran miogeosynclinal belt had been the site of almost continuous deposition throughout late Precambrian times, all of the Paleozoic, and into early Mesozoic time. This deposition had begun in the miogeoclinal setting suggested by Stewart (1972) for the continental margin in late Precambrian and early Paleozoic times, and had continued during the Antler and Sonoma orogenies farther west. Throughout, the Las Vegas-Wasatch Line was perhaps the most important lineament exerting tectonic control, and acted as the hinge line between the Cordilleran geosyncline and the craton to the east. It is true that at times the transition from the miogeosynclinal belt to the craton was not sharply defined, a point stressed by Armstrong (1968a, p. 39). Sediments which accumulated on the platform adjacent to the miogeoclinal belt on the east are proportionately much thinner than rocks of comparable age to the west, and contain numerous unconformities that are absent in the miogeosynclinal belt. Sediment derived from the craton was transported southerly, from as far as Alberta, and thence westerly toward the continental margin. The eugeosynclinal belt to the west was characterized by sedimentation in deeper water, volcanic activity, and intermittent orogenesis. The Antler, Sonoma, and ancestral Sevier orogenic belts formed just to the west near the continental margin. Such an interpretation for the evolution of the Cordilleran region accords rather well with the explanation of the Appalachian and other orogens (Dietz, 1963, 1966).

Beginning in about mid-Triassic time, the Las Vegas-Wasatch Line behaved like a fulcrum; what had been the shelf and craton subsided, whereas the area formerly occupied by the geosyncline was uplifted to become a provenance for sediments of the Rocky Mountain geosyncline to the east during Mesozoic time. Crosby (1968) discussed the Paleozoic and Mesozoic miogeosynclinal sedimentation and later vertical uplift in terms of isostatic adjustment. It has been pointed out herein that various slow vertical uplifts developed irregularly in time and space as intrageosynclinal positive features adjacent to sinking depocenters. Armstrong (1968a, p. 39) remarked: "No complete explanation for the slow gentle vertical movements in the miogeosyncline has been proposed. The positive and negative paleogeographic features are ephemeral and, apparently, unrelated to igneous activity or to the structure of the underlying Precombrian basement. Consequently, most conceivable mechanisms, particularly those which might arise from within, or might be due to some characteristic of the crust itself are eliminated."

In the light of the plate tectonic model, however, perhaps a causal mechanism does exist in the upper mantle and crust. Assuming that the Las Vegas-Wasatch Line was a significant tectonic hinge along the western continental margin of the United States during the long interval from the late Precambrian, through the Paleozoic, and into the early Mesozoic, and that uncouplings of the North American plate from other plates in the Pacific region occurred at times, perhaps plate-tectonic theory offers promising insights. As Pacific plates to the west were subducted beneath North America to trigger the Antler, Sonoma, and Sevier orogenies with attendant igneous activity, waves of diastrophism spread eastward across the Cordilleran miogeosynclinal belt. At times and places during development of the Antler, Sonoma, and Sevier orogenic belts, obduction also occurred and upper plates from the west were displaced easterly on the former miogeocline and later on the shelf. In middle and late Mesozoic time, the Las Vegas-Wasatch Line again separated two contrasting tectonic elements, except that the area east of it then became the negative element as the area west of it was uplifted.

REFERENCES CITED

ARMSTRONG, R. L., 1968a, The Cordilleran miogeosyncline in Nevada and Utah: Utah Geol. and Mineralog. Survey Bull. 78, 58 p.

———, 1968b, Sevier orogenic belt in Nevada and Utah: Geol. Soc. America Bull., v. 79, p. 429–458.

BISSELL, H. J., 1959, Silica in sediments of the upper Paleozoic of the Cordilleran area, *in* Ireland, H. A.

(ed.), Silica in sediments—a symposium: Soc. Econ. Paleontologists and Mineralogists Special Pub. 7, p. 150–185.
———, 1960, Eastern Great Basin Permo-Pennsylvanian strata—preliminary statement: Am. Assoc. Petroleum Geologists Bull., v. 44, p. 1424–1435.
———, 1962a, Pennsylvanian and Permian rocks of Cordilleran area, *in* Pennsylvanian System in the United States—a symposium: Am. Assoc. Petroleum Geologists, Tulsa, p. 188–262.
———, 1962b, Permian rocks of parts of Nevada, Utah, and Idaho: Geol. Soc. America Bull., v. 73, p. 1083–1100.
———, 1962c, Pennsylvanian-Permian Oquirrh basin of Utah, *in* Geology of the southern Wasatch Mountains and vicinity, Utah: Brigham Young Univ. Geology Studies, v. 9, p. 26–49.
———, 1964, Ely, Arcturus, and Park City Groups (Pennsylvanian-Permian) in eastern Nevada and western Utah: Am. Assoc. Petroleum Geologists Bull., v. 48, p. 565–636.
———, 1967, Pennsylvanian and Permian basins in northwestern Utah—discussion: *ibid.,* v. 51, p. 791–802.
———, 1969, Permian and Lower Triassic transition from the shelf to basin (Grand Canyon, Arizona, to Spring Mountains, Nevada), *in* Baars, D. L. (ed.), Geology and natural history of the Grand Canyon region: Four Corners Geol. Soc. Guidebook, p. 135–169.
———, 1970, Realms of Permian tectonism and sedimentation in western Utah and eastern Nevada: Am. Assoc. Petroleum Geologists Bull., v. 54, p. 285–312.
———, 1972, Permian-Triassic boundary in the eastern Great Basin area: Bull. Can. Petroleum Geology, v. 20, p. 700–726.
——— AND CHILINGAR, G. V., 1968, Shelf-to-basin Permian sediments of southern Nevada, U.S.A.: 23rd Internat. Geol. Cong. Proc., Sec. 8, p. 155–167.
BREW, D. A., 1971, Mississippian stratigraphy of the Diamond Peak area, Eureka County, Nevada: U.S. Geol. Survey Prof. Paper 661, 84 p.
BRILL, K. G., JR., 1963, Permo-Pennsylvanian stratigraphy of western Colorado Plateau and eastern Great Basin regions: Geol. Soc. America Bull., v. 74, p. 307–330.
CHAMBERLAIN, C. K. AND CLARK, D. L., 1973, Trace fossils and conodonts as evidence for deep-water deposits in the Oquirrh Basin of central Utah: Jour. Paleontology, v. 47, p. 663–682.
CRAMER, H. R., 1971, Permian rocks from Sublett Range, southern Idaho: Am. Assoc. Petroleum Geologists Bull., v. 55, p. 1787–1801.
CROSBY, G. W., 1968, Vertical movements and isostasy in western Wyoming overthrust belt: *ibid.,* v. 52, p. 2000–2015.
DIETZ, R. S., 1963, Collapsing continental rises: an actualistic concept of geosynclines and mountain building: Jour. Geology, v. 71, p. 314–333.
———, AND HOLDEN, J. C., 1966, Miogeoclines (miogeosynclines) in space and time: *ibid.,* v. 74, p. 566–583.
DOTT, R. H., JR., 1955, Pennsylvanian stratigraphy of Elko and northern Diamond Ranges, northeastern Nevada: Am. Assoc. Petroleum Geologists Bull., v. 39, p. 2211–2305.
EARDLEY, A. J., 1947, Paleozoic Cordilleran geosyncline and related orogeny: Jour. Geology, v. 55, p. 309–342.
———, 1951, Structural geology of North America: Harpers, N.Y., 624 p.
HOSE, R. K., AND REPENNING, C. A., 1959, Stratigraphy of Pennsylvanian, Permian and Lower Triassic rocks of Confusion Range, west-central Utah: Am. Assoc. Petroleum Geologists Bull., v. 43, p. 2167–2196.
KAMEN-KAYE, MAURICE, 1967, Basin subsidence and hypersubsidence: *ibid.,* v. 51, p. 1833–1842.
KAY, MARSHALL, 1947, Geosynclinal nomenclature and the craton: *ibid.,* v. 31, p. 1289–1293.
LONGWELL, C. R., PAMPEYAN, E. H., BOWYER, BEN, AND ROBERTS, R. J., 1965, Geology and mineral deposits of Clark County, Nevada: Nevada Bur. Mines Bull. 62, 218 p.
MCKEE, E. D., ORIEL, S. S., KETNER, K. B., MACLACHLAN, M. E., GOLDSMITH, J. W., MACLAGHLAN, J. C., AND MUDGE, M. R., 1959, Paleotectonic maps of the Triassic System: U.S. Geol. Survey Misc. Geol. Inv. Map I-300.
MCKEE, E. D., AND ORIEL, S. S., 1967a, Paleotectonic maps of the Permian System: *ibid.,* Map I-450.
———, 1967b, Paleotectonic investigations of the Permian System in the United States: *ibid.,* Prof. Paper 515, 271 p.
ORGILL, J. R., 1971, The Permian-Triassic unconformity and its relationship to the Moenkopi, Kaibab, and White Rim formations in and near the San Rafael Swell, Utah: Brigham Young Univ. Geology Studies, v. 18, p. 131–179.
PAULL, R. A., WOLBRINK, M. A., VOLKMANN, R. G., AND GROVER, R. L., 1972, Stratigraphy of Copper Basin Group, Pioneer Mountains, south-central Idaho: Am. Assoc. Petroleum Geologists Bull., v. 56, p. 1370–1401.
ROBERTS, R. J., 1964, Stratigraphy and structure of the Antler Peak Quadrangle, Humboldt and Lander Counties, Nevada: U.S. Geol. Survey Prof. Paper 459-A, 93 p.
———, 1968, Tectonic framework of the Great Basin: UMR (Univ. Missouri, Rolla) Jour. No. 1, p. 101–119.
———, HOTZ, R. J., GILLULY, JAMES, AND FERGUSON, H. G., 1958, Paleozoic rocks of north-central Nevada: Am. Assoc. Petroleum Geologists Bull., v. 42, p. 2813–2857.
———, CRITTENDEN, M. D., JR., TOOKER, E. W., MORRIS, H. T., HOSE, P. K., AND CHENEY, T. M., 1965, Pennsylvanian and Permian basins in northwestern Utah northeastern Nevada, and south-central Idaho: *ibid.,* v. 49, p. 1926–1956.
———, AND THOMASSON, M. R., 1964, Comparison of Late Paleozoic depositional history of northern Nevada and central Idaho: U.S. Geol. Survey Prof. Paper 475-D, p. D1–D6.
SCHUCHERT, CHARLES, 1923, Sites and nature of North American geosynclines: Geol. Soc. America Bull., v. 34, p. 151–230.

SILBERLING, N. J., 1971, Geological events during Permian-Triassic time along the Pacific margin of the United States (abs.): Internat. Permian-Triassic Conf., Calgary, Canada, p. 355.
———, AND ROBERTS, R. J., 1962, Pre-Tertiary stratigraphy and structure of northwestern Nevada: Geol. Soc. America Special Paper 72, 58 p.
SKIPP, B. A. L., 1961, Interpretation of sedimentary features in Brazer Limestone (Mississippian) near Mackay, Custer County, Idaho: Am. Assoc. Petroleum Geologists Bull., v. 45, p. 376–389.
STANLEY, L. O., JORDAN, W. M., AND DOTT, R. H., JR., 1971, New hypothesis of Early Jurassic paleogeography and sediment dispersal for western United States: *ibid.*, v. 55, p. 10–19.
STEELE, GRANT, 1960, Pennsylvanian-Permian stratigraphy of east-central Nevada and adjacent Utah, *in* Geology of east-central Nevada: Intermountain Assoc. Petroleum Geologists 11th. Ann. Field Conf. Guidebook, p. 91–113.
STEVENS, C. H., 1965, Pre-Kaibab Permian stratigraphy and history of Butte Basin, Nevada and Utah: Am. Assoc. Petroleum Geologists Bull., v. 49, p. 139–156.
———, 1966, Paleoecologic implications of Early Permian fossil communities in eastern Nevada and western Utah: Geol. Soc. America Bull., v. 77, p. 1121–1130.
———, 1967, Leonardian (Permian) compound corals of Nevada: Jour. Paleontology, v. 41, p. 423–431.
STEWART, J. H., 1972, Initial deposits in the Cordilleran geosyncline: evidence of a Late Precambrian (< 850 m.y.) continental separation: Geol. Soc. America Bull., v. 55, p. 593–620.
———, POOLE, F. G., AND WILSON, R. F., 1972, Stratigraphy and origin of the Triassic Moenkopi Formation and related strata in the Colorado Plateau Region: U.S. Geol. Survey Prof. Paper 691, 195 p.
STOKES, W. L., AND HEYLMUN, E. B., 1958, Outline of the geologic history and stratigraphy of Utah: Utah Geol. and Mineralog. Survey, 37 p.
WALCOTT, C. D., 1893, Geologic time: as indicated by the sedimentary rocks of North America: The American Geologist, v. 12, p. 343–368.
WILLIS, BAILEY, 1909, Paleogeographic maps of North America: Jour. Geology, v. 17, p. 203–208, 253–256.
YOCHELSON, E. L., AND FRASER, G. D., 1973, Interpretation of depositional environment in the Plympton Formation (Permian), southern Pequop Mountains, Nevada, from physical stratigraphy and a faunule: Jour. Research U.S. Geol. Survey, v. 1, p. 19–32.
ZABRISKIE, W. E., 1970, Petrology and petrography of Permian carbonate rocks, Arcturus Basin, Nevada and Utah: Brigham Young Univ. Geology Studies, v. 17, p. 83–160.

RELATIONSHIPS OF CRATONIC AND CONTINENTAL-MARGIN TECTONIC EPISODES[1]

L. L. SLOSS AND ROBERT C. SPEED
Northwestern University, Evanston, Illinois

ABSTRACT

The Phanerozoic history of continental cratons is marked by repeated global episodes of three types: (1) *Oscillatory*—generally elevated or oscillating with respect to sea level; marginal and submarginal areas subject to highly differentiated uplift and subsidence; periodicity of oscillations and uplifts 10^5–10^6 years; wave lengths of intracratonic tectonic elements 10^1–10^2 km; duration of episodes 10^7–10^8 years. (2) *Emergent*—progressively elevated in time; without significant topographic relief; tectonically undifferentiated below wave lengths of 10^3 km; duration 10^6–10^7 years. (3) *Submergent*—progressively depressed below sea level to form widespread epicontinental seas; sub-episodes (10^6–10^7 years) of differential subsidence to form basins and arches ($\lambda = 10^2$–10^3 km); duration 10^7–10^8 years.

The time spans of cratonic episodes were: (1) *Oscillatory*—much of the Cenozoic, including the present, and the period from Pennsylvanian to Early Jurassic (time spans of Appalachian-Hercynian, Laramide, and late Alpine orogenies); (2) *Emergent*—latest Precambrian, early Middle Ordovician, and Early Devonian (lacunal intervals between accumulations of cratonic sedimentary sequences); and (3) *Submergent*—time spans of Caledonian, Antler-Acadian, and Nevadan orogenies.

In plate-tectonic terms, the present is characterized by convergent boundaries of oceanic and continental plates relatively remote from cratonic margins. It is postulated that these were the prevailing conditions during times of *oscillatory* cratonic behavior. *Emergent* cratons, by historical analysis, appear to be related to quiescent episodes at continental margins, and possibly reflect minima in spreading rates at sea. *Submergent* cratons seems to coincide with times of active plate convergence expressed by obduction and subduction at the oceanic margins of cratons.

It is our postulate that cratonal tectonic states are related to the rates at which melt, produced by radial heat flow, is extracted from subcontinental asthenospheres. A high degree of retention of such melt beneath cratons results in asthenospheric inflation, accompanied by elevation of the continental lithosphere and cratonic emergence; expulsion of melt, or bulk flow of melt and solid, to suboceanic asthenospheres would accelerate sea-floor spreading rates while causing cratonic submergence through asthenospheric deflation below continents. Oscillatory states represent times of episodicity in asthenospheric inflation and deflation, leading to fracturing on cratonic margins and flanks, to cratonic block-faulting and volcanism, and to the development of spreading centers and interarc basins at cratonic margins.

INTRODUCTION

The term *craton* and derived adjectives (cratonic, cratonal) figure prominently in the following discussion and yet there is difficulty in framing an adequate definition. The original usage (Stille 1936; Kay 1947, 1951) defines [continental] cratons as the relatively immobile shields and surrounding regions of shallow crystalline basement circumscribed by peripheral orthogeosynclines. Emerging plate-tectonic data and concepts have placed geosynclinal theory in an as yet unresolved state of flux such that orthogeosynclines, and thus the limits of cratons, cannot easily be delineated. Further, it has become increasingly clear that cratons, rather than being "immobile," "stable," or "undeformed," have been subject to episodes of deformation of varying degree, including localized vertical displacements of several kilometers. Indeed, our thesis is largely based on analysis of the deformation history of cratons.

Nevertheless, the interiors of continents, underlain by thick "granitic" crust and tectonically stable over long spans of time, are recognizably distinct from continent-margin terranes where deformation, volcanism, and sedimentation are direct responses to plate interactions. Therefore, the term craton remains useful in identifying the sedimentary-tectonic behavior of a particular region during a particular time span, irrespective of antecedent or subsequent sedimentary-tectonic states. Kansas, in our view, has been cratonic since middle Precambrian, and the Appalachian Piedmont, since at least Permian; either area may revert to a noncratonic condition at some future time but such a change would not affect classification for the spans of time indicated. We exclude from consideration as cratonic the original sites of accumulation of "eugeosynclinal" assemblages since we believe these to be largely allochthonous oceanic, arc-trench, or interarc materials.

[1] The manuscript of this paper has been reviewed by our colleague, Norman Sleep; his comments and criticisms have been most helpful. The figures are the handiwork of Judi Wind. Much of the research has been supported by National Science Foundation Grant GA-22844.

However, we include within our cratons those "miogeosynclinal" terranes which, during those specific time spans, accumulated sediments under conditions which extended relatively unchanged from undisputed cratonic interiors. In the Ordovician of the Cordilleran region, for example, we would place the cratonic margin 200 km or more west of the "Wasatch line," to include much of the Millard miogeosynclinal belt defined by Kay (1947). With varying degrees of difficulty and uncertainty, we believe that we can identify cratons and their boundaries for the past 600–700 million years—the latest Precambrian and Phanerozoic time span from which our data and concepts are drawn.

Cratons and continental margins have shared dual roles as sediment sources and sediment traps. The relatively undeformed and unmetamorphosed late Precambrian-Phanerozoic cratonic sedimentary record is abundantly documented, by generations of outcrop observation and by hundreds of thousands of drill holes, such that acceptable models relating tectonics, sedimentation, and sediment preservation can be derived, and reasonably detailed historical reconstructions can be achieved. In spite of the wealth of data, cratonic studies have not led to a comprehensive tectonic theory applicable on a global, or even continental, scale. On the other hand, observation and acquisition of readily interpretable data are much more difficult at continental margins subject to recent or ancient plate interactions. Here, successive events tend to obscure if not erase the records of precursors; historical reconstruction is dependent on structural relationships of long time resolution or on the dating of intrusive or thermal events that are widely spaced in comparison with the small time increments identifiable for many cratonic terranes. Where some of these problems may be absent, as on long-dormant, submerged margins, there is a heavy dependence on remotely-sensed seismic, gravimetric, and magnetic data which either lack time significance or do not provide adequate temporal detail. Nevertheless, in the face of these impediments, or perhaps because of them, investigations of continental margins have led to the development and exposition of a large number of global-scale models which, in latter years, have related continental margins to plate-tectonic concepts.

Analysis of the well-documented stratigraphic record on cratons clearly reveals that here, too, there has been a complex tectonic history, largely expressed in the changing geometry, vector properties, and facies responses of preserved sediments, but including extrusive, intrusive, and metamorphic events as well as vertical movements ranging from epeirogenic to orogenic scales. Further, it is now clear that certain tectonic events and phases are synchronous on more than one craton (Sloss 1972a, 1972b); that is, episodes of cratonic submergence and basin subsidence and other episodes of uplift and fracturing (as today) are simultaneous on separate cratons, leading to the suggestion of global controls on cratonic tectonism. In this light, it is natural to consider whether there is a relationship between the tectonic history of cratons and the more vigorous history of deformation and orogeny at continental margins. One of us (Sloss, 1966) has denied any such relationship, but new syntheses and evolving concepts force reconsideration. Our approach to a fresh evaluation follows three steps: (1) classification and time placement of modes of tectonic behavior of cratons, (2) identification of continental-margin modes and their correlation with cratonic events, and (3) consideration of theoretical bases for the proposed correlation.

TECTONIC MODES OF CRATONS

The following discussion leans heavily on North American data but, as has been stated, it is implicit that several or all cratons behaved similarly and synchronously, at least at the large scale of the phenomena considered. At this gross scale we identify three cratonic modes that characterize episodes a few million to more than a hundred million years long, and recur asystematically over the span of Phanerozoic time.

Emergent Mode

Definition.—Emergent episodes are characterized by slow elevation of cratons with respect to sea level to maxima greater than a kilometer over time spans of 10^6–10^7 years followed by equally slow submergence. Emergence is maximal in degree of elevation and in duration in craton-interior shield areas, and is minimal on both counts at craton margins. During emergence, cratons are smoothly uparched without major inflections below wavelengths of 1000 to 2000 km; therefore, cratonic surfaces lack significant structural relief during emergent episodes.

Sedimentary response.—Figure 1 is our view of a portion of a craton at the climax of an emergent episode. Sediments that filled the subsiding basins of a preceding submergent episode and covered much of the cratonic interior are shown in the process of being stripped away. Strata of varying resistance to erosion are exposed, and cover rocks have been removed from

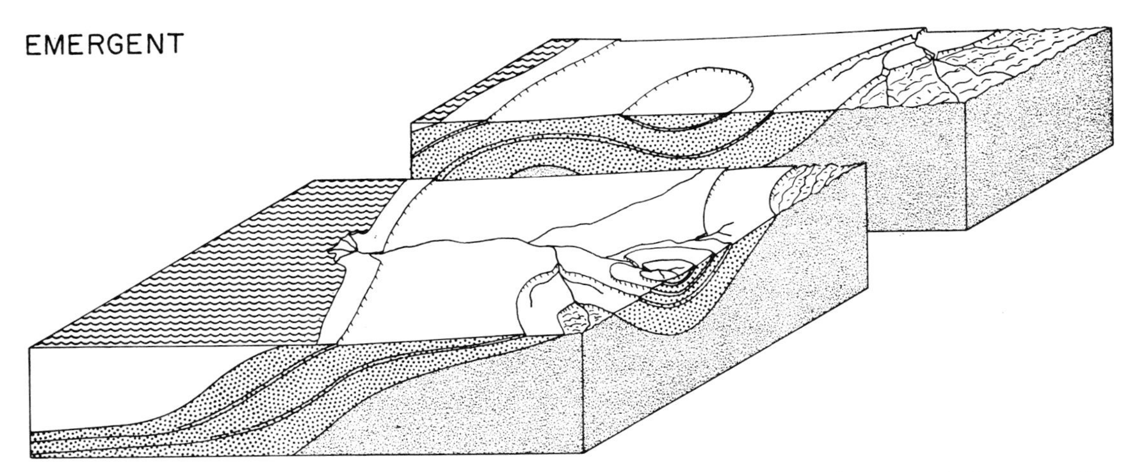

Fig. 1.—Diagrammatic view of a portion of a craton during an emergent episode; vertical exaggeration approximately 500:1. At time illustrated, craton is an uparched slab without significant local intracratonic vertical dislocations. Entire craton interior is above base level and lacks marked topographic relief except that due to differences in resistance to erosion of exposed shield rocks (right) and of sediments of preceding submergent episodes. Products of erosion are carried to cratonic margins where, in the absence of subsiding sediment traps, accumulation is limited to continental rises and abyssal plains.

shield areas. In the absence of cratonic sediment traps, the products of erosion are transported to the craton edge. As developed on later pages, we believe emergent episodes to have been times of minimal plate convergence; as a consequnce, emergent-craton margins were largely passive, lacking silled marginal or interarc basins, arc-trench systems, or other closed sediment traps. Under these postulated conditions, in combination with emergent, or at least non-subsiding, continental shelves, there would be small opportunity for the growth of major delta systems and the detrital materials delivered to the cratonic margin would be contributed to deep-water accumulations. Thus, the cratonic stratigraphic record of an emergent episode is a major craton-wide unconformity—a lacuna marked by increase of both degradational (vacuity) and non-depositional (hiatus) dimensions from margin to interior.

Inasmuch as the major process active during an emergent episode is erosion, there are no sediments of such an episode preserved in cratonic interiors. Rather, the corresponding sediments must have been deposited at cratonic margins as slope-rise accumulations and beyond on deep-sea floors where they perhaps mingled with materials of extracratonic source. Erosion of carbonates and shales from emergent cratons produces pelagic sediments that bear no source identity at their ultimate sites of deposition. Shield rocks, through extensive weathering on gentle slopes, long transport, and repeated deposition and reworking enroute to continental margins, are transported as dissolved solids, clays, quartz silt, and a sand-sized quartz fraction of high textural and mineralogic maturity. Dissolved and suspended components are dispersed to the pelagic realm; the quartzose sands join like material eroded from mature sediments of older cycles and may ultimately form apparently anomalous intercalations in extracratonic "deep-water" successions.

In sum, sediments of emergent episodes are confined to extracratonic sinks. Here, the identifiable sediments of cratonic derivation are quartzose sands which contrast with the immature sands delivered and deposited at continental margins from interior sources during oscillatory episodes as discussed in succeeding paragraphs.

Time relationships.—The time spans and chronology of Phanerozoic cratonic modes are shown graphically on figure 2. Emergent episodes are represented by the sub-Sauk, sub-Tippecanoe, and sub-Kaskaskia unconformities in North America and by equivalent stratigraphic discontinuities on other cratons. The duration of each episode of emergence is difficult to estimate because, although the close of the hiatus is recorded in the age of the first transgressive sediments of the succeeding submergence, intrahiatal erosion strips away the youngest sediments of the preceding submergent episode, stretching the stratigraphic lacuna backward in time, and obscuring the date of transition from submergence to emergence.

The greater longevity ascribed to the pre-Sauk emergent episode may be genuine but uncertainty is introduced by at least two factors: (1) Estimates of the absolute age of the earliest

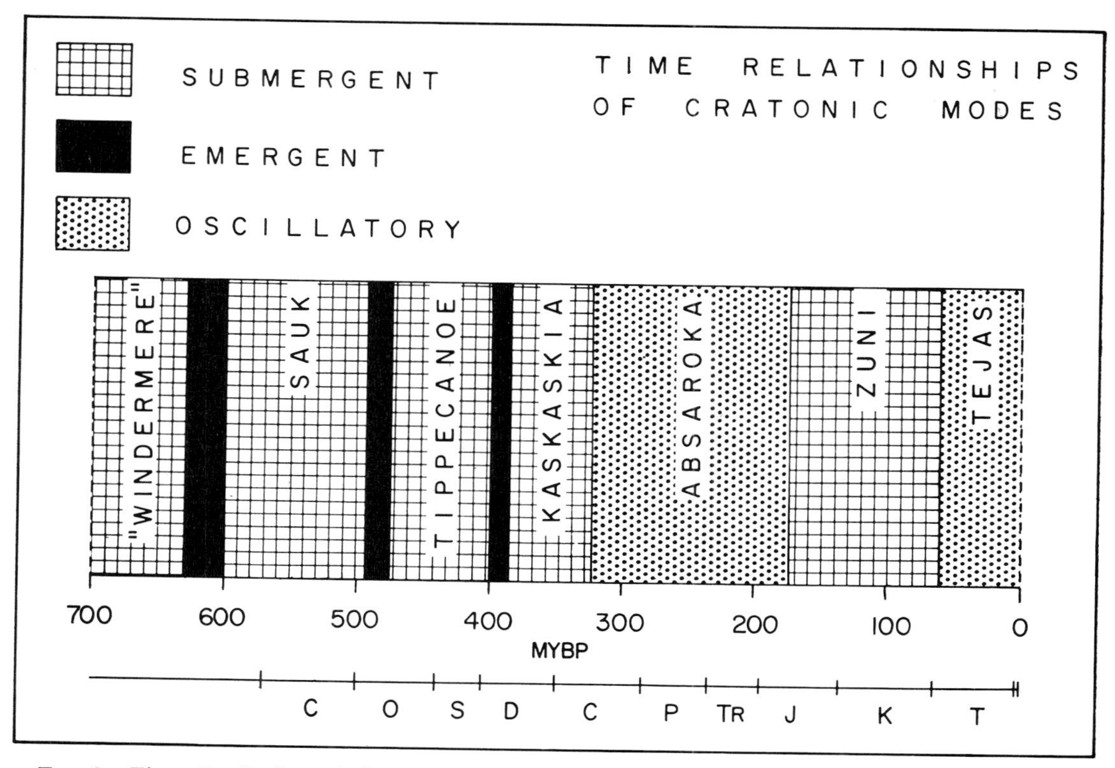

FIG. 2.—Time distribution of the three identified cratonic modes. Relationships to late Precambrian-Phanerozoic absolute time, to the Phanerozoic periods, and to the episodes of deposition of the North American stratigraphic sequences are shown.

Cambrian biostratigraphic zones cluster near 570 million years, but considerable thicknesses of strata that occur near cratonic margins below the positions of such zones, and above the bounding interregional unconformity, suggest the passage of a significant span of time; we have rather arbitrarily chosen 600 my as the time of initiation of Sauk transgression of the craton but this estimate could easily involve an error of greater than 10 million years. (2) The youngest strata preserved below the sub-Sauk unconformity, variously termed Eocambrian, Late Riphean, Adelaidean, Hadrynian, or Precambrian Z, define the maximum age that can be ascribed to the beginning of pre-Sauk emergence. These units are also imprecisely dated—the Windermere "System" of western Canada, for example, may be younger than 675 my (Douglas and others, 1970, p. 371), but how much younger is unknown. Although the two subsequent emergent episodes are bracketed by better biostratigraphic and radiometric resolution, no high degree of accuracy is claimed; however, as indicated later by Figure 5, it is clear that the duration of emergent episodes is short in comparison to the lengths of time occupied by other cratonic modes.

Cratonic uplift and the development of interregional unconformities are also characteristic of the oscillatory mode. The problem of discrimination between unconformities related to emergent and oscillatory states is more logically treated after discussion of the latter mode.

Oscillatory Mode[2]

Definition.—We define as *oscillatory* those episodes of cratonic history marked by pulsatory

[2] In earlier papers (Sloss 1964, 1966) the stratigraphic products of what are here termed oscillatory episodes were referred to "orogenic sequences," in recognition of the cratonic mountains characteristic of the episodes represented, and the term "epeirogenic" was used for what are here considered as products of submergent episodes. Many geologists use "orogeny" to cover processes responsible for structures, plutonism, and metamorphism *within* mountain chains, and especially for structures and other features thought to be related to lithospheric-plate convergences. The previous usage thus becomes a source of confusion. Hence, we substitute descriptive terms without necessary genetic implications.

vertical movements leading to general net elevation of cratons with respect to sea level and thus characterized by a sedimentary record of marine regression punctuated by repeated transgressions with a periodicity ranging from less than a million to five million years. Oscillatory episodes are further characterized by abrupt, localized vertical movements within cratons, commonly accompanied by high-angle faulting to form mountainous blocks and complementary yoked basins variously identified as grabens, half-grabens, intermontane basins, and rift valleys. High-angle, basement-involved faulting may occur at any cratonic site, but, in the general case, the amplitude of vertical displacement declines toward cratonic interiors and becomes more widely separated in space and time at distances greater than 1000 km from cratonic margins. Some oscillatory-mode vertical movements, such as those that created the Cenozoic fault-block mountains and intermontane basins of Colorado, are partially controlled by ancient basement lineaments. Some may appear as modifications of precursor arches and basins; for example, Pennsylvanian fault movement along the Nemaha Ridge-Forest City Basin couple followed the trend of gentle flexures active in early Paleozoic time. In general, however, vertical movements by flexure or faulting within oscillatory cratons transform the tectonic and environmental geography of areas affected, creating new erosional and depositional sites and significantly altering the compositional and textural parameters of material contributed to the sedimentary record.

Fault-block couples typical of the oscillatory mode may be from a few tens to several hundreds of kilometers in length, commonly in discontinuous subparallel or en echelon trends, but including blocks disposed at any angle to the major trends. Individual blocks range from a few kilometers to 200 km in width, are typically asymmetric, and have vertical amplitudes as great as seven kilometers, decreasing, as previously noted, with approach toward cratonic interiors. Periods of vertical movements of blocks range from 10^5 to 10^6 years. Intracratonic igneous activity and high-T/low-P metamorphism are relatively common, but far from universal, accompaniments of oscillatory behavior.

Sedimentary response.—Figure 3 is a diagram of our view of a portion of a craton during an oscillatory episode. The craton is shown as elevated with respect to sea level, and significant relief is developed in an interior shield area. Streams draining the craton carry relatively immature sediments to the cratonic margin or to traps formed by intracratonic basins. Where the latter are yoked to adjacent mountainous uplifts the basin fill is dominated by very immature sediment derived from the stripping of older sediments and from exposed basement crystallines.

Thus the stratigraphic record of the initiation of an oscillatory episode is commonly represented by an interregional unconformity marking cratonic emergence, by the displacement of carbonates and mature clastics at craton margins by immature clastics of internal derivation, and by thick wedges of arkose or coarse lithic wacke, accompanied in places by evaporites, in craton-interior positions that had been occupied

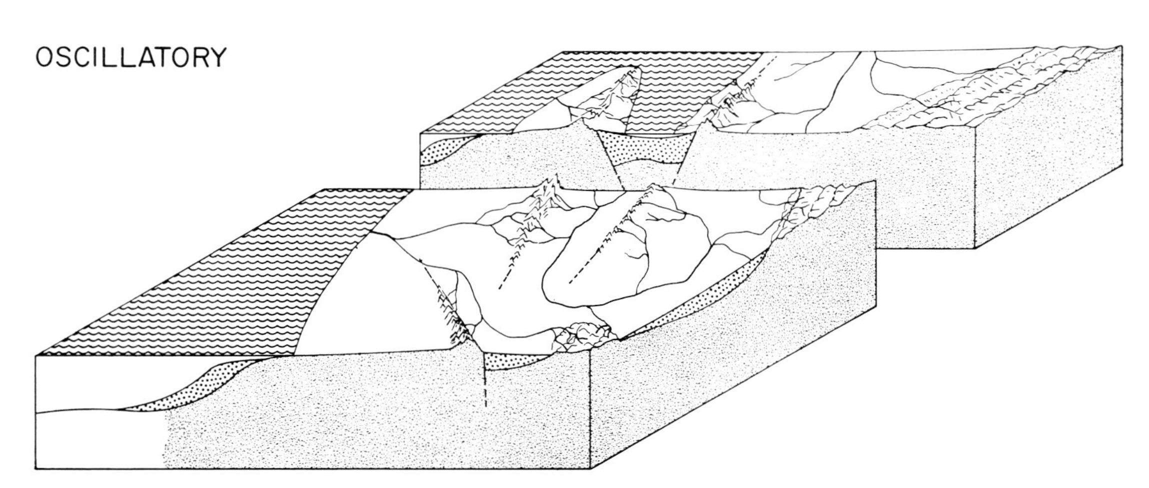

FIG. 3.—Diagram of a portion of a craton during an oscillatory episode. Time shown is one of extreme uplift involving a strongly emergent, sediment-producing shield and a numerous local vertical dislocations represented by mountainous blocks and adjoining grabens, half-grabens, and synclinal, sediment-trapping basins.

previously by quite different depositional conditions.

Cratonic emergence and intracratonic mountains during oscillatory episodes lead to a sediment flux of a magnitude sufficient to keep interior sediment traps filled to and above base level. One result is abundant bypassing of sediment to prograding cratonic margins; another product is periodic or long-continued exclusion of the sea from subsiding interior basins and the accumulation of thick successions of nonmarine sediments or of cyclically alternating marine and nonmarine deposits.

The rapid fluctuation of parts of cratons with respect to sea level during oscillatory phases is indicated by many regional and interregional unconformities; these represent rapid alternations of transgression and regression leading to cyclic accumulations, including coal cyclothems, which may be modulated at still higher frequencies by continental glaciation. In general, the record indicates net regression culminating in broad cratonic uplift above base level as in emergent modes. A major distinction lies in the occurrence of numerous local unconformities marked by angular discordances not commonly associated with cratonic conditions. Such local unconformities are the products of short-lived and geographically variable vertical displacements typical of the oscillatory phase.

Seismic profiling of Cenozoic sediments at submerged continental margins suggests numerous shifts in the loci of slope/rise accumulations in apparent correlation with cratonic transgressions and regressions. It would seem that during emergent pulsations the sediment flux from interior sources is increased, continental shelves are exposed to wave-base or subaerial erosion, and the sedimentary prisms at continental slopes and rises prograde distally while undergoing proximal truncation, perhaps by contour currents. Conversely, it appears that renewed submergence leads to a reduced sediment flux, sediment trapping on continental-shelf sites, and proximal onlap of slope-rise deposits.

Time relationships.—Phanerozoic history includes two major oscillatory episodes, one extending from Late Carboniferous to Early Jurassic, the other from late Paleocene to the present. The earlier is recorded in the rocks of the Absaroka Sequence (Sloss, 1963) in North America and much of the interpretation of oscillatory cratonal tectonics derives from analysis of these rocks and their structural relationships. Contemporaneous rocks and structures on other cratons corroborate the interpretations. In this connection special attention is called to the late Paleozoic and early Mesozoic sediments and associated structures of Southern Hemisphere cratons where these sedimentary- tectonic assemblages are widely ascribed to unique conditions prevailing during the existence and fragmentation of the Gondwana continent. As developed on later pages, we believe that cratonic rifting is common during oscillatory episodes but we see nothing unique in Gondwana sediments or demanding of special circumstances other than those typical of Absaroka strata on this continent. Rather, postglacial Gondwana rocks and the tectonic milieu of their deposition in peninsular India (Chaterji and Ghosh, 1967), South Africa (Ryan 1967, Theron 1967, Winter and Venter 1970), Australia (Veevers 1967, Connolly 1969), and Brazil (Fulfaro and Landim 1972) bear striking parallels with Absaroka sediments and their conditions of deposition in North America.

The second oscillatory episode produced the rocks and structures of the Tejas Sequence of North America and its equivalents elsewhere. Since this episode has not yet run its course, tectonic phenomena and their sedimentary responses can be directly observed or, at worst, are subject to interpretation without superposition of the effects of younger events. Differences in the time perspective through which events responsible for Absaroka and Tejas sediments and structures are viewed, the marked differences in present-day exposures and topographic expressions, and perhaps most importantly, the divergencies between concepts applied to interpretation of the two episodes tend to obscure the fundamental kinship of late Paleozoic to early Mesozoic cratons with those of Cenozoic time. There is insufficient space here to detail the many points supporting such a kinship but it is noteworthy that descriptions of the oscillatory state in preceding paragraphs apply equally to the sedimentary-tectonic responses of both episodes.

Because of the large vertical displacement of fault basins within cratons during oscillatory episodes, their sediments and subjacent rocks buried under them have a high probability of at least local preservation in the face of subsequent periods of erosion. Thus, the only late Precambrian rocks preserved in the North American cratonic interior are Keweenawan, deposited during what we would interpret as a pre-Phanerozoic oscillatory episode.

Submergent Mode

Definition.—Submergent cratons are characterized by progressive depression relative to sea level, subsidence of cratonic-interior basins, and

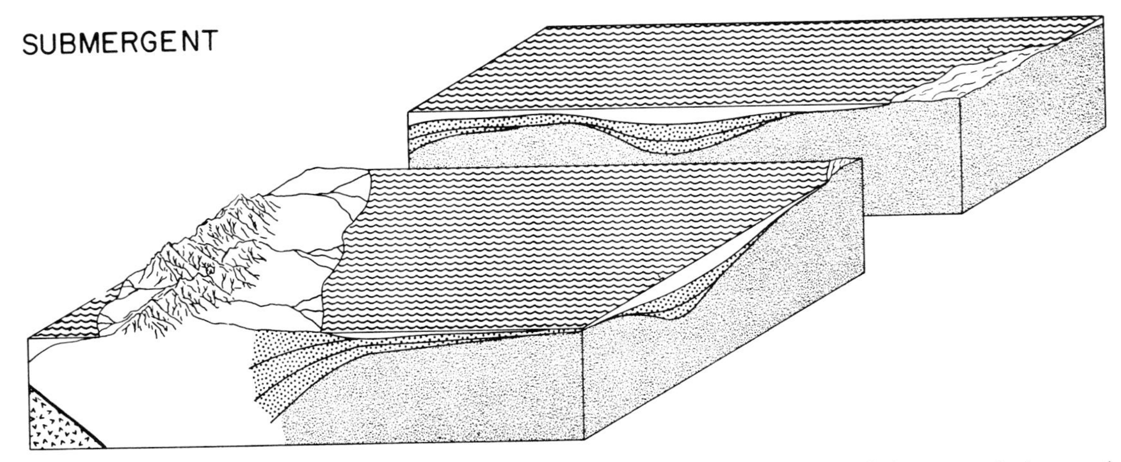

FIG. 4.—Diagram of a portion of a craton during a submergent episode. Most of the craton is depressed below sea level and marked by interior basins where rates of subsidence and concomitant sedimentation exceed those of adjoining shelf terranes. Segments of the periphery of the craton are characterized by mountain belts, the sources of clastic wedges which prograde toward the cratonic flanks and interior.

encroachment at one or more margins by clastic wedges (fig. 4). The four submergent episodes identified in the Phanerozoic record of cratons range in duration from 70 to 110 million years. The greater part of each episode involves marine transgression, in pulses of a few millions to ten million years, which, although interrupted by shorter regressive phases, results in the spread of epicontinental seas from cratonic margins to and across interior shield regions until less than ten percent of cratonic areas may remain above sea level.

Stratigraphic evidence indicates that cratonic submergence is not purely eustatic but is accompanied, even controlled, by differential subsidence of basins which tend to recur at the same sites and which are separated by more slowly subsiding "positive" elements identified as arches and domes. Analyses of data on preserved sedimentary thicknesses show that, over the course of a submergent episode, rates of subsidence range from greater than 200 meters per million years in major basins, miogeosynclines of the Stille-Kay classification, near cratonic margins to an order of magnitude less in smaller interior basins (autogeosynclines). Even the latter, however, are subject to short pulses of subsidence at rates exceeding 100 meters per million years for 1 to 10 million years. Figure 5 illustrates the contrasting geometries and rates of subsidence of typical basins developed during oscillatory and submergent episodes.

Sedimentary response.—Figure 4 is a diagram of part of a craton during a submergent episode. As the diagram suggests, sedimentation in such an episode is dominated by three factors: clastic sediment derived from erosion within the craton, nonclastic sediment derived from epicontinental seas, and clastic wedges of sediment derived from mountainous highlands presumed to exist at cratonic margins.

At submergent times such as that illustrated, transgression has covered the greater part of the craton and only very small volumes of craton-derived detritus are contributed to the accumulating sediment. During the early stages of transgression, as at the initiation of a submergent episode, and repeatedly during regressive phases, major cratonic areas are exposed above base level. Where such areas are underlain by older clastic sediments and by basement crystallines, large volumes of detrital material are delivered to the margin of the sea. Here, as a result of the very slow rate of migration of the strand line, and the general absence of rapid subsidence, the detritals are repeatedly deposited and reworked before ultimately coming to rest. The most obvious product is a blanket of highly mature quartz arenite representing the slow transgression and regression of strand environments. At the initiation of a submegent episode, when strand lines are confined to cratonic margins, any minor and temporary regressive phase will tend to sweep mature sands into the more rapidly subsiding basins of the craton margin where they may accumulate to great thickness. Alternatively, mature sands may be passed beyond the limits of the craton to form, in the manner of like sediments of emergent episodes, unexpected intercalations among pelagic clays, cherts, and volcanics of oceanic affinities.

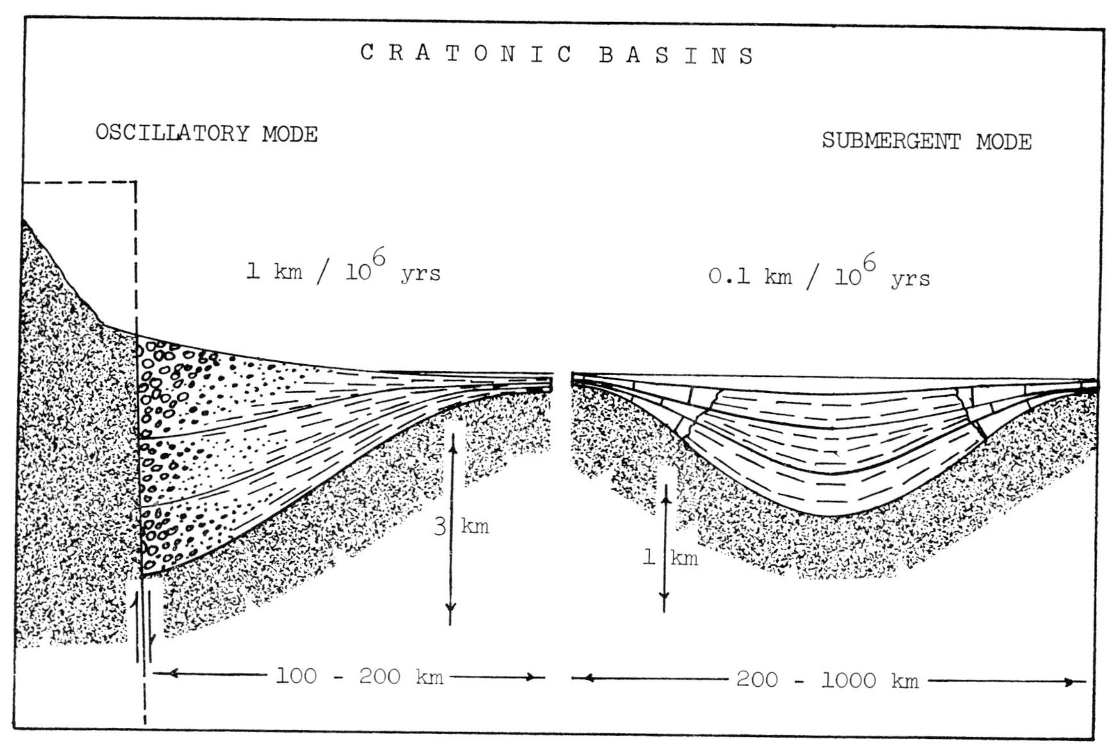

FIG. 5.—Diagrammatic cross sections of cratonic basins typical of the oscillatory mode (left) and the submergent mode (right), showing differences of scale and rates of subsidence.

Carbonate-evaporite deposition is the dominant mode of cratonic sedimentation during submergent episodes. Nonclastics prevail at a distance from strandlines, except when the epicontinental sea is loaded with distal clays of advancing extracratonic or craton-margin clastic wedges, and commonly extend to the supratidal zone in the absence of interior sources of detritals. Subsiding interior basins, previously discussed and diagrammatically illustrated on figure 5, are sites of the most commonly preserved submergent-mode sedimentation. Although the basins are typically filled to near base level with sediments at all times, the more rapid subsidence of the basin interiors finds a subdued response in submarine slopes. The gentle bathymetric relief, in turn, is reflected in concentrically disposed environmental zones: barrier banks and reefs, basin-interior micrites and evaporites, basin-margin lagoonal deposits, etc.

One of the major hallmarks (see table 1) of the submergent mode is the spread of clastic wedges from one or more cratonic margins toward cratonic interiors. Such wedges thicken and coarsen toward the presumed marginal highlands believed to constitute their sources, but the proximal roots of ancient clastic wedges are nowhere preserved and no present-day analog can be identified. Thus, the composition, geometry, and textural characteristics of clastic wedges are thoroughly documented while the nature of their source regions remains obscure. The typical sandstones of clastic wedges are lithic and feldspathic wackes which may grade distally to arenites. Clasts may include rock types of normal cratonic affinities, radiolarian cherts and spilitic volcanics, and metasediments and plutonic rocks of ages but slightly older than the wedge itself.

Regardless of the origin of clastic wedges, it is clear that they appear early in the history of a submergent episode along one or more segments of cratonic margins. The wedges prograde in the direction of the cratonic interior, progressively occupying sites of previous interior-source clastic or carbonate-evaporite deposition, and ultimately extending several hundred kilometers onto the craton. Mid-Cretaceous wedge deposits replacing earlier Cretaceous interior-source detritals in Wyoming, and Silurian clastic-wedge conglomerates and

TABLE 1.—NORTH AMERICAN EXAMPLES OF THE SUBMERGENT MODE

Time span of episode	Sedimentary sequence	Times of maximum cratonic subsidence	Examples of subsident basins	Examples of clastic wedges
Latest Precambrian-Early Ordovician	Sauk	Late Cambrian-Early Ordovician	Anadarko Illinois-Mississippi embayment	Grant Land Middle and Late Cambrian Franklinian margin
Middle Ordovician-Early Devonian	Tippecanoe	Late Middle and Late Ordovician, Middle and Late Silurian	Michigan Hudson Bay Williston	Queenston Late Ordovician Appalachian margin Lands Lokk Lt Ordovician Franklinian margin
Early Devonian-Early Carboniferous	Kaskaskia	Middle and Late Devonian, Early Carboniferous	Michigan Elk Point	Catskill Late Devonian Appalachian margin Diamond Peak Early Carboniferous Cordilleran margin Melville Island Middle and Late Devonian Franklinian margin
Middle Jurassic-Early Paleocene	Zuni	Late Cretaceous	Hanna* San Juan*	Mesaverde Late Cretaceous Cordilleran margin

* Hanna and San Juan basins are Tertiary features. The names are commonly applied to precursor Zuni basins which occupied parts of the same sites but with quite different shapes and boundaries.

sandstones succeeding carbonate banks of the Appalachian area are examples.

At the stage illustrated by figure 4 the clastic wedge, prograding from the left, has advanced across the cratonic margin; continued uplift of the source highlands would cause continued cratonward spread of supratidal delta-platform environments and sedimentation until, near the climax of the submergent episode, marine deposition would be confined to craton-interior positions between the clastic wedge and a barely emergent shield. Thus, during a submergent episode, those cratonic sectors affected by rising mountainous uplands at or near their margins may be characterized by marine regression at times of maximum cratonic subsidence.

Time relationships.—Three Phanerozoic episodes clearly fit the pattern of the submergent mode; a fourth episode is slightly less obvious. The time spans of the Tippecanoe (late Early Ordovician-Early Devonian), Kaskaskia (Early Devonian-Early Carboniferous), and Zuni (Early Jurassic-early Paleocene) sedimentary sequences are marked by cratonic depression and accompanying major transgressions of epicontinental seas, differential subsidence of interior basins, and the spread of clastic wedges from cratonic margins toward the interior (see table 1). Assignment of the Sauk Sequence (latest Precambrian-Early Ordovician) to accumulation under conditions of the submergent mode is less clear. The criterion of long-continued subsidence and marine transgression is met, although greater areas of most cratons appear to have remained above sea level at the climax of Sauk submergence than is the case for later submergent episodes. Further, the degree to which the tectonic pattern of the cratonic interior is differentiated into subsiding basins and less subsident shelves, domes, and arches is distinctly less than that typical of younger submergent episodes. Instead, differential subsidence was largely confined to basins on the cratonic margins and to embayments of the margins. The Anadarko Basin (Adler and others, 1971), the Mississippi Embayment-Illinois Basin trend (Bond and others, 1971), and several basins of the Arctic Interior Platform (Christie, 1972) are examples of marginal embayments which extended into craton-interior terranes. Major clastic wedges, if developed during Sauk deposition at the Appalachian and Cordilleran margins of North America, are not well preserved, but Ellesmere Island and adjacent areas of the Canadian Arctic expose evidence of Cambrian wedges at the Franklinian margin (Trettin, 1972).

In sum, there is a real difference between the tectonism expressed during Sauk deposition and that of the younger submergent episodes but we believe the difference to be one of degree rather than kind.

Cratonal Tectonic Modes and the Sequence Concept

Our classification of tectonic modes operating on cratons is heavily dependent on stratigraphic observation, synthesis, and interpretation, and has evolved from continuing consideration of the concept of stratigraphic sequences (Sloss and others, 1949; Sloss 1963) as an approach to analyzing stratigraphic data. Some of the problems, other than those inherent in any attempt to erect a taxonomy to codify a continuum of natural phenomena, encountered in application of the proposed tectonic classification are equally present as problems raised in application of the sequence concept, as the following discussion indicates.

Emergent vs. oscillatory modes.—It is our thesis that the tectonic history of cratons, at least during Phanerozoic time, is marked by periods of subsidence with respect to sea level; the submergent episodes are represented stratigraphically by the Sauk, Tippecanoe, Kaskaskia, and Zuni sequences of North America (fig. 2). Submergent episodes are separated by periods of cratonic uplift which may be (1) relatively brief, uncomplicated by extensive fracturing, and identified in the stratigraphic record as craton-wide unconformities such as the sub-Sauk, sub-Tippecanoe, and sub-Kaskaskia discontinuities; these are here termed emergent episodes; or, (2) extended periods, the oscillatory episodes represented by the Absaroka and Tejas sequences ,of alternating uplift and subsidence accompanied by cratonic fracturing and rifting, but culminating in broad cratonic exposure evidenced by interregional unconformity. The oscillatory mode is conceived to be a special, protracted, case of emergent behavior; either state may intervene between submergent episodes, and both produce extensive unconformities in cratonal stratigraphic successions. As a result, there are stratigraphic difficulties in identifying certain sequence boundaries and problems of tectonic classification arise in referring certain episodes to specific tectonic modes.

Submergent-to-oscillatory transitions.—There is little dispute over the identification and significance of sequence boundaries where successions deposited in submergent episodes are separated by unconformities representing emergent modes. The Sauk and Tippecanoe sequences, and the base of the Kaskaskia Sequence, are clearly delimited by craton-wide discontinuities. However, the Late Mississippian (Namurian) transition from the submergent episode represented by the Kaskaskia Sequence to the Absaroka oscillatory episode is more difficult to define, both stratigraphically and tectonically. The sub-Absaroka discontinuity is manifest over large areas of the North American craton as the easily recognized pre-Pennsylvanian unconformity, but it is obscure or lacking on the Appalachian, Ouachitan, and Cordilleran margins at positions where the stratigraphic lacunas of earlier emergent episodes remain obvious. Much the same may be said of the late Paleocene boundary between the Zuni Sequence, a product of the submergent mode, and the depositional record of the oscillatory Tejas interval. That is, the sub-Tejas unconformity is clear over major craton-interior areas but loses definition before reaching the presumed limits of the Paleocene craton. Thus, the observational data applied to identifying the point in time of submergent-to-oscillatory shifts in tectonic behavior do not permit a high degree of precision.

Stratigraphic gradation or transition, rather than precise stratigraphic resolution, at these sequence boundaies is matched by gradual transition in tectonic behavior as this is interpreted from facies analysis. In Late Mississippian time (late Viséan and early Namurian) toward the close of Kaskaskia deposition and near the initiation of the oscillatory Absaroka episode, cratons began to emerge from the submerged state. Simultaneously, craton-interior sources of detrital sediment, presumably local uplifts, appeared, and relatively high-frequency vertical oscillations of parts of cratons began, leading to the development of Chester-age (early Namurian) coal cycles in the Illinois Basin and elsewhere. In recognition of these observations the Kaskaskia-Absaroka stratigraphic boundary was originally placed (Sloss and others, 1949) at the base of the Chester Series, but later revised (Sloss, 1963) when it became apparent that a much more significant break is present near the Mississippian-Pennsylvanian boundary. Similarly, the late Paleocene Zuni-Tejas submergent-oscillatory transition is anticipated by quite localized but well-defined evidences of fault-bounded basins and sharp vertical uplifts of latest Cretaceous and early Paleocene age. Such relative lack of resolution and definition is, in the light of our concept, the predictable result of intermittent, protracted, and locally variable cratonic uplift at the initiation of an oscillatory episode. Identification of the time of changeover from submergent to oscillatory behavior may well be more or less arbitarary within a transition period of a few million years while selection of a stratigraphic surface to represent the base of a sequence deposited in an oscillatory episode may be equally arbitrary within well-defined limits.

Oscillatory-to-submergent transition.—The lone example of an oscillatory-to-submergent transition available in the Phanerozoic record suggests that boundary problems exist here, too, both in time and in stratigraphic definition, but to a lesser degree. Perhaps fortunately, Zuni transgression was relatively slow on most cratons and is marked in many areas by Cretaceous strata which overstep a major unconformity. However, where the interval from Late Triassic through Early Jurassic is represented by a significant preserved record, as on parts of the Colorado Plateau, there is difficulty in identifying the time of onset of the submergent state and selection of a stratigraphic base for the Zuni Sequence is somewhat arbitrary. If, as we presume, an oscillatory episode ends with gradual diminution of the amplitude of geographically variable vertical displacements accompanying net cratonic depression, it is logical to anticipate the need for choices that are arbitrary within well-defined limits in selecting the chronologic and stratigraphic boundary between the Absaroka and Zuni episodes.

In sum, the proposed tectonic classification is in harmony with the older sequence concept, and consideration of cratonal tectonic states aids in understanding certain of the difficulties encountered in identification of sequence and episode boundaries involving the initiation or termination of oscillatory phases. The stratigraphic record of such phases, the Absaroka and Tejas sequences, remain as distinct entities, clearly different from other sequence assemblages in petrology, geometry and distribution of stratal units, and in tectonic interpretation, but with more diffuse stratigraphic and temporal limits than are associated with successions bounded by unconformities representing emergent modes.

CONTINENTAL MARGINS AND CRATONIC MODES

In the context of the present paper, certain significant implications arise from our interpretation of cratonal tectonism and its sedimentary-stratigraphic responses:

(1) If the tectonic behavior of cratons has changed in recurring patterns through Phanerozoic time, then the causative mechanisms must also have undergone recurring variations.

(2) If the tectonic modes discussed in the preceding section have been synchronously operative on all cratons, then the causative mechanisms are not confined to an individual continent but must be manifestations of global tectonics.

Acceptance of these two points in the light of current global-tectonic concepts leads, unavoidably, to two further consequences:

(3) Vertical movements of cratons including upwarps and differential subsidence of basins must involve continental lithospheres.

(4) Vertical movements of continental lithospheres relate to the behavior of oceanic lithospheres.

Ocean-continent relationships are explored in the following section in terms of their responses at continental margins during episodes of dominance of each of the cratonal tectonic modes. Before proceeding, however, it is necessary to define a few terms.

Definitions

Cratonic margins.—As noted in our introductory statement, cratonic interiors are easily identified as terranes of exposed or shallowly buried thick "granitic" crust. By extension, cratonic margins are defined for specific spans in geologic time at those farthest peripheral positions where, at the time in question, sedimentary-tectonic behavior may be interpreted to have been more craton-like than not. Beyond these positions, again for the time span in question, either abruptly or transitionally, noncratonic sedimentation and tectonics are the logical interpretations.

Continental margins.—The modern joins of continents and ocean basins are identified chiefly by topographic, tectonic, and seismic criteria, and with less certainty, by petrographic differences. The positions of such boundaries today are little disputed except for certain minor regions. Modern continental margins of the following types are identified for purposes of discussion:

(1) *Passive margins,* the apparently welded joins of continental and oceanic lithospheres; may be sites of long-term strain but without rapid slip.

(2) *Active margins* are plate boundaries, sites of differential motions of plates bearing continental and oceanic terranes, and subdivisible into:

(a) *Convergent boundaries,* involving subduction and plate consumption,

(b) *Conservative boundaries,* involving tangential motions at plate boundaries along which no plate is consumed.

The delineation of ancient continental margins is by no means as straightforward because the only available criteria are petrographic and structural, the latter taken to be the product of active-margin tectonism.

Cratonic vs. continental margins.—Cratonic margins may coincide with continental margins, as on the Atlantic Coast of North America today; or cratonic margins may lie several hundred kilometers within continental margins, with intervening island arcs and active or dormant interarc basins as on the present Pacific coast of Asia. Ancient cratonic and non-cratonic continental terranes are generally distinguishable by their sedimentary-tectonic records, but the rocks of ancient island arcs and interarc basins would be readily confused with tectonically mixed or superposed oceanic materials. Thus, ancient continental margins are generally poorly resolved, and it is the ancient cratonic margins that are more commonly recognized.

Continental Margins of Oscillatory Episodes

Conditions during oscillatory episodes are exemplified by the present characteristics of continents and their oceanic margins. All of the types of margins noted above exist today but only consideration of the two sub-types of active margins adds significantly to resolution of the problem at hand.

Convergent boundaries.—The model (fig. 6A) for present-day convergent plate boundaries involving continental margins, and, by extension, for similar boundaries during all times of oscillatory behavior, is drawn from the western Pacific area. Here, plate convergence takes place along island arc systems that are convex toward the ocean and separated by several hundred kilometers from the adjoining craton. Marginal basins that lie between arc and craton are either tectonically inactive or actively extensional interarc basins (Karig, 1971) characterized by spreading axes along which new oceanic-type crust is formed. The volume of detrital and volcanogenic sediment supplied from the island arcs is currently small relative to the width and depth of the marginal basins, as is the sediment volume deriving from the craton and encroaching on the basins as prograding delta complexes or as slope-rise accumulations. As a result, marginal basins of the present, and presumably throughout the course of oscillatory episodes, are maintained at near-oceanic depths and there is no progradation of detritus from mountainous terranes at continen-

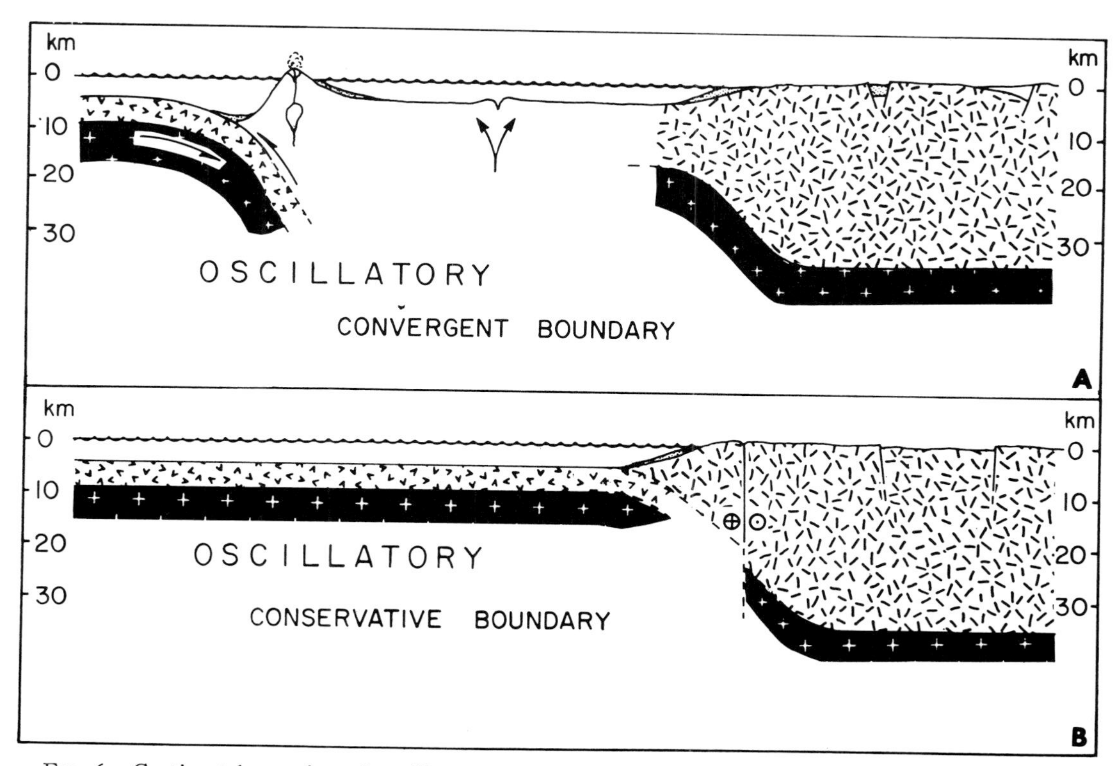

FIG. 6.—Continental margins of oscillatory episodes; oceanic and continental crust indicated by v's and dashes, respectively; upper mantle by white crosses; vertical exaggeration approximately 500:1. In the upper diagram, plate convergence and the resulting arc-trench system is separated from the cratonic margin by an interarc basin and spreading axis driven in part by the expulsion of subcontinental melt. The lower figure illustrates the less commonly anticipated situation characterized by tangential, plate-conserving motions of opposed plates.

tal margins to cratons in the form of clastic wedges.

Conservative margins.—It is implicit in plate-tectonic theory that both convergent, plate-consuming, and conservative plate boundaries characterized by tangential motions of adjacent plates will exist whenever the relative velocities of such plates are nonzero. For the sake of completeness and because conservative boundaries are believed to be active during the current oscillatory episode, we include a cartoon of a conservative continental margin (fig. 6B). The model is California where the opposed motions of the Pacific and American plates is presumed to be resolved by right-lateral slip along the San Andreas fault system, a putative ridge-ridge transform (Wilson, 1965; Morgan, 1968).

Vertical displacements near cratonic margins.—As has been noted on earlier pages, oscillatory episodes are the times of marked structural and topographic relief owing chiefly to high-angle faulting; the further point has been made that such relief tends to damp out irregularly from margins of cratons toward their interiors, making it appear to be a manifestation of ocean-continent relationships. In North America the late Paleozoic and early Mesozoic oscillatory episode coincided with the Arbuckle, Wichita, "Ancestral Rockies," and Palisades "orogenies" which were exclusively cratonic but which occupied broad, irregularly discontinuous areas or belts adjacent to earlier convergent plate boundaries at cratonic margins. Very similar tectonic and geomorphic conditions, modified locally by glaciation, are evidenced in the Gondwana cratons by Karroo, Parana, and related sequences. The same time span in Europe was characterized by the many spasms of the Hercynian orogeny which are interpreted (e.g., Zwart, 1967, 1969) to be dominated by vertical displacements and to have involved broad, interrupted areas of the sub-interiors of the craton as well as its margins in patterns strongly reminiscent of oscillatory-mode conditions on other continents. It should be noted that, while there is wide recognition of marked differences between Hercynian tectonic and plutonic patterns and those expressed in Alpine or Cordilleran models, there are numerous workers (e.g., Laurent 1972) who perceive Hercynian events as the consequences of plate-boundary interactions. In our view, such interpretations fail to explain the idiosyncratic mode of Hercynian tectonism, its spatial distribution, or its coincidence in time with equivalent vertical displacements on other continents.

Significant surface relief has again been imposed on all cratons during the current Cenozoic oscillatory episode. Once more, the most marked coupling of uplift and downwarp is concentrated in broad belts not far removed from continental margins but lacking evidence of direct relationship to plate boundaries. The most obvious product in North America is Laramide mountain-building as expressed in the Central and Southern Rockies.

Continental Margins of Submergent Episodes

The characteristics of continental margins of an oscillatory episode such as the present can be described in terms of direct observations of present-day sedimentational and geomorphic processes and responses; further, an abundance of geophysical data contributes to our understanding of currently operative crustal and deep-seated phenomena. Consideration of the continental margins of submergent episodes, on the other hand, is wholly dependent on interpretation of ancient rocks and structures. In this light it is useful to review the pertinent obser-

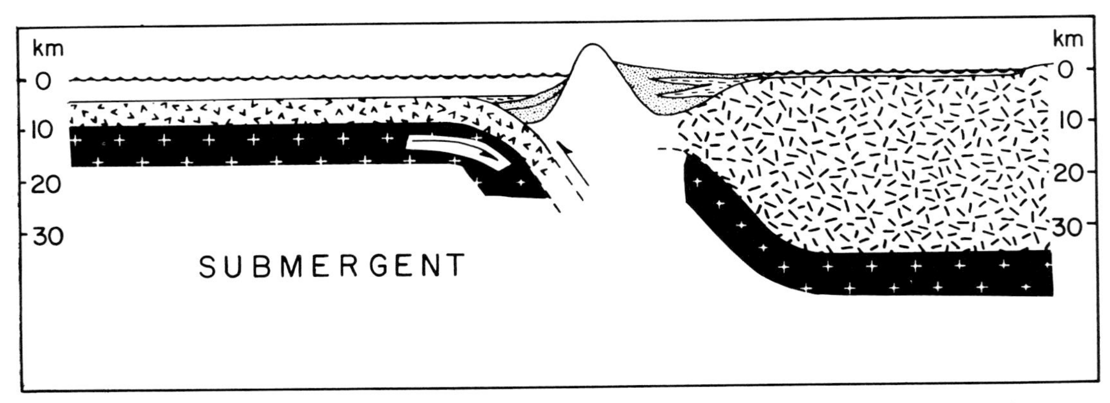

FIG. 7.—Active continental margin of a submergent episode, showing plate convergence and consequent mountains at or near the cratonic margin.

vations and their first-order interpretations before proceeding to deductive reconstruction.

Observed phenomena of submergent-episode margins.—The most significant observations can be outlined as follows:

(1) Maximal marine transgression of cratons is accompanied by accelerated downwarp of craton-margin basins as well as craton-interior basins.

(2) Development of clastic wedges prograding from craton margins toward interiors is synchronous with marine transgression, as correctly noted by Johnson (1971).

(3) Clastic wedges contain materials derived from older cratonic rocks, both sedimentary and crystalline; from extracratonic, probably oceanic, sediments and volcanics; and from extrusive, intrusive, and metamorphic rocks formed almost concurrently with deposition of the clastic wedge.

(4) Volcanics and volcanogenic sediments are common intercalations in clastic wedges where they increase in number and thickness with approach toward presumed continental margins.

(5) Major thrust faults dated as contemporaneous with clastic wedges involve tectonic transport of extracratonic and craton-margin rocks toward the interiors of cratons.

Reconstruction of submergent-episode margins.—The picture that emerges (fig. 7) from consideration of these observations is one of a submergent craton at least partially ringed by active orogenic belts which, at places and at times, reach mountainous elevations. As currently interpreted, such orogenic belts represent responses to plate convergence and accompanying subduction and obduction. Thus, the reconstruction is modeled after present-day (oscillatory-mode) boundaries, but with important differences:

(1) The convergent boundary closely coincides with the cratonic margin such that orogeny is localized at or very near the edge of the craton. This is in contrast to conditions in an oscillatory episode when convergent boundaries lie at a distance oceanward from craton margins, resulting in island arcs separated from the craton by marginal or interarc basins.

(2) Where and when, as shown on figure 7, convergence-induced orogenesis results in the maintenance above base level of a mountain chain near the margin of the craton, accelerated erosion leads to a high rate of sediment supply that exceeds the receptor capacity of the subsiding craton margin. The product is a clastic wedge including an aggrading piedmont plain and a prograding delta complex that advance across the craton margin toward the interior. Thus the major distinction between sedimentational characteristics at convergent continental margins of oscillatory and submergent episodes is in the processes of cratonward transport and environments of deposition of detritus derived from growing mountain chains. Continued turbidity-current transport and deep-water sedimentation characterize the oscillatory state. Rapid transition to stream and littoral transport, and to subaerial, intertidal, and shallow-water deposition, occurs under submergent-mode conditions.

Continental Margins of Emergent Episodes

Emergent episodes, as discussed in earlier paragraphs, are conceived to be times of essentially uniform vertical displacement, both at cratonic interiors and continental margins. That is, cratonic blocks during emergent episodes lack the relatively high-frequency, high-amplitude vertical displacements typical of the oscillatory mode and the slower, more subdued undations of the submergent mode; at the same time we

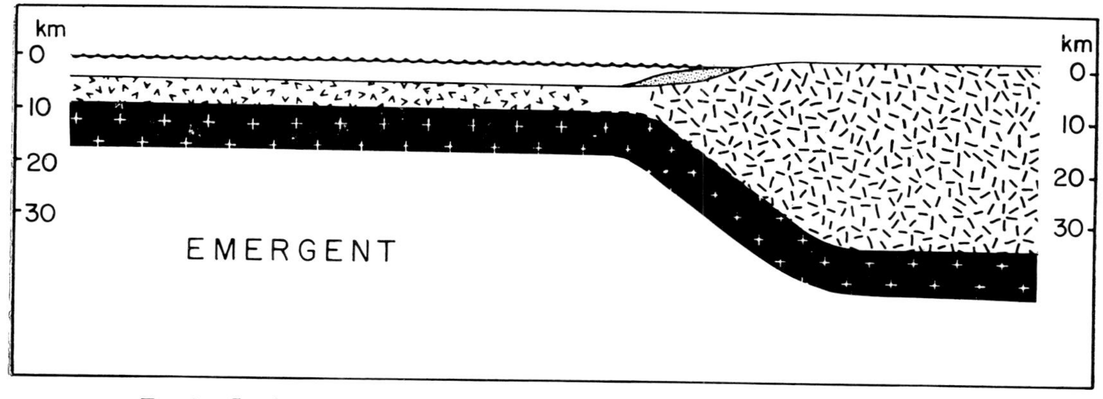

FIG. 8.—Continental margin during an emergent episode, showing the passive mode believed to be typical of such episodes.

see no compelling evidence of major tectonic activity at cratonic or continental margins during emergent episodes, suggesting an absence of plate convergences. Largely for the sake of completeness we include (fig. 8) a diagrammatic representation of an emergent-mode continental margin shown as the passive site of deposition of sediments derived from erosion of a featureless emergent craton.

THEORY OF CRATONAL TECTONICS

Review of the interrelationships in space and time between tectonic states at continental margins and in cratonic interiors, as described and interpreted in the foregoing parts of this paper, leads naturally to consideration of possible mechanisms. We have developed a theory of the generation and sequence of cratonal vertical motions of the types described in preceding paragraphs. The theory is presented here in qualitative form and will be given more rigorous treatment elsewhere (Speed and Sloss in prep., 1974).

The propositions on which the theory is constructed are:

(1) Modes of vertical displacement of cratons occur repetitively through time and such modes are synchronous among cratons.

(2) Cratonic modes (submergent, emergent, and oscillatory) are coupled with distinct types of tectonism at continental margins.

(3) Cratonic modes are tectonic phenomena not directly related to changes in the capacity of ocean basins, glacial events, or other eustatic responses; similarly, cratonic modes are not isostatic responses to loading by ice or mountains. Thus, there is implicit rejection of the view introduced by Hallam (1963), and repeated in numerous subsequent papers by other workers, relating transgression and regression to the enlargement and attenuation of oceanic ridge systems causing repeated advance and withdrawal of marine waters with respect to essentially inert or passive cratons. Rather, it is one of our fundamental postulates that cratonal tectonics control land/sea relationships and that eustacy is a second-order phenomenon.

(4) Cratonic modes are responses to global tectonics and are not the results of localized forces acting on a single continent or part of a continent.

(5) Cratonic modes are products of distributed vertical forces. Elastic or viscous buckling from horizontal forces on the edges of continents can be dismissed because the stress required for such buckling in the most susceptible mode exceeds the fracture strength of the materials involved by an order of magnitude. Moreover, the vertical forces must be distributed in fields that are virtually coextensive with cratons because the observed flexural rigidity of the continental lithosphere (Walcott, 1970) is too small for broad warping by laterally restricted stresses.

Hypothesis

Mechanism.—Simultaneous vertical displacements of all cratons must involve at least part of the lithosphere of which the continents form the uppermost zone. The concept of concurrent displacements of continental lithosphere by distributed vertical forces is immediately suggestive of the mechanism of radial phase transformation at some level within or below the lithosphere. Solid-solid transformations as would occur in the lithosphere have long been invoked as tectonic sources (*e.g.,* MacDonald and Ness, 1960). A solid-melt transformation in the asthenosphere, however, is more attractive as a generator of global vertical displacements because it can account in principal for the associated volcanism, faulting, and sea-floor spreading. The existence of an asthenospheric zone of partial melt seems well established by observation (Anderson and others, 1972) whereas the presence of a solid-solid transformation is not indicated above the low-velocity zone. Thus, we propose a mechanism for global vertical tectonic motions by episodic changes in the proportion of melt in the asthenosphere below the continents. The boundaries of the asthenosphere are assumed to be solidi (McKenzie 1967; Press 1972), and the radial flux of heat into the asthenosphere is taken to be constant. According to the hypothesis proposed, the thickness of the subcontinental asthenosphere varies through time in proportion to the difference between the rate of partial-melt production and the rate of extraction of such melt from the asthenosphere. We assume that the continental lithosphere acts as a passive elastic member of constant thickness above the asthenosphere; therefore, changes in asthenosphere thickness create changes in the elevation of the upper surface of the lithosphere. Jacoby (1972) also proposed that elevation of continental lithosphere depends on asthenospheric changes, although the objective and mechanics of his model differ from ours.

Concepts of plate generation and motion do not require decoupling of continental lithosphere from subjacent asthenosphere. In fact, the two zones must necessarily move together over a subducted slab of oceanic lithosphere at convergent boundaries, if the Benioff zone is steeper than, say, 20 degrees and maintains modest planarity to depths exceeding a few hundred km. The slip boundary, if one exists, on which continental lithosphere translates, must therefore be below much of the continental asthenosphere whose base is roughly 300 km or less (Press, 1972). The slip boundary presumably steps up at the trailing edge of the continent to the base of the passively-joined oceanic lithosphere, assuming the oceanic lithosphere and oceanic asthenosphere are decoupled. Oceanic and continental asthenospheres are probably contiguous at passive margins, but the properties of the two asthenospheres differ (depths, amplitude of velocity reversal, attenuation), suggesting that they do not currently convect as a unified system. The point is that during plate motions, continental asthenosphere is cotransported with continents, at least in part, and such asthenosphere "stays continental" through time. Moreover, there is no reason to assume that circulation between different asthenospheres is constant. We present later a scheme of transient mass transfer between continental and oceanic asthenospheres, and the general theory will be expanded in Speed and Sloss (in prep.).

The mechanics of melt extraction from the subcontinental asthenosphere is by no means clear. In principle, one may appeal to separated flow of partial melt through a porous solid residue or to bulk flow when the fraction of melt is sufficient to reduce the mean viscosity to some critical value. Progressive separation of fluid during bulk flow may also occur. Regardless, return flow of solid is required to maintain mass balance. Episodicity in rates of melt extraction conceptually exists in both separated and bulk flows. In the flow of partial melt through solid residue, the flux is proportional to a power of the space occupied by the fluid. Thus, it is reasonable to assume that the outflux of melt is meager until the concentration reaches some threshold. Sleep (1974) calculates that, at 1% concentration with a melt viscosity of 10^8 poise as determined by seismic wave attenuation (Solomon, 1972), the flux would be about 1 cm/my. If the concentration were 10%, the flux would be increased 1000 times, and if the actual viscosity is somewhat lower than Solomon's value, fluxes of kilometers per million years might be achieved. At such concentrations, however, bulk flow may well have set in (N. H. Sleep, oral communication, 1973) such that the melt-rich asthenosphere is discharged without separation of fluids. Regardless, bulk flows as well as separated flows will be episodic and depend upon the attainment of a critical melt fraction.

Thus, as the steady heat input increases the melt fraction, the asthenosphere will thicken and the overlying craton will be lifted, but significant expulsion of melt will not occur until the critical concentration is achieved. When the outflow becomes significant, the asthenosphere will deflate and the craton will sink. Presumably, such events of inflation and deflation go on cyclically.

Cratonic emergence and submergence.—Translating the above hypothesis for application to the tectonic behavior of cratons, it becomes clear that the modes here termed emergent and submergent are expressions of variations in the rate of extraction of melt from the subcontinental asthenosphere where melt is continuously produced by radial heat flow. When extraction rates are low, the thickness of the asthenosphere expands, the overlying craton is uplifted as a rigid slab, and the conditions of the emergent state are achieved. Conversely, when melt extraction rates are high relative to the rate of melt production in the continental asthenosphere, asthenospheric deflation and accompanying cratonic submergence are the results. For example, a density decrease of 0.05 g/cc of a continental asthenospheric layer 140 km thick is equivalent to melting of roughly 10 percent assuming a solid-liquid density contrast of 0.5 g/cc, and would produce cratonic uplift of about two kilometers; thus, given a steady rate of melt production and perturbations in the rate of melt extraction, vertical motions of cratons of the amplitudes observed may be achieved.

Relationship to sea-floor spreading rates.—It is explicit in our discussion of cratonic modes that submergent episodes are characterized by plate convergence at craton margins whereas such activity is subdued or absent during emergent episodes. It would appear, therefore, that spreading rates are high in submergent episodes when the rate of melt expulsion from continental asthenospheres is accelerated, but low during emergent episodes when continental asthenospheres are inflated by retention of melt. The apparent correlation of spreading rates and cratonic modes requires explanation.

At the climax of an emergent episode, just before initiation of asthenospheric deflation leading to submergence, the area of highest uplift, presumably underlain by the maximum concentration of melt, is in the cratonic interior. At the initiation of deflation by melt extraction the maximum potential melt flux would be below the cratonic interior but there are no evidences of great volumes of volcanic rocks of appropriate ages at such interior positions to represent the volume of asthenospheric melt expelled. Rather, we must presume that the greater part of the melt extracted during deflation is transferred to the oceanic asthenosphere.

The subcontinental asthenosphere must be continuous with the suboceanic asthenosphere.

Observations indicate, however, that the two asthenospheres differ in depth and, probably, in composition or proportion of melt (Press, 1972). We assume that a steady convection occurs within the oceanic lithosphere-asthenosphere system, and that interchange of mass between oceanic and continental asthenospheres is possible. Therefore, variation in the rate of mass transfer from the continental to the oceanic asthenosphere will add or subtract from the steady oceanic rate and be manifested by variation in sea-floor spreading rates.

In principle, if continental asthenospheric melt is pumped across the boundary to the oceanic asthenosphere, with appropriate return flow of solid of equivalent mass, there are two possible results: (1) the oceanic asthenosphere will inflate, causing uplift of the oceanic lithosphere, or (2) the discharge of melt at oceanic ridges will increase proportionately. Assuming the latter is the operative effect, sea-floor spreading rates would be maximum during early phases of cratonic submergence when the melt fraction, hence permeability, is presumed to be highest. Figure 9 portrays relationships diagramatically based on the condition that rate of outflux of melt to the oceanic asthenosphere is proportional to melt content of the continental asthenosphere during deflation. During such intervals, convergent boundaries at continental margins would be marked by extreme activity, and the orogenesis and sedimentary responses typical of the submergent mode would be achieved.

Submergence and intracratonal tectonics.—Thus far, while expounding a theoretical basis for the emergence of cratons, we have not accounted for the development of subsiding basins and less subsident arches and domes that characterize cratonic interiors during submergent episodes. During cratonal emergence, the up-arched continental lithosphere is presumably strained extensionally. At the maximum of emergence, the lithosphere might be considered as a halfwave in which faulting and viscous decay have eliminated extensional stress owing to up-arching. With the beginning of asthenospheric deflation, the lithosphere would flatten under gravitational loading. If the oceanic lithosphere supplies restraining forces at the continental margins, the flattening during submergence would create internal compression in the continental lithosphere. Buckling of an elastic lithosphere above a dense viscous substrate during flattening could conceivably create the basins and arches of the dimensions observed in cratonic interiors. In fact, a realistic application of buckling theory to the internal deformation of cratons demands consideration of certain other factors:

(1) Loci of first buckling would be strongly influenced by inhomogeneities in the thickness and strength of the continental lithosphere and by non-horizontality of surfaces of constant strength such as bedding surfaces and other discontinuities. The latter effect, for example, is equivalent to the localization of first buckles at positions of maximum initial dip during the folding of layered rocks. As a result, once cratonic basins and arches are developed because of some primordial predilection of the underlying lithosphere, we believe that cratonic buckling will occur at the positions of precursor basins and arches formed during a previous cycle of cratonic submergence. The stratigraphic record amply documents the recurrent nature of cratonic sedimentary basins.

(2) In addition to the effects of inherited lithospheric inhomogeneities, basin-arch wave forms and their dimensions are not directly derivable from theory because of interference among wave forms whose loci basically depend on lithospheric inhomogeneities and on the positions and attitudes of confining continental margins. Actual sedimentary basins and intervening arches are not truly sinusoidal but their geometries and diameters approach the predicted values.

(3) Viscoelastic behavior of a continental lithosphere (Nadai, 1967; Walcott, 1970) permits significant changes in basin configuration by time-dependent viscous deformation of the initial elastic buckle. Basins would deepen while shrinking in areas of maximum subsidence. Analyses of the evolution of basin geometries (Scherer, 1973) and indications of synchronous maximum subsidence events in widely-separated basins (Sloss, 1972b) support the applicability of the theory.

Cratonic oscillatory episodes.—The preceding theoretical treatment avoids consideration of the oscillatory mode characterized by high-frequency cratonic oscillation with respect to sea level, abundant cratonic rifting and block faulting, cratonic volcanism, and plate convergence at island arcs. We conceive these peculiar conditions of the oscillatory state to represent a special case of protracted transition from submergent to emergent states. No oscillatory episodes are detected (fig. 2) following the early Paleozoic submergences, and one tentative conclusion is that these submergent-to-emergent transitions were sufficiently rapid that there is no sedimentational or interpretable structural record of what may have been brief and transient oscillatory states. Alternatively, it is noteworthy that the degree of cratonic submergence, as measured by the volume of sediment pre-

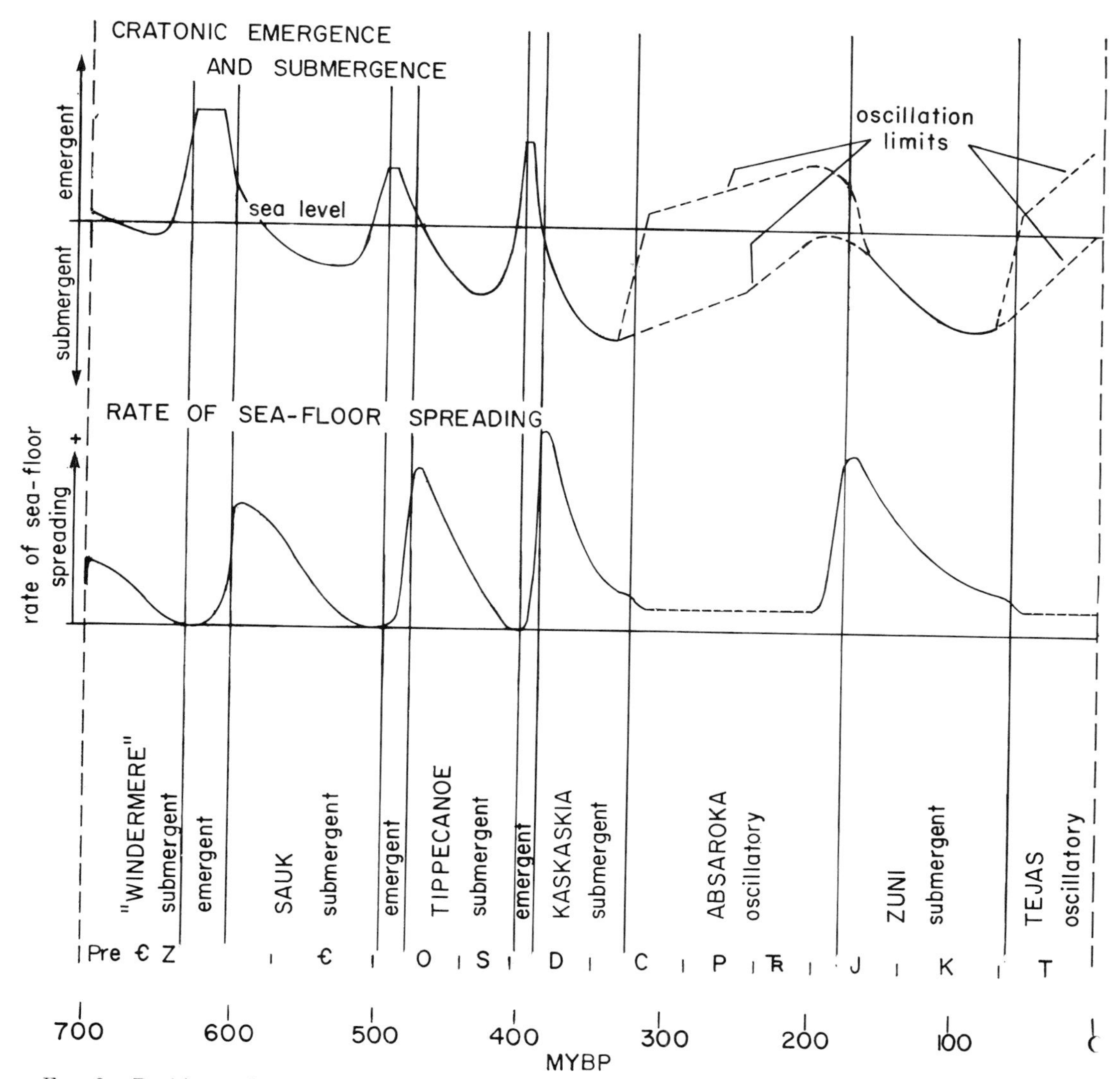

FIG. 9.—Positions of cratons with respect to sea level (upper graph) and inferred relative rates of oceanic sea-floor spreading (lower graph) in terms of geologic time, cratonal tectonic modes, and the periods of deposition of North American stratigraphic sequences. Dashed lines (upper graph) indicate positive and negative limits of high-frequency fluctuations of cratons relative to sea level during oscillatory episodes; sea-floor spreading rates in the same episodes are assumed to have similar short-period perturbations that are not shown on our generalized figure.

served, was relatively small during deposition of the Sauk and Tippecanoe sequences representing submergent episodes followed by well-defined emergences, whereas cratonic subsidence was relatively great during Kaskaskia and Zuni deposition during submergent episodes that were succeeded by oscillatory states. Our current thinking favors the latter relationship and suggests that extremes of asthenospheric deflation by melt expulsion and concomitant extremes of cratonic submergence set the stage for oscillatory behavior.

As noted in earlier discussion, it is our position that deflation and submergence result from expulsion and transfer of melt from continental to oceanic asthenospheres where it is added to convection currents in the suboceanic asthenosphere to emerge at midoceanic ridges. Both the emergent and oscillatory states of cratonic behavior represent modes of recovery from asthenospheic deflation and cratonic subsidence toward inflation and uplift. The relatively short-lived emergent state is achieved by rapid uplift of a moderately subsident continental lithosphere, presumably by a high ratio of melt retention to melt extraction in the continental asthenosphere. The result is a broadly up-arched, slab-like lithosphere, largely unfaulted except at the extreme continental margins and lacking conduits for the emplacement of cratonal volcanic and plutonic rocks.

According to our scenario, the oscillatory condition initiates from an episode of extreme asthenospheric deflation. Reversal toward ultimate lithospheric uplift is accompanied by erratic perturbations in the ratio of melt retention to extraction on either side of unity but favoring net inflation of the continental asthenosphere. The result is vertical oscillation of the craton and net uplift toward the highly elevated condition typical of the climaxes of oscillatory episodes. We presume that continuity of the continental and oceanic lithospheres is maintained during uplift such that extension occurs in the continental block as manifested by high-angle faulting with rotation on the flanks of the upwarp. This tendency, in combination with long-continued pulsatory uplift, we identify as the explanation for repetitive block faulting and rifting on the flanks of cratons during oscillatory episodes. Further, the same fractures may become, in some instances, the flow paths of melt expulsion to form the extensive volcanic fields typical of late Paleozoic-early Mesozoic and Cenozoic time.

Intermittent melt expulsion through extensional fractures on the flanks of cratons is an attractive mechanism to explain vertical oscillations of cratons. That is, if uplift is caused by an excess of asthenospheric melting over melt extraction, any such excess would be drained off by volcanism, leading to a check in uplift and, perhaps, to minor deflation and downwarp. With fractures filled and flow immobilized by cooling magma, the cycle of inflation and uplift followed by deflation and downwarp would be renewed. However, certain parts of oscillatory episodes, as exemplified by the Pennsylvanian cratonic record of western North America and the southern Midcontinent, although marked by abundant block-fault mountains and half-graben basins, are characterized by a general absence of igneous activity, either volcanic or shallow plutonic. Inasmuch as this was also a time of relatively high-frequency oscillation of the craton with respect to sea level, there is a presumption that intermittent melt expulsion from the continental asthenosphere prevailed, but not by cratonic volcanism.

We suggest that there are occasions, perhaps induced by the concentration of bending moments and tensional stress at the edge of the uplifting lithosphere, when the principal arteries of melt expulsion are confined closely to the ocean-continent lithosphere join. Under these hypothetical conditions of continent-margin tension, as contrasted to the compressional margins envisaged during deflation and submergence, melt expelled from the continental asthenosphere would not join the main stream of oceanic mantle convection but would be extruded along fractures at or close to the continental margin. Among the results would be the emergence of spreading axes at continental margins, rifting of continental slivers to island-arc positions, and the development of the interarc basins that we hold to be typical of oscillatory episodes.

Interplay between flow paths of melt expulsion onto cratons and to continental margins would explain not only the creation of spreading centers such as the Tasman Sea axis recently confirmed by Hayes and Ringis (1973) but also their abandonment and subsequent genesis at new positions, differentiating active interarc basins from inactive marginal basins (Karig, 1971).

Summary

Figure 9 is a highly diagrammatic view of the application of our theory to the Phanerozoic history of cratons and oceans. The upper curve of the diagram shows the relative elevation of cratons with respect to sea level as a function of time. Three emergent episodes are identified; as estimated from the apparent depth of erosion at each of the defining unconformities, we judge the late Precambrian (pre-Sauk) episode to have involved the greatest cratonic uplift and the early Middle Ordovician (pre-Tippecanoe) emergence to represent the least emergence. The relative degree of cratonic depression indicated for the four submergent episodes derives from a highly subjective integration of data on the volume of cratonic sediment preserved, the amplitudes of basins and arches, and the apparent magnitude of craton-margin orogenesis. There is a danger of circularity of reasoning in such an exercise—the late Precambrian "Windermere" submergence is shown as very slight because sediments of the episode are confined to cratonic margins; yet, we assign extreme uplift to the pre-Sauk emergence because of the degree to which older rocks were stripped by erosion. However, facies analysis strongly suggests that "Windermere" deposition did not encroach far onto cratons and evidence is lacking to indicate the activity of either intracratonic basins or major orogenesis at continental margins. Although the record of Sauk deposition is much reduced by subsequent erosion on all cratons, it is not difficult to demonstrate that cratonic submergence was not as extensive areally, and was not accompanied by the degree of differentiation of basins and arches as that indicated for younger submergent episodes which are shown to involve greater depths of cratonic depression.

Our emergence-submergence curve is smooth relative to second-order events of higher frequency and lower amplitude. There is no question that such events occurred, but neither the scale of the diagram nor the present state of knowledge is sufficient to permit delineation of details. Similarly, we cannot illustrate the many oscillations that characterized the time spans of Absaroka and Tejas deposition. Instead, we have shown, by estimating the upper and lower limits of mean cratonic elevation with time, the broad-band nature of the vertical amplitudes of cratonic uplift and depression during these two oscillatory episodes. The late Paleozoic and early Mesozoic episode appears to have been marked by a general decrease in the amplitude of oscillations from Late Carboniferous to Early Jurassic, concomitant with an increase in the mean elevation of the craton. The Cenozoic episode has not run its course and comparisons are hazardous. Although it is clear that net cratonic emergence has increased from Eocene time to the present, in accordance with the pattern set during Absaroka deposition, there are, as yet, no indications of the damping out of vertical oscillations.

The baseline of the lower plot of figure 9 represents the assumed rate of sea-floor spreading at midoceanic ridges from steady convection in the oceanic system involving essentially no contributions from the continental asthenospheres. Cratonic emergence, according to our hypothesis, is caused by retention of melt in continental asthenospheres. Retention would deprive oceanic convection systems of the discharge produced by the subcontinental heat flux and, according to our previous assumption that the spreading rate is proportional to discharge, there would be a minimum rate during emergent episodes. Conversely, the initiation of cratonic submergence is thought to result from the expulsion of subcontinental melt, chiefly into suboceanic convection systems, thus increasing the volume extruded at midoceanic ridges and raising the rate of sea-floor spreading. The maximum in the outflux of subcontinental melt would be at the incidence of deflation, the time of maximum asthenospheric permeability, and maximum spreading rates would be achieved early in a submergent episode.

It is also a presumption that the rate of melt expulsion, and spreading rates, would decay gradually as the proportion of melt in the asthenosphere decreases during a submergent episode to approach the steady oceanic condition. The history of continental margins can be interpreted to indicate variations in spreading rates during cratonic submergence, but our simplified diagram excludes such second-order perturbations.

We claim our oscillatory state of cratonic behavior represents a special variant of the emergent mode, differing by intermittency of retention and expulsion of subcontinental melt as opposed to the more complete melt storage believed typical of the emergent state. We further suggest that, in times of asthenospheric deflation during oscillatory episodes, the expelled melt may follow any or all of three paths: to volcanic conduits on the flanks or margins of cratons, to interarc spreading axes, or to oceanic convection systems. At times of melt retention, oceanic sea-floor spreading rates would be depressed, as during emergent episodes; when there is melt discharge via interarc basins, oceanic spreading rates would continue to be low, but the rate of plate convergence at active island arcs would be high; during times of maximum diversion of melt to oceanic systems, spreading rates would be temporarily high. Although, in the interests of simplicity, our diagram (fig. 9) does not attempt to show variations in spreading rates during oscillatory episodes, it is possible that Cenozoic spreading-rate changes (*e.g.*, Rona, 1973) are consequences of the interplay between melt retention and expulsion and among paths of melt outflow.

PROBLEMS

Segments of the Pacific border of the Americas appear to have been convergent plate boundaries, at least at intervals, during the oscillatory episode that occupies all but the first few million years of Cenozoic time. Plate convergences of oscillatory episodes are confined to island arc positions remote from continental margins according to our tenet; thus, the Andes, the Peru-Chile trench, and our understanding of their recent histories constitute a minor embarrassment. Our suggestion would be that plate boundaries may not be uniquely remote from cratons during oscillatory states, but only that such boundaries dominate the geographies of oscillatory episodes. Similarly, Andean margins prevail in submergent states, but remote island arc boundaries may have occurred here and there at such times.

A further complication is introduced by apparent interruptions of submergent episodes by brief periods of emergent or oscillatory behavior. Late Cambrian stratigraphy, for instance, bears many evidences of a significant marine regression within the time span of the Sauk Sequence (the sub-Franconia unconformity of the North American interior; the absence of Middle Cambrian below a Late Cambrian unconformity in much of arctic Canada). Wheeler (1963) has applied regional unconformities in the latest Ordovician and latest Devonian of parts of North America to define subdivisions of the Tippecanoe and Kaskaskia Sequences. One of us (Sloss, 1964, 1966) has identified these discontinuities, plus others in the Late Jurassic and Early Cretaceous, as characteristic of the median time position of what we now would call submergent episodes. The Late Devonian and Late Jurassic-Early Cretaceous events are accompanied in a number of areas near craton flanks and margins by high-angle faulting and, in some localities, by volcanism and/or the emplacement of sills and dikes. According to our current scenario, we would invoke temporary cessations of asthenospheric deflation and transient melt retention below continents to explain short-lived reversions to emergent or oscillatory states within submergent episodes.

We have ignored the question of continental collisions and their role in affecting tectonic and sedimentary conditions at craton margins. Collision, a seemingly inevitable consequence of plate convergence (*e.g.*, Dickinson, 1971), can be considered a relatively rare and short-term event that, if craton margins are involved, may be more likely to occur during submergent episodes than at other times.

REFERENCES CITED

Adler, F. J., 1971, Anadarko Basin and central Oklahoma area: Am. Assoc. Petroleum Geologists Mem. 15, p. 1061–1070.

Anderson, D. L., Sammis, C. G. and Jordan, T. N., 1972, Composition of the mantle and core, *in* Robertson, E. C. (ed.), The nature of the solid earth: McGraw-Hill, N.Y., p. 41–66.

Bond, D. C., 1971, Possible future petroleum potential of region 9—Illinois Basin, Cincinnati Arch, and Northern Mississippi Embayment: Am. Assoc. Petroleum Geologists Mem. 15, p. 1165–1218.

Chaterji, G. C., and Ghosh, P. K., 1967, Tectonic frame-work of the peninsular Gondwanas of India, *in* Gondwana stratigraphy, International Gondwana Symposium: UNESCO, Paris, p. 923–925.

Christie, R. L., 1972, Central stable region, *in* Christie, R. L., Cook, D. G., Nassichuk, W. W., Trettin, H. P., and Yodath, C. J. (eds.), Guidebook to the Canadian Arctic Islands and the MacKenzie Region: 24th Intern. Geol. Cong., Harpell's Press Co-operative, Quebec, p. 41–87.

Connolly, J. R., 1969, Models for Triassic deposition in the Sydney Basin: Geol. Soc. Australia Special Pub., v. 2, p. 209–223.

Dickinson, W. R., 1971, Plate tectonics in geologic history, new global tectonic theory leads to revised concepts of geosynclinal deposition and orogenic deformation: Science, v. 174, p. 107–113.

Douglas, R. J. W., Gabrielse, Hubert, Wheeler, J. O., Stott, D. F., and Belyea, H. R., 1970, Geology of Western Canada, *in* Douglas, R. J. W. (ed.), Geology and economic minerals of Canada: Geol. Surv. Canada, Economic Geology Report No. 1, p. 367–490.

Fulfaro, V. J., and Landim, P. M. B., 1972, Tectonic and paleogeographic evolution of the Paraná sedimentary basin by trend surface analysis: 24th Internat. Geol. Cong. Rept., Sec. 6, p. 379–388.

Hallam, Anthony, 1963, Major epeirogenic and eustatic changes since the Cretaceous, and their possible relationship to crustal structure: Am. Jour. Sci., v. 261, p. 397–423.

Hayes, D. E., and Ringis, John, 1973, Seafloor spreading in the Tasman Sea: Nature, v. 243, p. 454–458.

Jacoby, W. R., 1972, Plate theory, epeirogenesis, and eustatic sea-level changes: Tectonophysics, v. 15, p. 187–196.

Johnson, J. G., 1971, Timing and coordination of orogenic, epeirogenic and eustatic events: Geol. Soc. America Bull., v. 82, p. 3263–3298.

Karig, D. E., 1971, Origin and development of marginal basins in the Western Pacific: Jour. Geophys. Research, v. 76, p. 2542–2561.

Kay, Marshall, 1947, Geosynclinal nomenclature and the craton: Am. Assoc. Petroleum Geologists Bull., v. 31, p. 1289–1293.

———, 1951, North American geosynclines: Geol. Soc. America Mem. 48, 143 p.

Laurent, Roger, 1972, The Hercynides of South Europe, a model: 24th Internat. Geol. Cong. Rept., Sec. 3, p. 363–370.

MacDonald, G. J. F., and Ness, N. F., 1960, Stability of phase transitions within the Earth; Jour. Geophys. Research, v. 65, no. 7, p. 2173–2190.

McKenzie, D. P., 1967, Some remarks on heat flow and gravity anomalies: *ibid.*, v. 72, p. 6261–6273.

Morgan, J. W., 1968, Rises, trenches, great faults, and crustal blocks: *ibid.*, v. 73, p. 1959–1982.

Nadai, Arpad, 1967, Theory of flow and fracture of solids, v. II.: McGraw-Hill, N.Y., 705 p.

Press, Frank, 1972, The earth's interior as inferred from a family of models, *in* Robertson, E. C. (ed.), The nature of the solid earth: McGraw-Hill, N.Y., p. 147–171.

Rona, P. A., 1973, Relationships between rates of sediment accumulation on continental shelves, sea-floor spreading, and eustacy inferred from central North Atlantic: Geol. Soc. America Bull., v. 84, p. 2851–2872.

RYAN, P. J., 1967, Stratigraphy of the Ecca series and lowermost Beaufort beds (Permian) in the Great Karroo Basin of South Africa, *in* Gondwana stratigraphy, International Gondwana Symposium; UNESCO, Paris, p. 945–967.

SCHERER, WOLFGANG, 1973, A mathematical model for the differential subsidence of intra-cratonic basins (Ph.D. thesis): Northwestern Univ., Evanston, 60 p.

SLEEP, N. H., 1974, Segregation of magma from a mostly crystalline mush: Geol. Soc. America Bull., v. 85 (in press).

SLOSS, L. L., 1963, Sequences in the cratonic interior of North America: *ibid.*, v. 74, p. 93–114.

———, 1964, Tectonic cycles of the North American craton, *in* Merriam, D. F. (ed.), Symposium on cyclic sedimentation: Kansas Geol. Survey Bull. 169, p. 449–460.

———, 1966, Orogeny and epeirogeny: the view from the craton: N.Y. Acad. Sciences Trans., Ser. II, v. 28, p. 579–587.

———, 1972, Concurrent subsidence of widely separated cratonic basins (abs.): Geol. Soc. America Abstracts with Programs, v. 4, p. 668–669.

———, 1972, Synchrony of Phanerozoic sedimentary-tectonic events of the North American craton and the Russian platform: 24th Internat. Geol. Cong. Rept., Sec. 6, p. 24–32.

———, KRUMBEIN, W. C., AND DAPPLES, E. C., 1949, Integrated facies analysis, *in* Longwell, C. R. (chm.), Sedimentary facies in geologic history: Geol. Soc. America Mem. 39, p. 91–123.

SOLOMON, S. C., 1972, Seismic-wave attenuation and partial melting in the upper mantle of North America: Jour. Geophys. Research, v. 77, p. 1483–1501.

STILLE, HANS, 1936, Die Entwicklung des amerikanischen Kordillerensystems in Zeit und Raum: Preuss. Akad. Wiss. Phys.-Math. Kl. Sitzungsber., v. 15, p. 134–155.

THERON, J. C., 1967, A palaeocurrent analysis of a portion of the Beaufort series, Karroo System, *in* Gondwana stratigraphy, International Gondwana Symposium: UNESCO, Paris, p. 725–741.

TRETTIN, H. P., 1972, The Franklinian geosyncline, *in* Christie, R. L., Cook, D. G., Nassichuk, W. W., Trettin, H. P., and Yorath, C. J. (eds.), Guidebook to the Canadian Arctic Islands and the MacKenzie Region: 24th Internat. Geol. Cong., Harpell's Press Co-operative, Quebec, p. 87–109.

VEEVERS, J. J., 1967, The Phanerozoic geological history of north-west Australia: Jour. Geol. Soc. Australia, v. 14, p. 253–271.

WALCOTT, R. I., 1970, Flexural rigidity, thickness, and viscosity of the lithosphere: Jour. Geophys. Research, v. 75, p. 3941–3954.

WHEELER, H. E., 1963, Post-Sauk and pre-Absaroka Paleozoic stratigraphic patterns in North America: Am. Assoc. Petroleum Geologists Bull., v. 47, p. 1497–1526.

WILSON, J. T., 1965, A new class of faults and their bearing on continental drift: Nature, v. 207, p. 343.

WINTER, H. DE LA R., AND VENTER, J. J., 1970, Lithostratigraphic correlation of recent deep boreholes in the Karroo-Cape sequence, *in* Second Gondwana Symposium (proceedings and papers): The Natal Witness Ltd., Natal, p. 395–409.

ZWART, H. J., 1967, The duality of orogenic belts: Geologie en Mijnbouw, v. 46, p. 283–309.

———, 1969, Metamorphic facies series in the European orogenic belts and their bearing on the causes of orogeny: Geol. Assoc. Canada Special paper 5, p. 7–16.

SEDIMENTARY AND TECTONIC HISTORY OF THE OUACHITA MOUNTAINS[1]

ROBERT C. MORRIS
Northern Illinois University, DeKalb, Illinois

ABSTRACT

Ordovician through Mississippian strata of the Ouachita Mountains developed in a classical "starved trough," and consist mostly of dark slates and cherts with only minor intercalations of coarser clastic sediment. Mature, well-sorted quartz and rounded carbonate sands were derived from the North American craton whereas immature, fine-grained sands came from the south or east. The succeeding Carboniferous section, almost 12,250 m (40,000 ft) thick, consists of proximal and distal turbidite sandstones, black shales, and minor interlayered wildflysch and volcanic ash. Sedimentary structures indicate predominant westward sand dispersal; compositions suggest a quartz-rich, cratonic provenance as well as an extracontinental source of feldspathic and lithic clasts.

South of the Ouachita trough, during the early Paleozoic, oceanic crust plunged under a plate including North America to create an island arc-trench-subduction zone. Northwest of the trench, a slope, rise, and abyssal plain formed along the southern margin of the North American continent. East of the Ouachitas, continent-to-continent collision began in the late Devonian, creating growing source areas from which detrital materials were subsequently shed westward to build a subsea fan during the Carboniferous. South of the Ouachita trough continued subduction by the late Pennsylvanian ultimately created a series of uplifted tectonic lands and resultant sliding of the sedimentary succession northward as folds and thrust sheets against North America. Mesozoic spreading then disrupted the Paleozoic fold belt.

INTRODUCTION

The Ouachita Mountains of Oklahoma and Arkansas are formed by a thick succession of folded and faulted Paleozoic strata that constitutes one of the important geosynclinal belts of the North American continent. The object of this paper is to describe the major characteristics of these rocks and to interpret their origin. The concept of sea-floor spreading and plate tectonics provides a framework that best explains the sedimentary and tectonic history of this region.

The Ouachita terrane has been deeply eroded since its deformation and is overlapped by Mesozoic and Cenozoic clastic rocks except along an outcrop belt (fig. 1) that extends for only 350 km (220 mi). From considerable well data (Flawn and others, 1961), we know that the Ouachita fold belt continues in the subsurface. To the southeast, it makes a poorly understood connection with the folded Appalachians. To the southwest, it passes around the southern margin of the Llano uplift and connects with the Marathon area of west Texas, where folded rocks of the belt are again exposed in a small area. Gilluly and others (1970) have made a very conservative estimate that 230,000 km^3 of Paleozoic rock comprise the Ouachita belt.

The major physiographic and tectonic provinces near the Ouachita Mountains are more clearly outlined in figure 2. Some explanation is needed for the term Frontal Ouachitas. In Oklahoma this term is reserved for a narrow belt between the Ti Valley and Choctaw faults, but because the Choctaw fault disappears in Arkansas this belt becomes simply part of the Arkoma basin along strike. The Frontal Ouachitas of Arkansas compare somewhat with the Central Ouachitas of Oklahoma, and consist of a succession of folded and imbricated blocks of Carboniferous rocks. The Southern Ouachitas provide an opportunity to examine rocks deposited farthest seaward from the North American craton. The Broken Bow and Benton uplifts and Potato Hills locate the Ordovician, Silurian, and Devonian rocks in the area. A more detailed geologic map of the Ouachita Mountains is provided by figure 3, although all faults and many small outcrops are omitted.

[1] Much of the field and laboratory work on Carboniferous rocks was supported by a grant from the National Science Foundation (GA-844). Several timely grants from the Council of Academic Deans, Northern Illinois University, allowed me and several graduate students to work on various aspects of Ouachita geology. The Arkansas Geological Commission, under the direction of N. F. Williams, also provided geological and logistical support. C. G. Stone aided in the study of the pre-Stanley rocks. M. P. Weiss read the manuscript and offered suggestions for its improvement. The writer is especially grateful to W. R. Dickinson for his constructive comments.

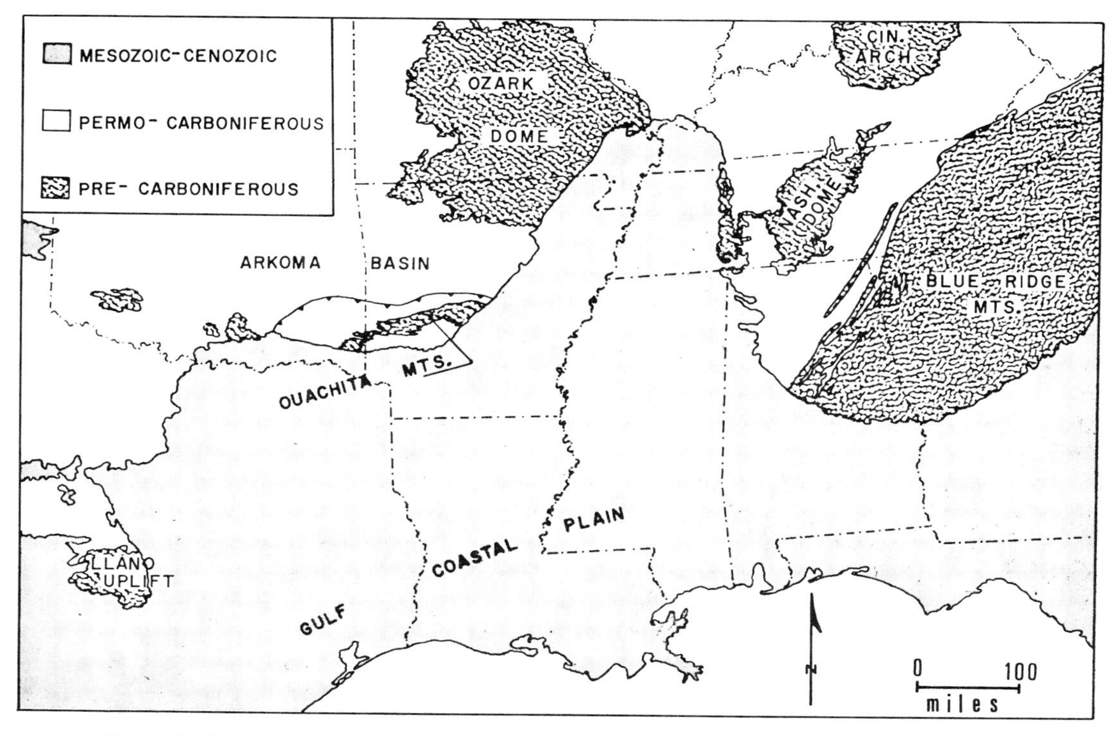

FIG. 1.—Generalized geologic map of south-central United States showing location of Ouachita Mountains.

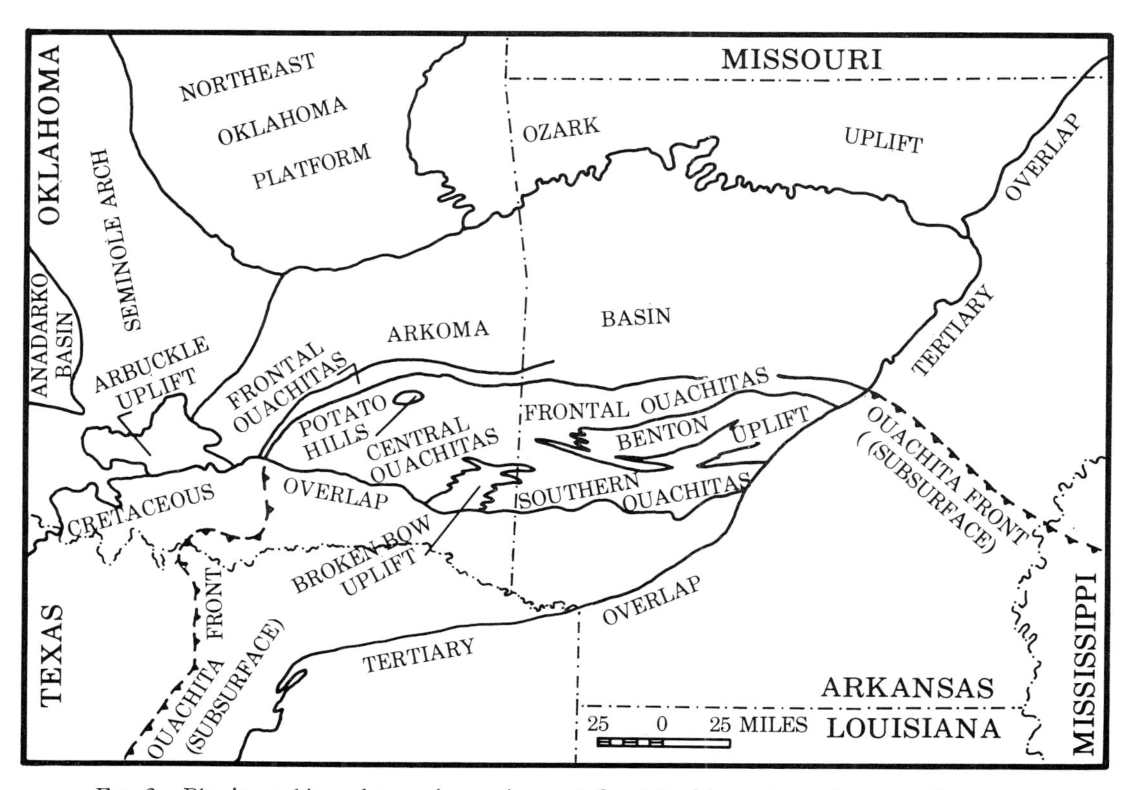

FIG. 2.—Physiographic and tectonic provinces of Ouachita Mountains and surrounding areas.

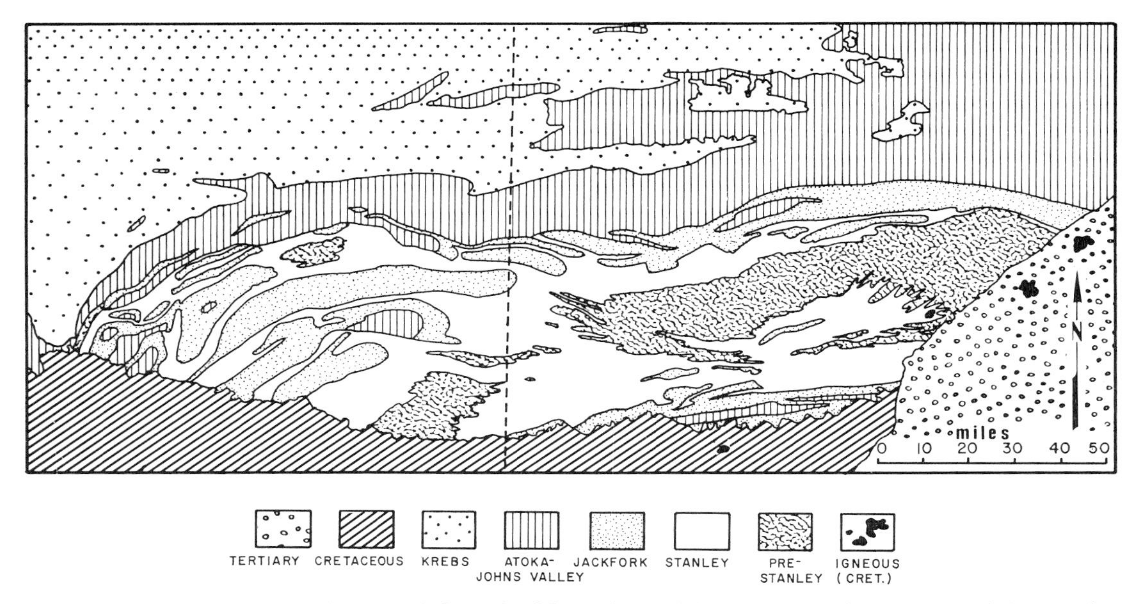

FIG. 3.—Generalized geologic map of Ouachita Mountains and Arkoma basin. See table 1 and figure 4 for stratigraphy. All faults and minor outcrops are cmitted. The Ordovician, Silurian, and Devonian units are grouped together as pre-Stanley strata because of their structural complexity and limited outcrop widths individually. Modified from Miser (1959).

STRATIGRAPHY

Correlation

The strata of the Ouachita Mountains form a thick, deep-water succession and have a limited number of geologic names (table 1). Figure 4 shows the currently accepted correlations of these units with adjacent, thinner shelfal or foreland rocks of the Ozark and Arbuckle Mountains. The Stanley and Jackfork have additional subdivisions as proposed by Harlton (1938, 1959) and Morris (1971a). As faulting or sedimentary cover prohibits actual tracing of shelfal stratigraphic units laterally into the Ouachita units, and lithologic characteristics are completely different, sparse fossil evidence is our only means of correlation. The Ordovician and Silurian formations can be accurately correlated by the presence of both graptolite-bearing black shales (Miser and Purdue, 1929; Decker, 1959) and turbidite limestones with conodonts (Repetski and Ethington, 1973). Conodonts were used by Hass (1950, 1951) to zone the Arkansas Novaculite and also to date the contact between the Arkansas Novaculite and the Stanley. Plant fossils of Carboniferous age from the Stanley and Jackfork were described by White (1934, 1937). An understanding of how the Stanley-Jackfork-Johns Valley succession is related to shelfal rocks of the same age has been hampered because geologists were not sure whether Mississippian (Chesterian) goniatites found within the lower Johns Valley of Oklahoma were allochthonous or autochthonous. Cline (1960) and Miser and Hendricks (1960) concluded that these fossils had not been transported and that the Johns Valley spans the Mississippian-Pennsylvanian boundary. However, evidence offered by Gordon and Stone (1973) from a study of numerous Stanley and Jackfork sites of resedimented fossils suggests that the upper Stanley is younger than any of the Chesterian rocks of the Ozark uplift, hence that the Jackfork and Johns Valley must be Pennsylvanian (Morrowan). This age determination is very important to the reconstruction of the paleogeography at that time. Trace fossils from the Carboniferous strata (Chamberlain, 1971) also provide important clues to bathymetry and paleoecology in the Ouachita trough.

Thicknesses and Rates of Sedimentation

By using the maximum thickness of each Ouachita unit as estimated by Stone and others (1973), and taking the absolute ages given on the correlation chart (fig. 4), rates of deposition have been calculated (table 1). In general, the pre-Stanley units are thinner and were deposited more slowly than the remainder. Also,

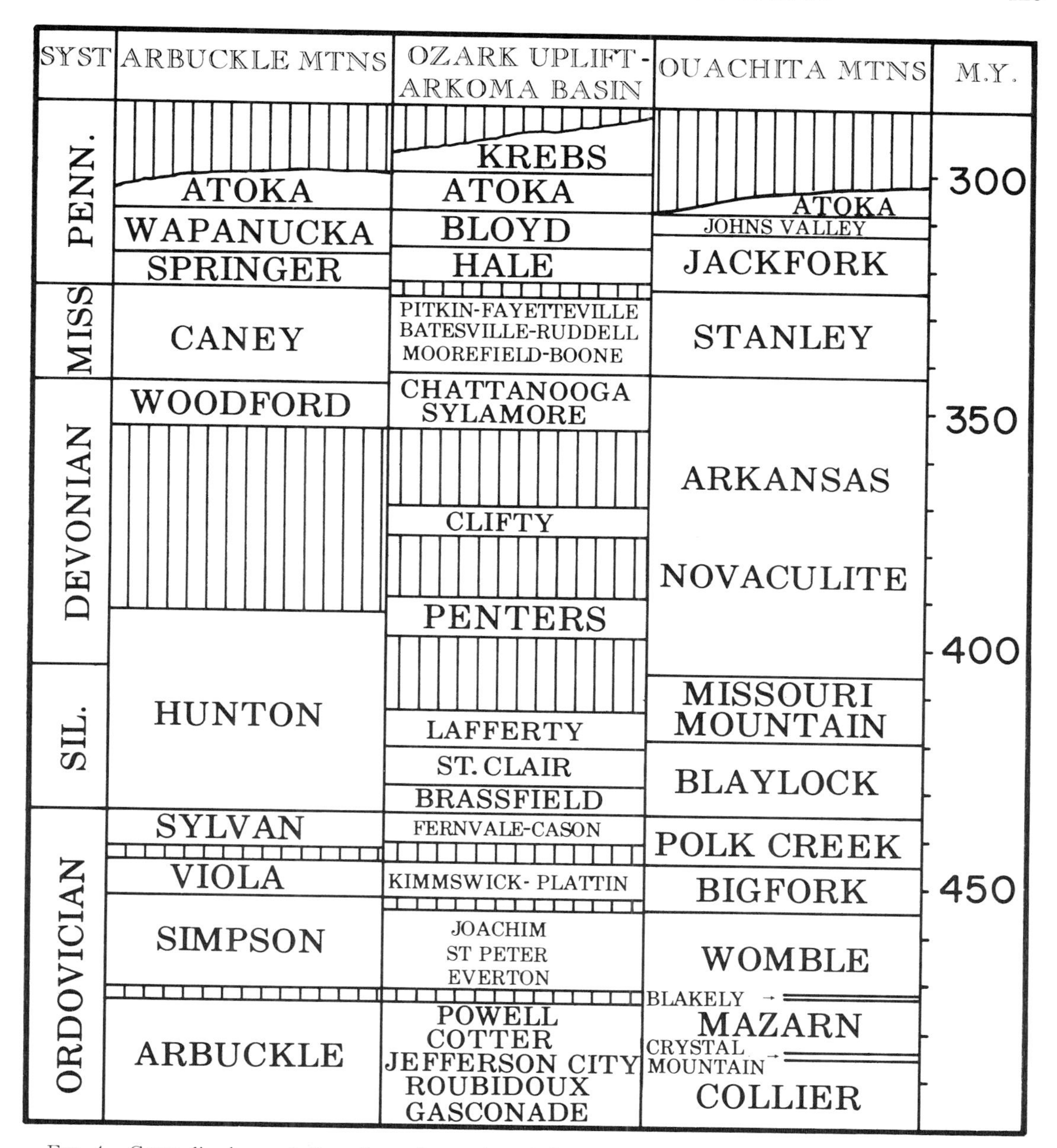

FIG. 4.—Generalized correlation chart for rocks of Ouachita Mountains and adjacent areas using key stratigraphic names only. No major breaks in the stratigraphic section have been discovered in the Ouachita Mountains. The rocks of the Ozark area probably contain more unconformities than actually shown (Frezon and Glick, 1959). Numbers are millions of years before present.

according to Ham (1959) and Cline (1960), the pre-Carboniferous shelfal rocks now present in the Arbuckle Mountains are thicker than those of an equivalent age in the Ouachita Mountains. Thus, the Ouachita area may have been "starved" for coarser clastic sediment during the early Paleozoic. Figure 5 is a cross-section, modified from Ham (1959), showing facies relations restored to the end of Arkansas Novaculite time. It clearly shows a thinner Ouachita column. However, if the maximum thicknesses of Stone and others (1973) are more nearly correct, the Ouachita and Arbuckle sections would have the same approxi-

TABLE 1.—RATES OF DEPOSITION, OUACHITA FACIES, ARKANSAS

Age	Rock unit	Maximum thickness[1]	Rate of deposition[2]
Pennsylvanian	Atoka Formation	27,500	3,000
	Johns Valley Shale	1,500	375
	Jackfork Sandstone	6,000	500
Mississippian	Stanley Shale	8,500	550
Devonian	Arkansas Novaculite	950	14.6
Silurian	Missouri Mountain Shale	250	20
	Blaylock Sandstone	1,500	97
Ordovician	Polk Creek Shale	175	18
	Bigfork Chert	800	80
	Womble Shale	3,500	206
	Blakely Sandstone	400	500
	Mazarn Shale	3,000	300
	Crystal Mountain Sandstone	850	570
	Collier Shale	1,000+	50±

[1] in feet
[2] in feet per million years

mate thickness, although the lithologies are completely different. The section of the Ozark area is only about a third as thick (Frezon and Glick, 1959), and the Marathon area had only 1000 m (3000 ft) of sediments by Early Mississippian time (King, 1973).

The base of the Stanley marks a quickening of sedimentation rates that reached a maximum for the Atoka. Figure 6 shows selected, generalized stratigraphic columns in which Carboniferous Ouachita rocks are compared with upper slope or lower shelf rocks of the Frontal Ouachitas of Oklahoma and the eastern Arkoma basin. The southward thickening is most notable in the Stanley and Jackfork, mainly due to an increase in turbidite sandstones in this direction. Disturbed bedding (or wildflysch) has its thickest development in the Jackfork across the Frontal Ouachitas (Morris, 1971b).

Sedimentary Facies

The first geologist to consider the Stanley and Jackfork strata a flysch facies was van der Gracht (1931), although Cline (1960, 1970) was responsible for the widespread use of this term for Carboniferous rocks of the Ouachitas. Cline (1970) stated that flysch sedimentation began in middle Mississippian time and described 17 characteristic features of the Ouachita flysch, the most notable being a rapid alternation of fine- and coarse-grained strata. Morris (in press) has erected a formal classification of Carboniferous Ouachita flysch facies that include pelagic rocks, disturbed bedding, proximal turbidites, and distal turbidites, and these terms will be used throughout this report. Although more rapid sedimentation rates did occur during the Carboniferous, I suggest the use of the term flysch for the *entire* Phanerozoic succession of the Ouachitas because the pre-Stanley rocks also possess most of the characteristics of flysch as described by Cline (1970), Dzulynski and Walton (1965), McBride (1970), and others (Lajoie, 1970).

The pelagic facies include strata dominated by sediments that accumulated slowly by settling of suspended material, possibly altered by diagenesis. Such rocks probably formed in the deeper, more stable reaches of the Ouachita trough where incursions of turbidity flows and mass slumps were infrequent. Included in this category are thick, undisturbed shale or mudstone intervals, as well as siliceous shale, impure chert (porcelanite), and dense, even-textured, light-colored, cryptocrystalline siliceous rock (novaculite).

Disturbed bedding, a collective term for all types of soft-sediment deformation, occurs throughout the entire section. It is especially abundant in Jackfork rocks and has been described and classified in detail by Morris (1971b). When soft sediment moves down a depositional slope, the resulting deposit depends upon type of material involved, its state of consolidation, and its distance of transport down the slope. Rubble bedding, defined as thoroughly churned mudrocks with a complex assortment of angular or rounded, non-oriented blocks, is referred to as wildflysch by some

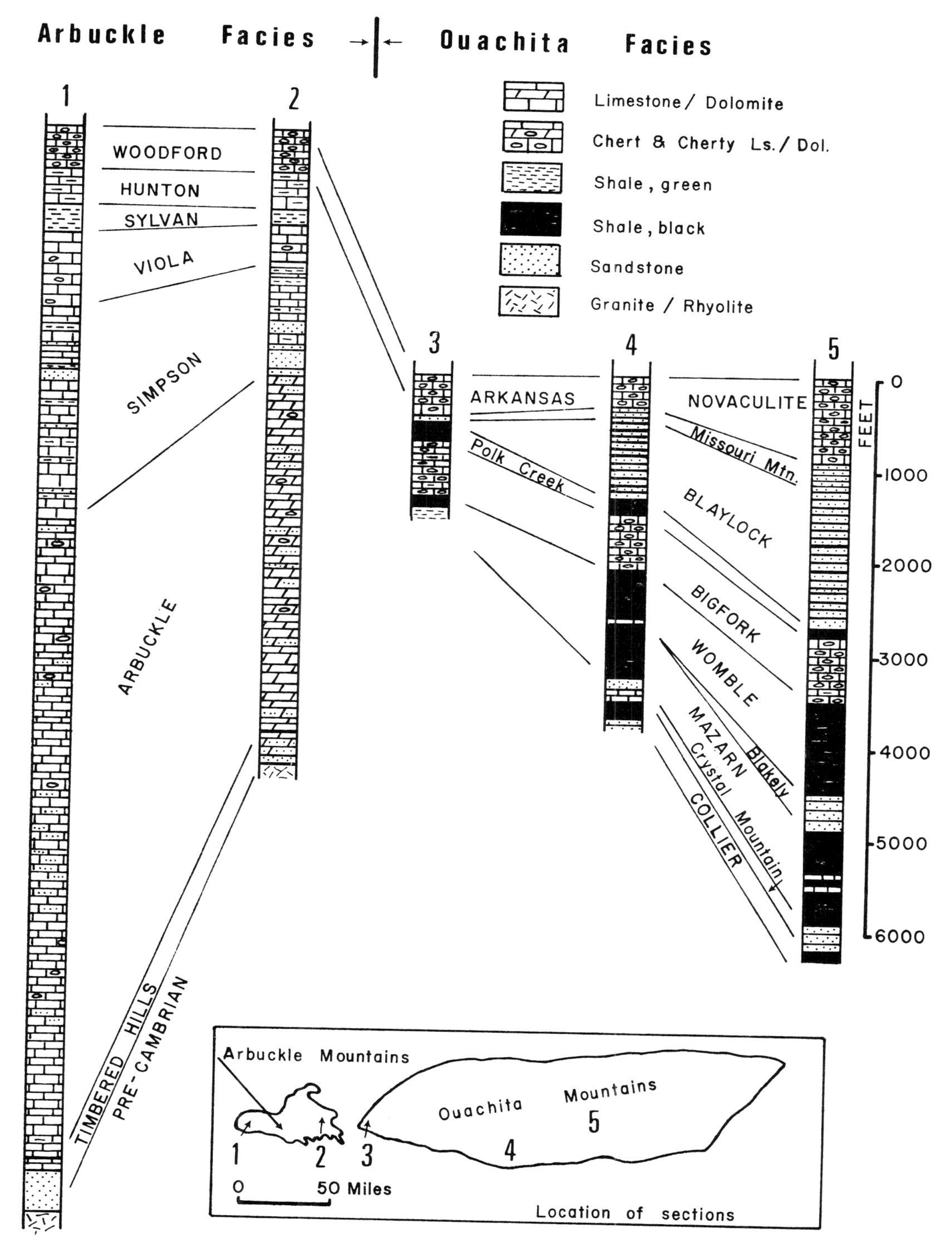

FIG. 5.—Correlation and facies of pre-Stanley strata in Ouachita and Arbuckle Mountains. Modified from Ham (1959).

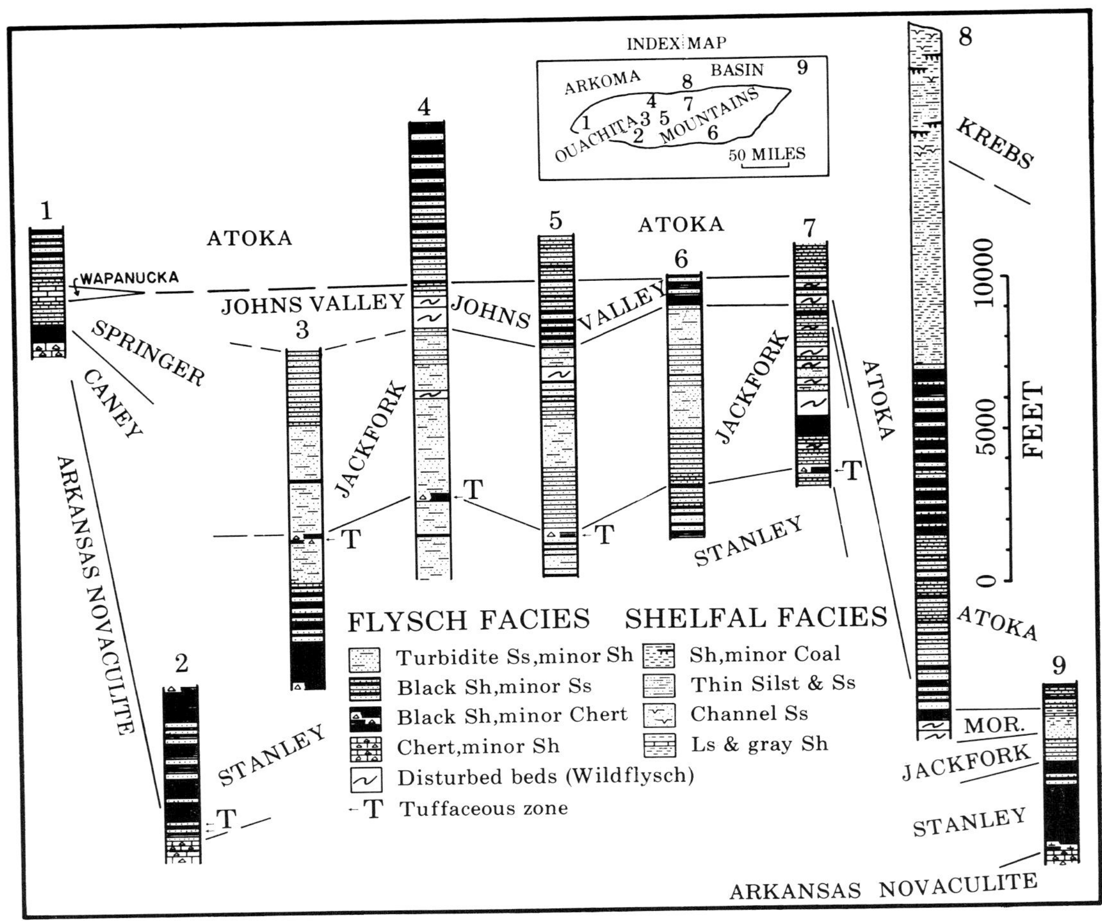

Fig. 6.—Correlation and facies of Carboniferous strata in Ouachita Mountains and Arkoma basin. Sources as follows: section 1 (Cline, 1960); sections 2 and 3 (Hill, 1967); section 4 (Hart, 1963); section 5 (Cline and Moretti, 1956); sections 6 and 7 (Morris, 1971a); section 8 (Reinemund and Danilchik, 1957); section 9 (Maher and Lantz, 1953).

geologists. This type of bedding occurs most commonly across the Frontal Ouachitas, suggesting that slumping was mainly north to south; however, some evidence of south-to-north slumping as well is present in upper Carboniferous rocks.

Thick-bedded Crystal Mountain and Blakely sandstones as well as repeated intervals within the Jackfork and Stanley form conspicuous ridges across the Ouachitas. These non-rhythmic, partially scoured, shale-poor units match Walker's (1967) descriptions of proximal turbidites so well that they must have a similar genesis. Close examination of these rocks permits subdivision of this facies into two subfacies, primarily based upon whether the contacts between beds are scoured or nonscoured (table 2).

Distal turbidites are characterized by rhythmically interbedded shales and turbidite sandstones which generally have hieroglyphs and internal lamination. A two-fold subdivision of this facies has been employed, depending upon whether turbidities (sandy flysch) or black shales (shaly flysch) dominate a succession (table 3). Figure 7 illustrates the gross characteristics of sandy flysch in outcrop. Most of the turbidities are quartzose sandstones but thin turbidite limestones are common in Ordovician rocks of Arkansas. The lucid descriptions of Cline (1970) of the late Paleozoic flysch succession in Oklahoma primarily focus upon the distal turbidite facies.

Major Lithologic Characteristics

Listed below are capsule descriptions of the

TABLE 2.—DISTINCTIVE BEDDING CHARACTERISTICS OF PROXIMAL TURBIDITES

Scoured contacts	Non-scoured contacts
Minimum bed thickness 25 cm, beds commonly exceed 200 cm	Minimum bed thickness 25 cm, beds rarely exceed 100 cm
Channel-fills locally	Channels never present
Scour-and-fill contacts	Sharp, even contacts with shales
Laminated sandstones absent	Laminated sandstones scarce
Bedding planes obscured	Rhythmic, distinct bedding
Absence of hieroglyphs	Hieroglyphs rare
No cross-bedding	No cross-bedding

major units (see table 1). More complete descriptions of the pre-Carboniferous rocks are given by Amsden and others (1967), Ham (1959), Miser and Purdue (1929), Pitt and others (1961), Sellars (1967), Sterling and others (1966), and Stone and others (1973). Some of the more recent descriptions for Stanley, Jackfork and Atoka rocks include those of Cline (1960), Cline and Moretti (1956), Fellows (1964), Hammes (1965), Hart (1963), Hill (1967), Maher and Lantz (1953), Morris

TABLE 3.—DISTINCTIVE BEDDING CHARACTERISTICS OF DISTAL TURBIDITES

Sandy flysch	Shaly flysch
Beds 3 to 100 cm thick, average 15 cm	Beds mostly less than 5 cm thick, average 2 cm
Broad scours rare, channels absent	No scours or channels
Sharp, even contacts upon shales	Sharp, even contacts upon shales
Subequal amounts of sandstones and mudrocks	Mudrocks dominant over sandstones
Laminated sandstones common	Laminated sandstones common
Rhythmic distinct bedding	Rhythmic, distinct bedding
Tops rippled, planar, or gradational	Tops rippled, planar, or gradational
Abundance of tool and flute casts, load structures	Small tool and flute casts, few load structures
Small-scale cross-bedding common	Small-scale cross-bedding common
Bouma sequences common	Bouma sequences common

(in press), Reinemund and Danilchik (1957), Seely (1963), Shelburne (1960), Shideler (1970), Sullivan (1966) and Walthall (1967).

Collier Shale.—The bulk of the formation is

FIG. 7.—Carboniferous flysch—a typical distal turbidite succession consisting of alternating turbidite sandstones and shales at DeGray Dam, Arkansas.

bluish-black, soft graphitic clay shale with lenses and beds of steel-gray or blue-gray limestone. The limestones are distal turbidites, probably derived from the carbonate shelf to the north and emplaced upon a muddy, abyssal area above the carbonate compensation depth.

Crystal Mountain Sandstone.—This unit is light-gray, well-sorted, fine- to medium-grained, quartz-cemented, structureless, thick-bedded, ridge-forming quartz arenite with minor shales. The irregular contacts of the sandstones suggests that they are proximal turbidites and/or watery slides (slurried bedding of Morris, 1971b) derived from the northern shelf via submarine canyons.

Mazarn Shale.—The Mazarn Shale is chiefly dark-gray laminated graptolitic shale with numerous thin interbedded distal turbidites, both quartzose as well as carbonate types.

Blakely Sandstone.—The formation consists of massive to thin-bedded, well sorted, fine-grained quartz arenite and varicolored silts and shales. Several zones of rubble bedding contain swarms of rounded, weathered meta-arkose and granitic blocks up to 15 m (45 ft) in diameter. The blocks may have been emplaced as submarine slides of material derived from cratonal exposures to the north (Stone and others, 1973). The structureless quartz arenites, pulled apart in places, were apparently deposited rapidly as proximal turbidites or watery slides upon an unstable slope.

Womble Shale.—The Womble Shale consists principally of gray-black graptolitic slaty shale. Numerous thin-bedded calcareous distal turbidites occur within the lower part, whereas the upper part contains some turbidite siltstones, phosphatic breccias, and gray-black, well sorted, pelletal turbidite limestones.

Bigfork Chert.—This resistant unit consists of gray and black, thin-bedded, highly fractured chert with minor interbedded clay shales, siliceous shales, and siliceous limestones. The limestones appear to be turbidites whereas the other rocks are pelagic in origin.

Polk Creek Shale.—This thin shale unit is black, highly fissile, graptolitic, and carbonaceous with a few thin beds of calcareous chert.

Blaylock Sandstone.—The formation consists of subequal proportions of olive-gray, thin-bedded, laminated, feldspathic, very fine-grained wacke-sandstones and siltstones as well as greenish gray shales that together comprise a shaly flysch facies. From their fine grain size, poor sorting, and feldspar content, it appears that these distal turbidites are the oldest significant accumulation of Paleozoic clastic sediment that apparently was not derived from the North American craton, for they are important only in the Southern Ouachitas.

Missouri Mountain Shale.—Varicolored shales (or slates) comprise the bulk of the unit. Near the top are thin interbedded turbidites composed of white, fine- to medium-grained well rounded, silica-cemented quartz arenite, almost certainly of cratonic origin.

Arkansas Novaculite.—This distinctive unit consists chiefly of novaculite, a light-colored, extremely fine-grained, homogeneous, highly fractured siliceous rock similar to chert but characterized by a dominance of quartz rather than chalcedony. The upper and lower members contain laminated beds of rounded and angular quartz grains, as well as intraformational breccias of black chert, phosphate, and coarse quartz grains. A middle member is characterized by thin graded laminated siliceous beds and black shales (Lowe, 1973a).

Stanley Shale.—The Stanley is dominantly olive-gray to black shale with subordinate proximal and distal turbidites (fig. 8A and 8B), which are very fine-grained and generally feldspathic wackes, at the base and near the top. Along the Southern Ouachitas, sandstones become thicker, more numerous, and tend to be matrix-rich (fig. 8D); tuffs and tuffaceous sandstones are prominent at the base; and there is a rarity of disturbed bedding, impure cherts, and siliceous shales. Along the Frontal Ouachitas, a thinner Stanley section has less turbidites at the base and only a minor tuff interval, which is near the top, but more intervals of impure chert and siliceous shale. In Arkansas, shelfal limestone blocks occur within thick slump masses that have rubble bedding and were apparently derived from the northern shelf by submarine mass-flow mechanisms. A fine-grained quartz arenite, the Hot Springs Sandstone, 30 m (100 ft) thick, is a local channel-fill deposit of a submarine fan which apparently originated from the outer continental shelf to the north.

Jackfork Sandstone.—The Jackfork consists of gray-black shale and rhythmically interbedded whitish-gray, very fine-grained, quartz-rich turbidite packets. Rubble bedding (fig. 8C) dominates in the Frontal Ouachitas of Arkansas, proximal turbidites in the Southern Ouachitas, and thinner bedded distal turbidites and black siliceous shales in the Ouachitas of Oklahoma. The proximal turbidites are massive, ridge-forming quartz-rich scour-and-fill units (fig. 8E) with few or no shale interbeds. Distal turbidites have partial to complete Bouma sequences, better developed sole markings, greater evidence of trace fossils, common

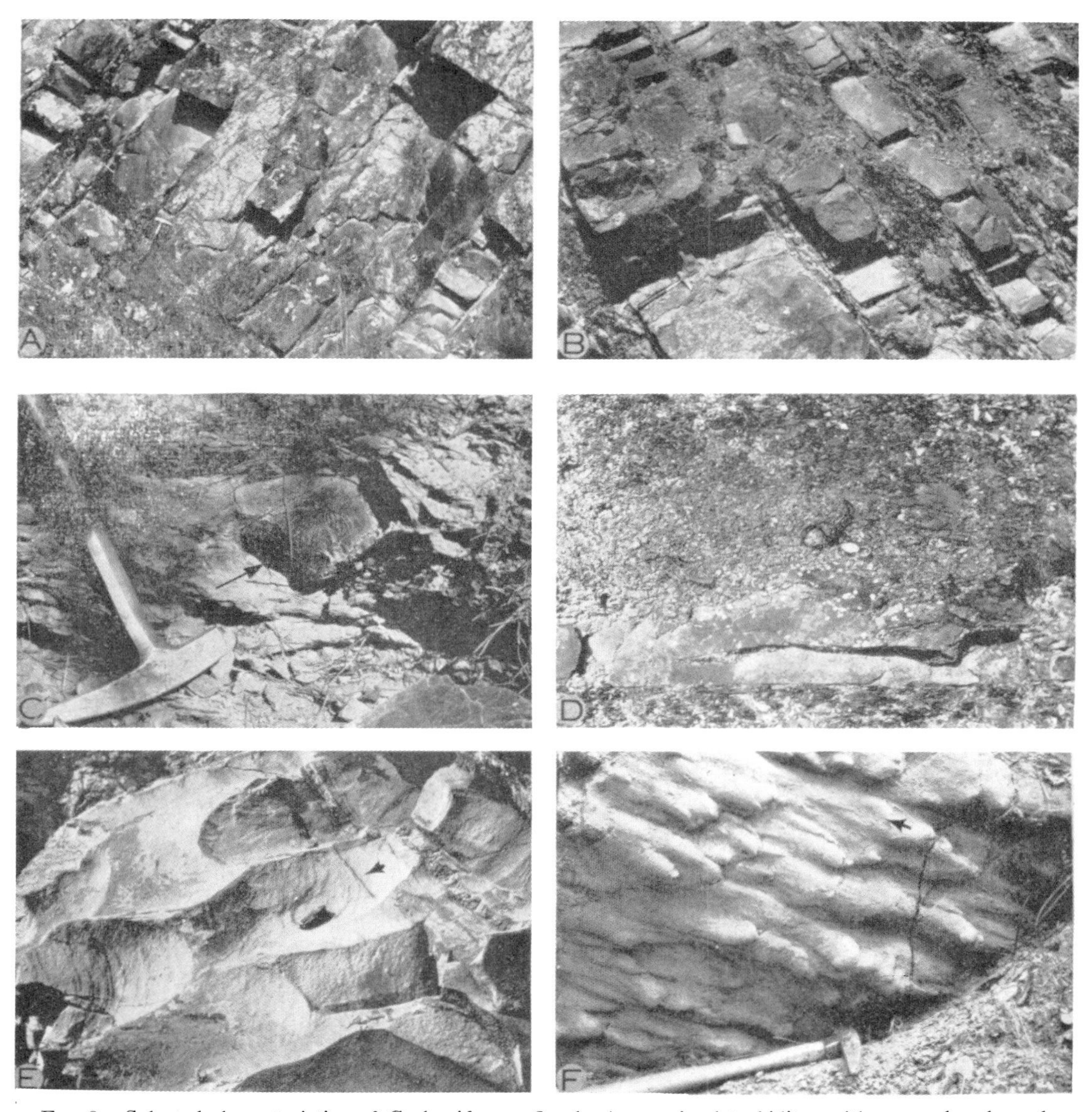

FIG. 8.—Selected characteristics of Carboniferous flysch. A, proximal turbidites with scoured and amalgamated bedding contacts, Moyers Formation of Stanley Shale; top is to the left. B, distal turbidites (sandy flysch) showing even-bedded characteristics, Moyers Formation of Stanley Shale. C, chert block (arrow) in rubble bedding, Irons Fork Mountain Formation of Jackfork Sandstone; note lack of fissility in mudrock. D, content grading showing a sharp base and gradational top; very characteristic of Stanley turbidites with high matrix contents. E, upper bedding plane of proximal turbidite with scour-and-fill bedding, Jackfork Sandstone; these sedimentary structures are useful as top and bottom indicators, and also show paleocurrent direction (arrow) for overlying bed. F, flute casts indicating paleocurrent direction (arrow) on sole of distal turbidite, lower Atoka Formation.

rippled tops caused by deep marine currents, and dewatering structures (LoPiccolo, 1973), as well as slope- or shock-induced disruptive structures. Several distinct zones consist of quartz granules, carbonate clasts, and shallow-water fossils that originated from the continental shelf.

Johns Valley Shale.—This unit is characterized by a thin distinctive zone of rubble bedding with enormous blocks of limestone, chert, and black shale in the Central Ouachitas of Oklahoma. Eastward into Arkansas, the deformational zones are replaced by gray-black shales and rhythmically interbedded very fine grained, matrix-rich, thin-bedded sandstones that constitute a distal turbidite facies. Some slump zones also occur in the Southern Ouachitas, where they apparently indicate sliding northward before consolidation.

Atoka Formation.—The lower part of the Atoka is probably the most nearly classic flysch sequence in the Ouachitas. It consists of a thick sequence of alternating interbeds of dark-gray shale and gray-brown, very fine-grained, laminated, graded subfeldspathic lithic wacke turbidites with common hieroglyphs (fig. 8F). Upward, shallow-water sedimentary structures and poor coal seams indicate basinal filling from the north.

Krebs Group.—This group consists dominantly of light-gray shales and sandstones that are medium- to fine-grained and have channel-fills, cross-bedding, upright tree trunks, and thin coals. They represent a deltaic and prodeltaic facies.

SEDIMENTARY PETROLOGY

A partial list of petrographic studies of pre-Stanley rocks includes those of Goldstein (1959), Goldstein and Hendricks (1953, 1962), Goldstein and Reno (1952), Park and Croneis (1969), and Sellars (1967). More recent petrographic studies of Carboniferous sandstones of the Ouachitas include those of Bokman (1953), Burkart (1969), Hill (1967), Howard (1963), Klein (1966), Laudon (1959), Luttrell (1965), Moretti (1958), Morris (1964), Russell (1969), Seely (1963) and Simonis (1967). Some of this work remains unpublished. Selected photomicrographs of some Ouachita rocks are grouped in figure 9. Quantitative data on grain-size distributions and principal-mineral compositions from thin-section studies by the writer and others have been compiled in table 4. All data in this table are from Arkansas except the fine work of Hill (1967).

The cratonic source area referred to in table 4 probably lay to the northeast of the Ouachita trough and provided well-sorted, rounded quartz grains which moved down south-facing submarine canyons to be emplaced upon the floor of the Ouachita trough as slumps, sand flows, and turbidity currents. Minor sandstones of the Mazarn and Womble Formations are too fine to provide provenance interpretations but the limestones of the Collier, Mazarn, and Womble were derived from a carbonate shelf and emplaced as turbidity flows upon a deep-sea floor. The rounded quartz sands and silts of the Arkansas Novaculite also were derived from a stable shelf. In the various Ordovician and Carboniferous units, exotic blocks of shallow-water fossiliferous limestones and black shales, cherts, meta-arkoses, and silicic plutonic rocks are clustered across the leading edge of the Frontal Ouachitas in a pattern strongly suggestive of a cratonic origin from fault scarps and submarine canyons of the outer shelf and upper slope.

The extracontinental source area is thought to have been folded orogens to the east and south of the Ouachita trough. The higher incidence of sedimentary and low-rank metamorphic clasts across the Southern Ouachitas is suggestive that they were derived from areas other than the craton.

Compositional data for the Carboniferous sandstones (figs. 10, 11, 12, 13) are plotted on the triangular diagrams of Williams and others (1954). No data from intraformational chert breccias are included. The quartz types include mostly common and strained quartz, with minor polycrystalline quartz and chert grains. The polycrystalline grains, generally more abundant in larger sizes, are branded, stretched, sutured, and attenuated in ways strongly suggestive of derivation from metamorphic sources. Feldspar consists of orthoclase, microcline, and plagioclase (oligoclase and andesine), both fresh as well as weathered, rarely calcitized, rounded to angular, and except for the tuffaceous sandstones, never in sufficient quantity to have been derived directly from igneous rocks. Rock fragments in Carboniferous strata most commonly are metamorphic (mica schists, quartz-mica schists, slates, and phyllites) and sedimentary (shales, siltstones, and cherts) in origin. Volcanic clasts have sporadic distribution except for two major intervals. Heavy minerals are badly worn and include zircon, tourmaline, rutile, garnet, and opaque minerals. In Oklahoma, Hill (1967) also reported apatite and staurolite. Other common accessories are detrital chlorite, biotite, and muscovite with authigenic calcite and muscovite. Clay minerals in the argillaceous Carboniferous units have been described by Davis (1968) and Weaver (1958). These units all contain illite, mixed layer material, chlorite, and kaolinite.

As seen from the textural data in table 4, the sandstones interpreted as solely shelf derived have slightly larger grains and less matrix than those apparently derived from extracontinental sources. Morris (in press) made a comparison of Stanley and Jackfork flysch subfacies with textures and concluded that (a) proximal turbidites with scoured bedding contacts have the least matrix and largest grains, and (b) distal turbidites, especially those in shaly flysch, have the smallest grains, the high-

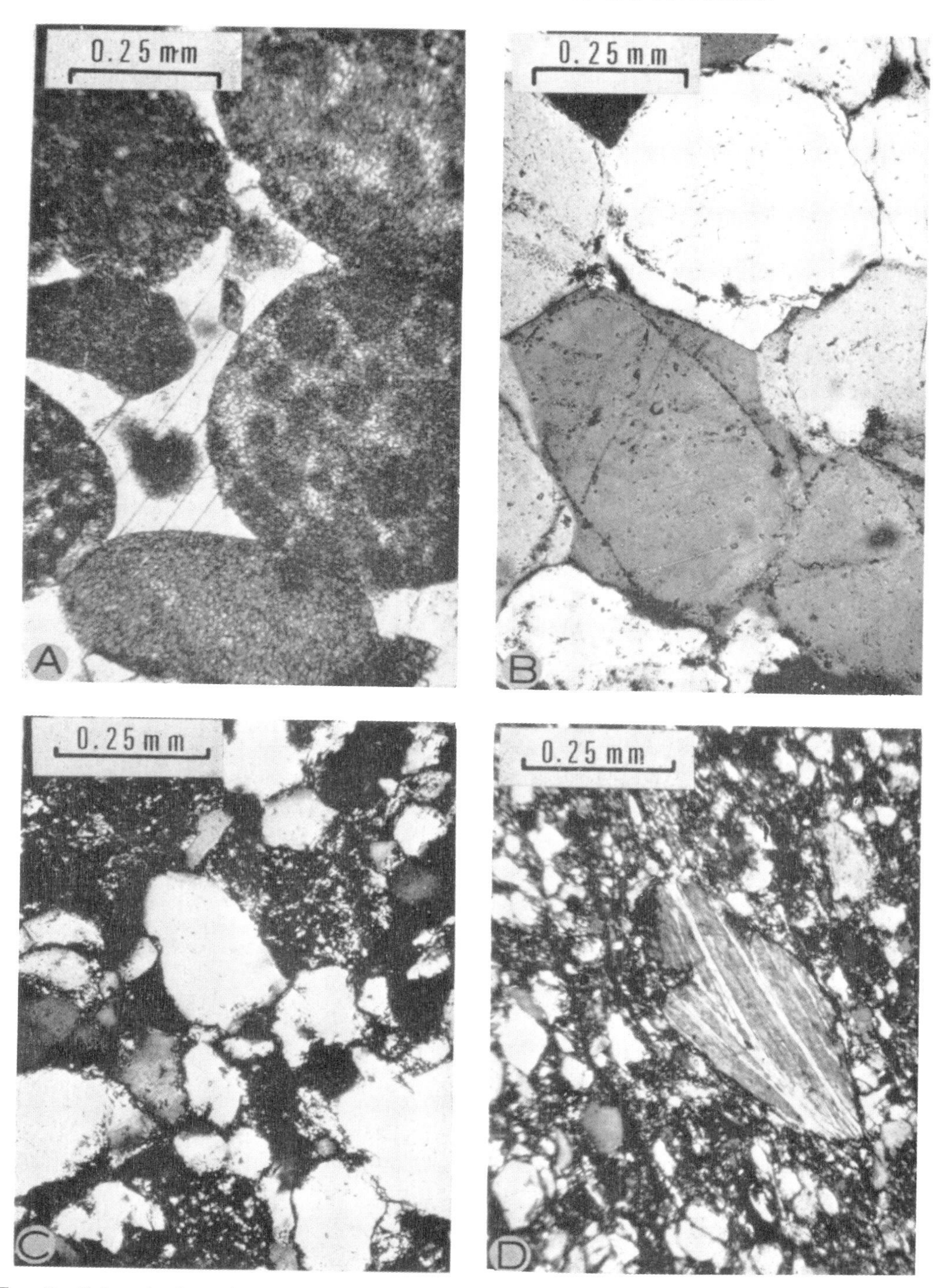

FIG. 9.—Selected photomicrographs from thin sections of Ouachita rocks. A, rounded intraclasts of pelletal and micritic limestone cemented by sparry calcite in turbidite limestone of Collier Formation (plain light); also in the thin section are rounded grains of quartz and some scattered feldspar. B, rounded quartz grains with well-developed overgrowths in quartz arenite derived from northern craton in Missouri Mountain Formation (crossed nicols). C, angular quartz and shale fragments in fine-grained lithic wacke, Atoka Formation (crossed nicols). D, detrital chlorite, quartz, trace feldspar, and shale fragments in quartz wacke, Tenmile Creek Formation of Stanley Shale (crossed nicols).

TABLE 4.—SUMMARY OF SANDSTONE PETROGRAPHY AND INTERPRETED SOURCE AREAS

Stratigraphic unit[1]	M_ϕ	σ_ϕ	Quartz[5]	Feldspar[5]	Rock fragments[5]	Matrix[5]	Other	Cratonic source	Extracratonic (orogenic) source
Atoka (18)	4.3	1.26	56.2	5.0	12.1	19.6	7.1	Yes	Yes
Jackfork[2] (201)	2.7	0.86	84.8	0.9	2.2	11.4	0.7	Yes	Yes
Stanley									
Arkansas[3] (73)	3.6	1.30	65.7	9.7	2.9	18.2	3.5	Yes	Yes
Oklahoma[4] (109)	3.4	1.77	66.9	11.5	1.0	17.6	3.0	No	Yes
Missouri Mountain (6)	2.1	0.62	68.9	0.0	0.0	2.4	28.7	Yes	No
Blaylock (8)	silt-sized	—	43.0	13.1	8.4	25.2	10.3	No	Yes
Blakely (3)	2.6	0.71	72.0	tr	1.0	2.0	25.0	Yes	No
Crystal Mountain (7)	2.3	0.80	75.0	2.4	tr	8.0	14.5	Yes	No

[1] Numbers in parenthesis refer to thin sections counted
[2] Some data from Burkart (1969) and Simonis (1967)
[3] Some data from Hamilton (1973), Russell (1969), and Simonis (1967)
[4] All from Hill (1967)
[5] Figures are percentage of whole rock

est matrix content, and the most abundant heavy minerals.

PALEOCURRENTS

No paleocurrent information is available for pre-Stanley rocks, even though sole markings are present in some of the turbidites. Abundant but scattered data exist for Carboniferous sandstones. Current-formed sole marks have been used almost exclusively for the determination of Stanley, Jackfork, and early Atoka paleocurrents. Tool marks include groove casts, bounce casts, prod casts, and brush casts; some give true current direction and some only current sense. Scour marks include large scours associated with proximal turbidites, flute casts, furrow casts, and internally located parting lineation. Almost all paleocurrent data have been gathered from distal turbidites. The sedimentary structures that provide paleocurrent information for upper Atoka and Krebs clastics include ripple marks (asymmetrical and sym-

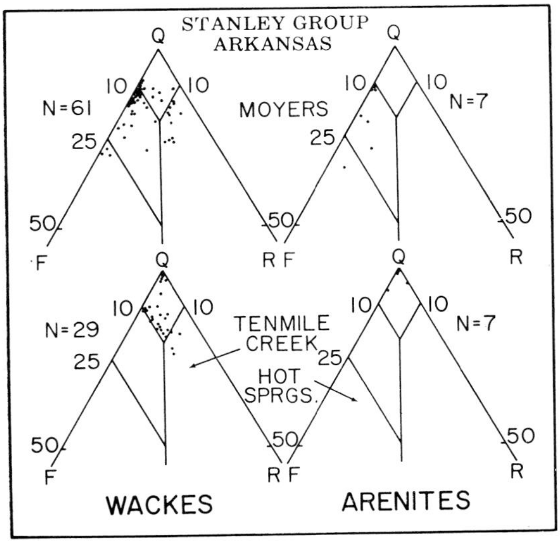

FIG. 10.—Compositional diagrams of Stanley sandstones in Arkansas as determined from thin-section studies. See table 4 for summary. Data from three stratigraphic members shown separately.

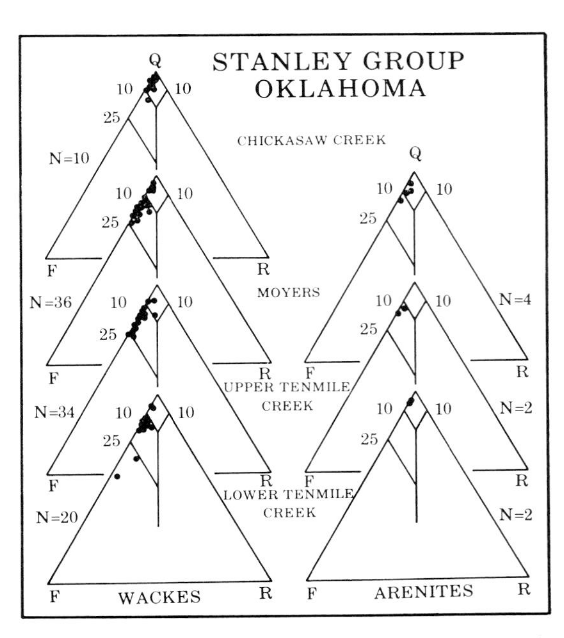

FIG. 11.—Compositional diagrams of Stanley sandstones in Oklahoma as determined from thin-section studies by Hill (1967). See table 4 for summary. Data from four stratigraphic members shown separately.

metrical), parting lineation, and cross lamination. These structures were formed in near-shore shallow waters of a southward prograding shore line.

The paleocurrent maps in this report were compiled from several sources. Johnson (1966) presented numerous maps summarizing 786 Stanley paleocurrent readings in Oklahoma. Briggs and Cline (1967) show the mean current directions from approximately 2400 readings from Jackfork, Johns Valley-Atoka, and Krebs rocks in Oklahoma. Raw data from 2365 paleocurrent readings of Jackfork, Atoka, and Krebs rocks in Arkansas were taken from a thesis by Sullivan (1966). In Arkansas, the Stanley and Jackfork paleocurrent maps also include data from 386 readings by Morris (in press) and Hamilton (1973).

Stanley paleocurrents (fig. 14) show a northwesterly flow that becomes more westerly towards Atoka, Oklahoma. As Stanley time progressed, there was a shift to more westerly current flow. Also, more thin-bedded siltstones were emplaced northwestward, whereas the majority of beds thicker than 2.5 cm were emplaced westward or southwestward. No data are available for the proximal turbidites. Jackfork paleocurrents (fig. 15) show westerly flow with components to the northwest and southwest but no overall change in current flow between lower and upper Jackfork rocks. Most of these readings were taken from distal turbidites. Johns Valley and lower Atoka paleocurrents (fig. 16) show west-northwesterly flow in Arkansas and southwesterly flow in Oklahoma, but the upper Atoka sandstones with shallow-water characteristics were emplaced almost due southward. Krebs paleocurrents (fig. 17) also show a southerly flow, a somewhat startling fact because subsurface evidence south of the present Ouachitas (Vernon, 1971) indicates uplift (orogeny) of the Ouachita fold belt at the same time that the Krebs deltas and near-shore clastics were moving southward. There is no evidence of northward sediment dispersal during Desmoinesian (Krebs) time.

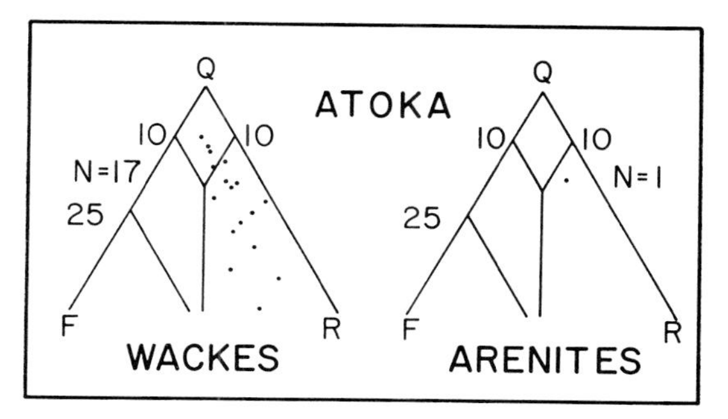

FIG. 13.—Compositional diagrams for lower Atoka sandstones in Arkansas as determined from thin-section studies. See table 4 for summary.

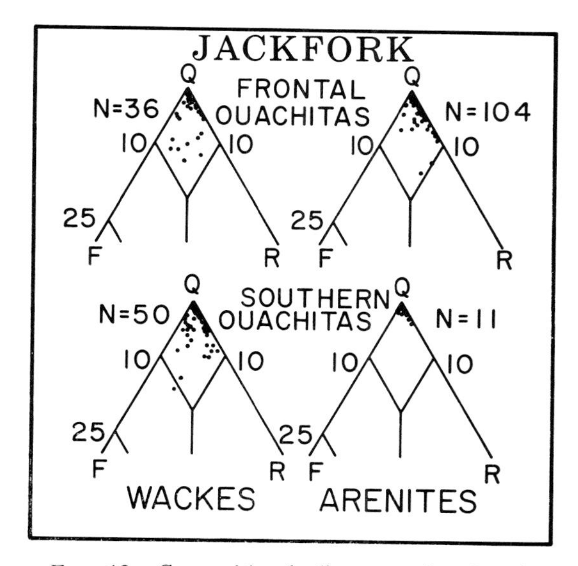

FIG. 12.—Compositional diagrams for Jackfork sandstones in Arkansas as determined from thin-section studies. See table 4 for summary.

DEPOSITIONAL MODEL

The Ouachita Mountains display a sedimentary, structural, and volcanic succession whose origin may have been related to convergence of lithospheric plates, as is the care for other folded mountain belts with thick geosynclinal sequences (Dewey and Bird, 1970; Bird and Dewey, 1970; Keller and Cebull, 1973). Critical to interpretation of the Ouachitas is the behavior of the southern Appalachians, where Hatcher (1972) postulated westward underflow of a proto-Atlantic plate which produced a rising tectonic sourceland by Late Devonian time. Continued compression generated large thrusts, folds, and mountain belts during the collision and suturing of Africa against southeastern North America during the middle to late Paleozoic. In the general area of the Ouachita trough, it seems certain that a shelf, slope, and abyssal plain extended southward from the North American craton during the early and middle Paleozoic. It is likely that an African-American plate boundary curved westward from the Appalachian region to constitute an arc-trench system at some distance beyond the deep-sea plain. This boundary was probably characterized by considerable strike-

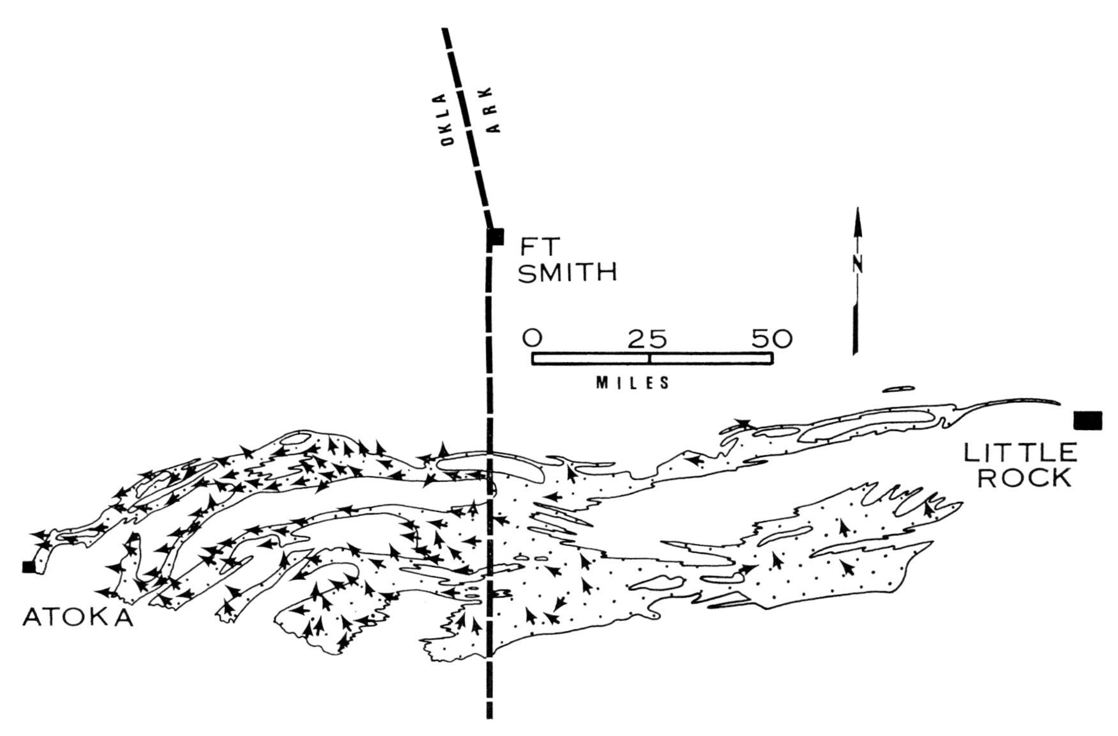

Fig. 14.—Outcrop map of Stanley Shale (stippled) showing mean paleocurrent directions at locations indicated by arrows.

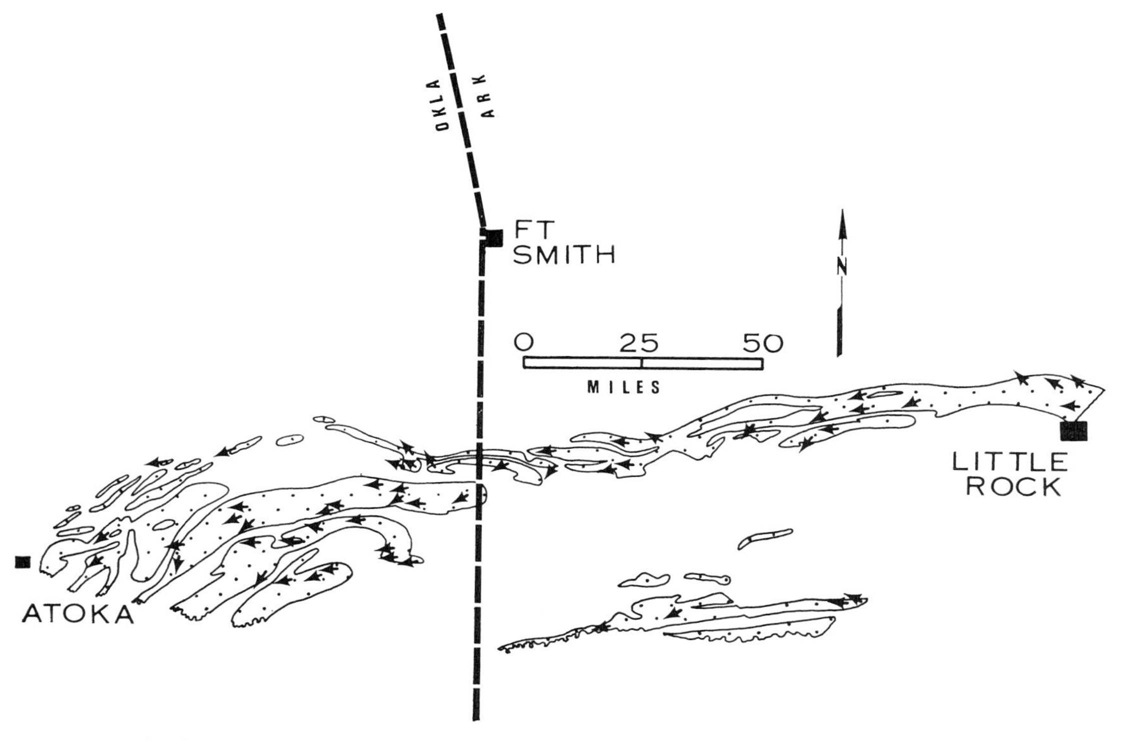

Fig. 15.—Outcrop map of Jackfork Sandstone (stippled) showing mean paleocurrent directions at locations indicated by arrows.

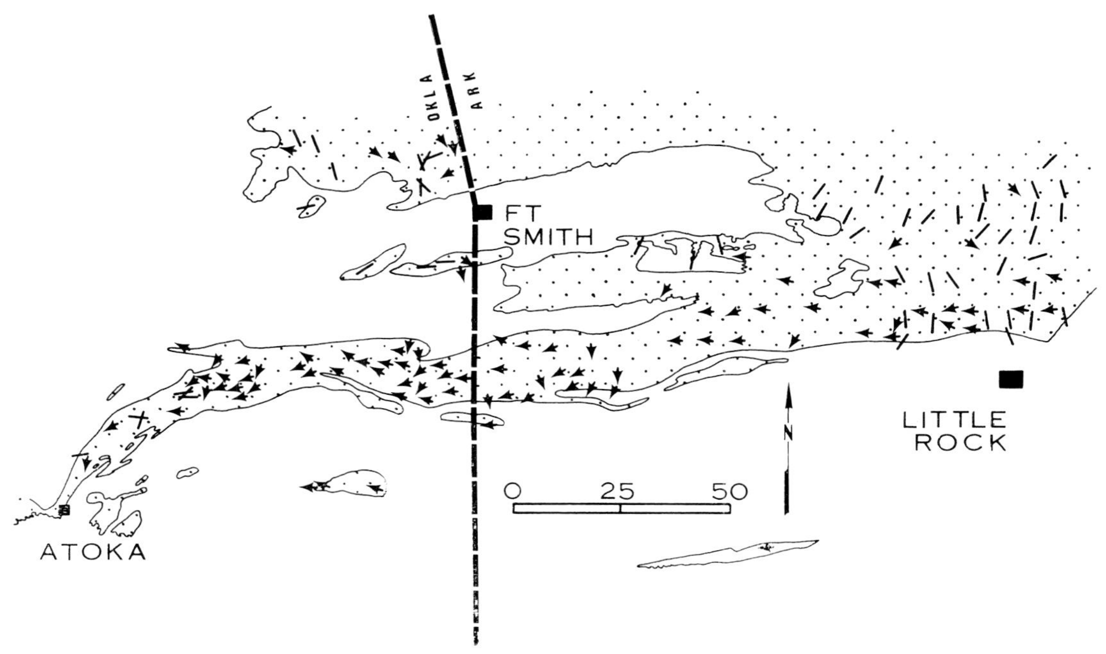

FIG. 16.—Outcrop map of Atoka Formation (stippled) showing mean paleocurrent directions at locations indicated by arrows. Short, heavy lines give average of readings at localities where only the bidirectional sense is known.

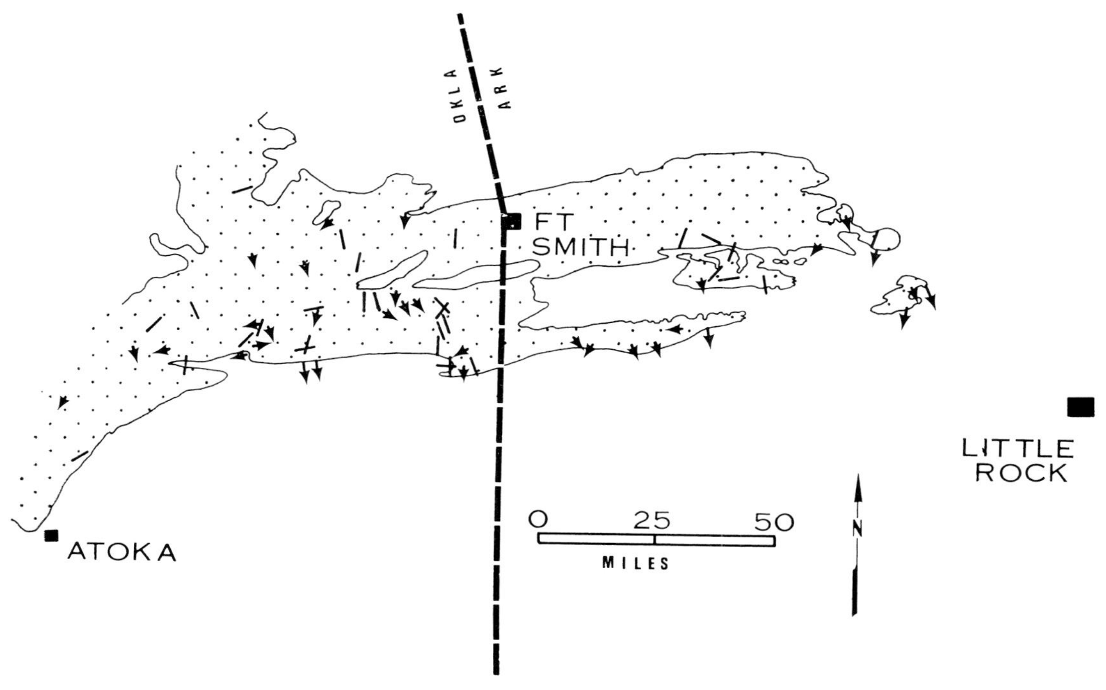

FIG. 17.—Outcrop map of Krebs Group (stippled) showing mean paleocurrent directions at locations indicated by arrows. Short, heavy lines give average of readings at localities where only the bidirectional sense is known.

slip movement and only minor subduction adjacent to the Ouachitas.

Peninsular Florida, Central America, Cuba, and possibly other present Caribbean islands are considered here to have been part of the plate including Africa and South America during the early Paleozoic. Only Florida (Rainwater, 1971) and Central America have definite Precambrian rocks, but the Paleozoic(?) of Cuba (Meyerhoff, 1971) may also overlie Precambrian rocks at depth. For the early Paleozoic, the connection of these areas to the plate including Africa and South America appears to be the key to the use of a plate model for the evolution of the Ouachita fold belt. There is a strong likelihood that the rocks of the Florida and Yucatan shelves have not been folded since the Precambrian; thus a plate boundary, if it existed, was located north of Florida and both south and west of the Yucatan shelf.

In the following paragraphs, an attempt is made to relate the known sequence of Paleozoic depositional events in the Ouachitas to a speculative series of plate-tectonic settings off the southern margin of North America.

Ordovician Period.—During the Ordovician, deposition was dominated by the accumulation of dark graptolitic muds and thin interbedded turbidite limestones derived from a northern shelf. On at least two separate occasions the northern shelf became sandy, allowing a generous quantity of mature, rounded quartz sand and carbonate grains to escape via turbidity currents, sand flows, and subaqueous slides to the deep-sea floor to the south. Minor earthquakes(?) caused slope collapses and development of rubbly zones with exotic blocks. One major cherty unit suggests high production of siliceous organisms that accumulated together with turbidite limestones and black muds.

Silurian Period.—The Silurian shaly flysch sequence contains turbidite sandstones derived from extracratonic sources, probably a large land mass east of the Ouachita trough (fig. 18). Another possible source area might have been uplifted tectonic lands along the arc to the south. Minor mature quartz sands in the

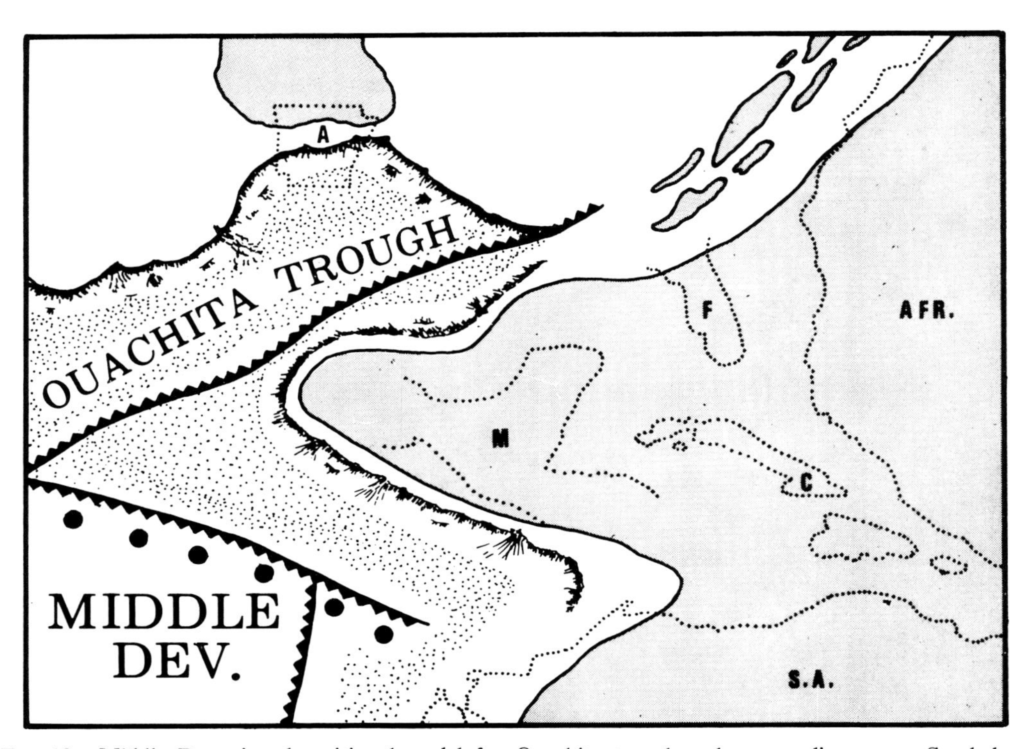

Fig. 18.—Middle Devonian depositional model for Ouachita trough and surrounding areas. Symbols used as follows (also apply to figs. 19, 20, and 21): A, Arkansas; Afr., African continent; C, Cuba; D, deltaic facies; F, Florida; G, graben basins (diagrammatic); M, "microcontinent" of southern Mexico, Guatemala, and Honduras; large solid dots, volcanoes (diagrammatic); fine pattern, land areas; no pattern, continental shelves; short irregular lines, continental slopes; irregular small dots, abyssal plains; sawteeth, trenches (pattern on overriding plate).

Upper Silurian entered as turbidity currents or sand flows from the northern shelf. Varicolored pelagic muds dominated the Ouachita trough, perhaps owing to a high influx of iron compounds from the southern Appalachian trough.

Devonian Period.—The Devonian was a time of major silica sedimentation and only minor clastic influx, chiefly pelagic. Lowe (1973b) suggested that closing of the Appalachian seaway might have reversed normal circulation patterns and brought about a large-scale influx of upwelling waters into the Ouachita trough where siliceous phytoplankton and sponge growth could have provided the silica. A shaly continental slope and a carbonate shelf, possibly emergent at times, lay to the north. Figure 18 is a conceptual model of major tectonic and physiographic features related to the Ouachita trough.

Mississippian Period.—Several major changes are inferred in the Ouachita trough during the Osagean Epoch as shown in figure 19. Conceivably, a large portion of the plate broke loose from the African and South American continents and newly formed volcanoes at the rift-trench junction perhaps supplied ashy detritus for north-flowing turbidity currents. The small spreading sea floor shown is intended to mark the beginning of a proto-Gulf of Mexico. Such a concept agrees with the conclusion of Wilhelm and Ewing (1972) that early simatic crust probably formed in the Gulf of Mexico during the late Paleozoic. The microcontinent, lying to the west and alternately emergent or submergent, probably should be called Llanoria, following Miser (1921); however, he conceived of a large stationary body that eventually became worn down and covered up. Collision of the microcontinent with the trench is inferred to have created rising welts of tectonic lands with their fine-grained sedimentary cover. In the Mississippian, the bulk of the sand-sized clastic material apparently entered the Ouachita trough from the southern Appalachian region at the eastern end of the trough as turbidity flows. Turbidity currents carried sediment northwestward to build a deep-

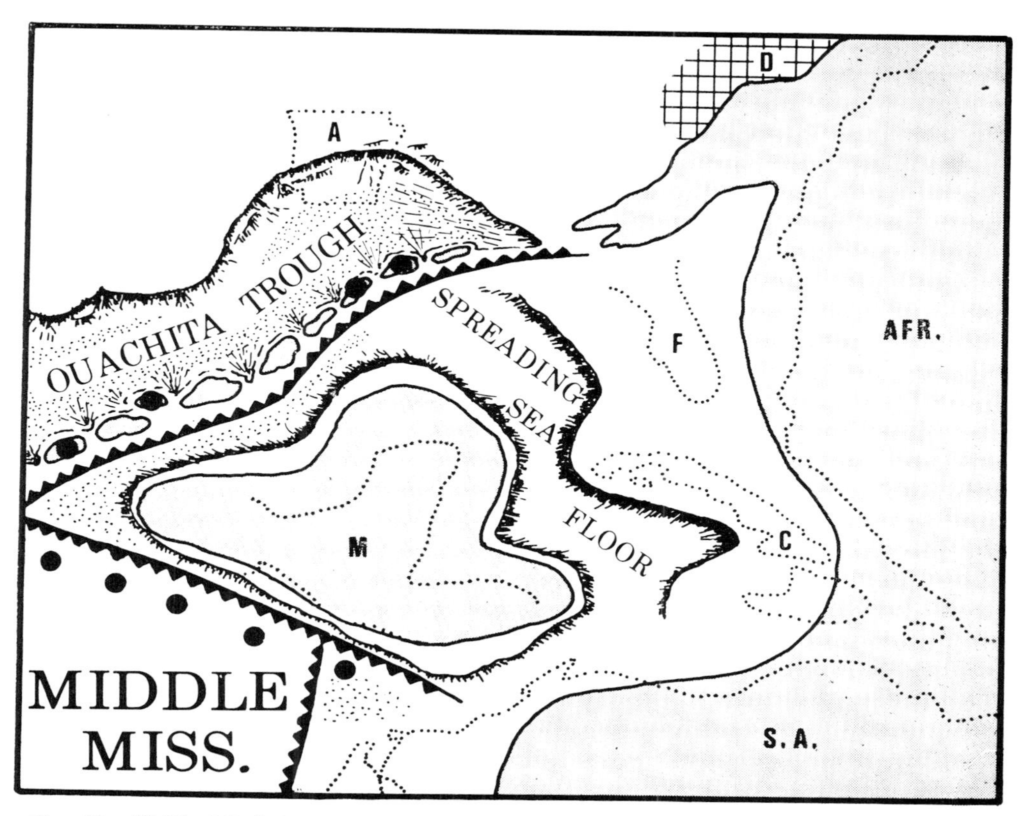

FIG. 19.—Middle Mississippian depositional model for Ouachita trough and surrounding areas. For symbols see figure 18.

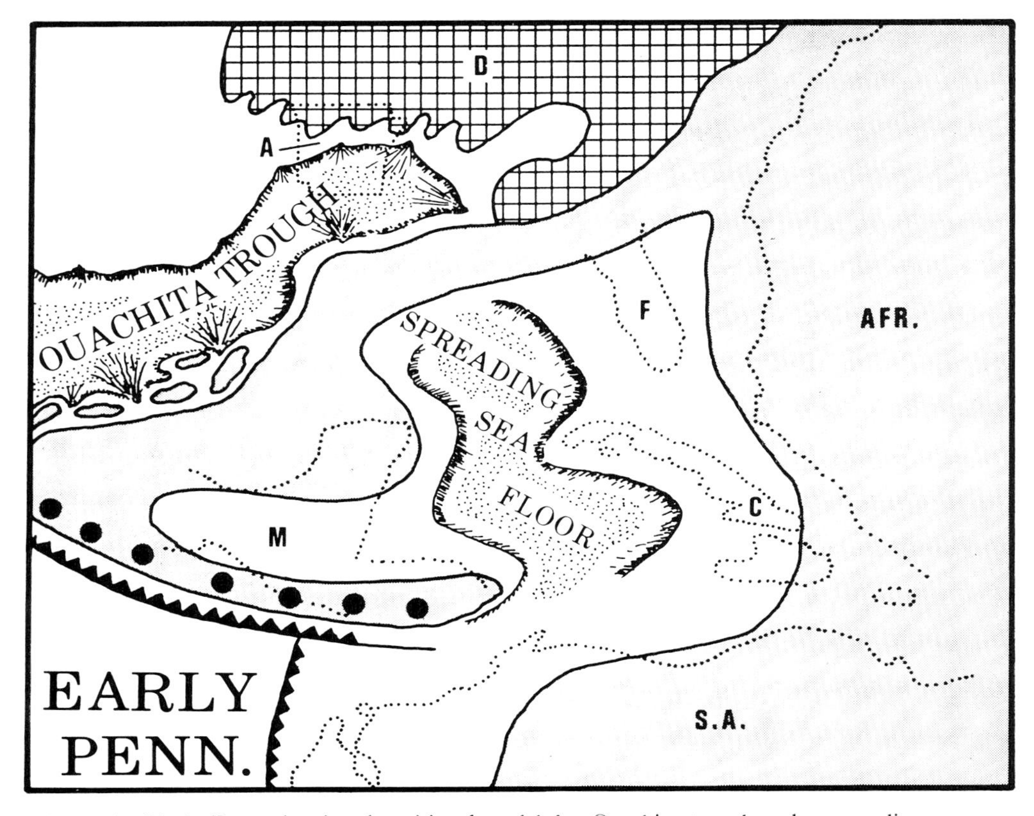

FIG. 20.—Early Pennsylvanian depositional model for Ouachita trough and surrounding areas. For symbols see figure 18.

sea fan and flat abyssal plain just south of the continental margin of North America. Minor silt-sized clastic material might have been derived from the tectonic lands or even the microcontinent itself. The widespread thin dark siliceous shales and impure cherts probably resulted from decreased clastic influx and/or increased pelagic organic activity due to updwelling waters. Minor zones of contorted bedding with limestone, chert, and other blocks are due to southward slumping along the unstable northern slope of the trough. The shelf break abruptly changed to a northwest-southeast trend to the east in Mississippi (Thomas, 1972). During the Chesterian Epoch, the deltas that prograded southward across Illinois apparently failed to provide clastic sediment to the Ouachita trough, for we don't know of any mature upper Stanley clastics in the Ouachitas. In what was to become Mexico, a flysch facies, partially volcanic in origin, developed to a considerable thickness in the late Paleozoic (Guzman and de Cserna, 1963).

Pennsylvanian Period.—At the onset of the Morrowan Epoch, lowered sea levels caused the northern shelf to be eroded, resulting in a flood of nonfeldspathic clastics from the northeast that found their way to the continental rise and abyssal plain of the Ouachita trough. The shelf edge was retreating by collapse, and the axis of the trough also seems to have migrated northward. Major slumps and slides moved southward to the edge of the continental rise, probably triggered by earthquakes from slopes located to the south. Filling with sparsely feldspathic, sand-sized clastic sediment began mainly from the eastern end of the trough. Down the axis of the trough the number of sand beds increased, the thickness of each sand bed decreased, and various point sources blended into a single westward building subsea fan of abyssal-plain deposits (fig. 20).

During the Atokan, the axis of the trough shifted northward, possibly due to several down-to-the-south growth faults, but the westward flow of turbidity currents continued. The highest depositional rates occurred during this

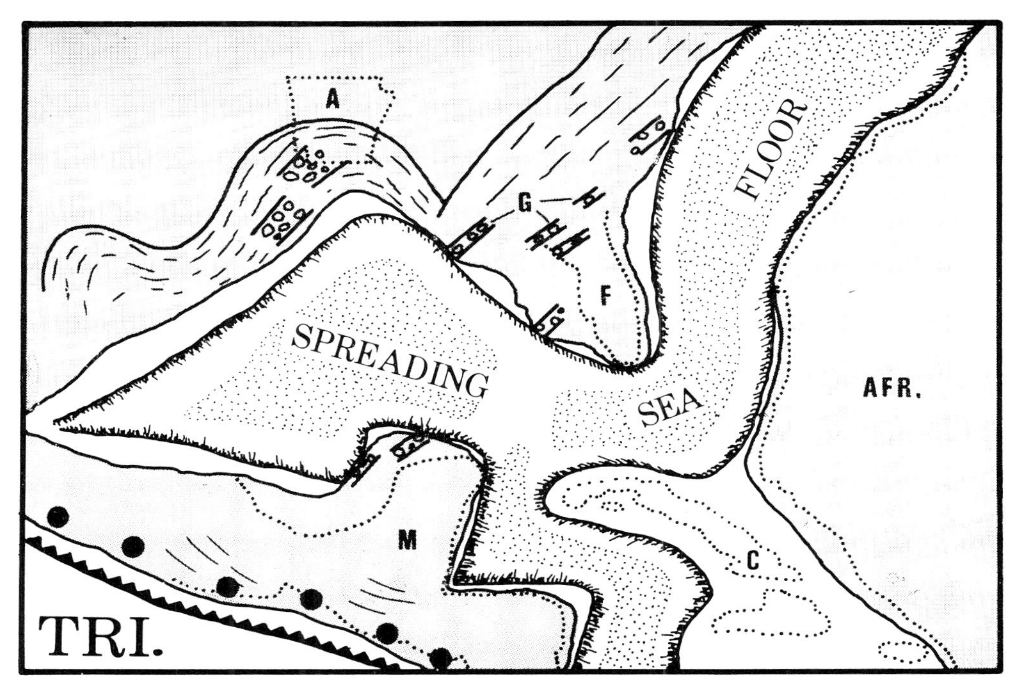

FIG. 21.—Triassic depositional model for Ouachita trough and surrounding areas. For symbols see figure 18.

time, so that the trough became nearly filled. During the late Atokan and Desmoinesian, all sediment to the basin was supplied by southward-building deltas from across the eastern interior of the continent.

Viele (1973) has suggested that synchronously with Carboniferous sedimentation, the pre-Carboniferous rocks may have been gliding northward as large, imbricate nappes. Certainly by the late Atokan or early Desmoinesian, uplift long the tectonic lands of the continental margin apparently caused the thick sedimentary wedge to slide northward along giant sole thrusts, faulting and folding all sedimentary rocks of the Ouachita and Arkoma basins. Subsurface evidence south of the Ouachitas suggests that part of the fold belt was leveled rapidly (or was never raised very high out of the water) before the end of the late Desmoinesian. In this area, shallow-water shales, minor sandstones, and fossiliferous limestones, including fusulinids (*Wedekindellina*, *Fusulina*), were deposited upon highly deformed Ouachita facies. However, in northeastern Texas during the Late Pennsylvanian and Permian, the Ouachitas were mountainous and supplying sediment for westward building deltas into north-central Texas (Brown, 1969). The Ardmore basin of southeastern Oklahoma also received cherty detritus from the Ouachitas in the Desmoinesian (Tomlinson and McBee, 1962).

Permian-Triassic-Jurassic.—No Permian history can be read from the Ouachita Mountains; presumably they were deeply eroded by the Triassic. Figure 21 illustrates my concept of major Triassic events: the opening of the Gulf of Mexico-Florida Straits-North Atlantic by sea-floor spreading, the development of numerous grabens, the associated arkose conglomerates and igneous dikes, and the vast evaporite deposits of the Gulf of Mexico. By Late Jurassic time the zone of rifting left the Gulf of Mexico area and began instead to split apart the continents of Africa and South America as postulated by Dietz and Holden (1970). This modification of spreading patterns may have created the deep-seated fractures that fed the basic dikes cutting the eastern Ouachitas.

REFERENCES CITED

AMSDEN, T. W., CAPLAN, W. M., HILPMAN, P. L., MCGLASSON, E. H., ROWLAND, T. L., AND WISE, O. A., JR., 1967, Devonian of the southern midcontinent area, United States *in* Oswald, D. H. (ed.), International symposium on the Devonian System: Alberta Soc. Petroleum Geologists, v. 1, p. 913–932.

BIRD, J. M., AND DEWEY, J. F., 1970, Lithosphere plate-continental margin tectonics and the evolution of the Appalachian orogen: Geol. Soc. America Bull., v. 81, p. 1030–1060.
BOKMAN, JOHN, 1953, Lithology and petrology of the Stanley and Jackfork Formations: Jour. Geology, v. 61, p. 152–170.
BRIGGS, GARRETT, AND CLINE, L. M., 1967, Paleocurrents and source areas of late Paleozoic sediments of the Ouachita Mountains, southeastern Oklahoma: Jour. Sed. Petrology, v. 37, p. 985–1000.
BROWN, L. F., JR., 1969, Geometry and distribution of fluvial and deltaic sandstones (Pennsylvanian and Permian), north-central Texas: Gulf Coast Assoc. Geol. Soc. Trans., v. 19, p. 23–47.
BURKHART, M. R., 1969, Stratigraphy, structure, and petrology of Carboniferous rocks from Crystal Mountain region, Arkansas (M. S. thesis): Northern Illinois Univ. DeKalb, 75 p.
CHAMBERLAIN, C. K., 1971, Bathymetry and paleoecology of Ouachita geosyncline of southeastern Oklahoma as determined from trace fossils: Am. Assoc. Petroleum Geologists Bull., v. 55, p. 34–50.
CLINE, L. M., 1960, Stratigraphy of the late Paleozoic rocks of the Ouachita Mountains, Oklahoma: Oklahoma Geol. Survey Bull. 85, 113 p.
———, 1970 Sedimentary features of late Paleozoic flysch, Ouachita Mountains, Oklahoma *in* Lajoie, J. (ed.), Flysch sedimentology of North America: Geol. Assoc. Canada Special Paper No. 7, p. 85–101.
———, AND MORETTI, F. J., 1956, Two measured sections of the Jackfork Group in southeastern Oklahoma: Oklahoma Geol. Survey Circ. 41, 20 p.
DAVIS, C. G., 1968. An investigation of the clay mineral suite in a type section of the Jackfork Group, Ouachita Mountains, Arkansas (M. S. thesis): Northern Illinois Univ., DeKalb, 91 p.
DECKER, C. E., 1959, Correlation of lower Paleozoic formations of the Arbuckle and Ouachita areas as indicated by graptolite zones *in* Cline, L. M., Hilseweck, W. J., and Feray, D. E. (eds.), The geology of the Ouachita Mountains, a symposium: Dallas and Ardmore Geol. Soc., p. 92–96.
DEWEY, J. F., AND BIRD, J. M., 1970, Mountain belts and the new global tectonics: Jour. Geophys. Research, v. 75, p. 2625–2647.
DIETZ, R. S., AND HOLDEN, J. C., 1970, Reconstruction of Pangaea: Breakup and dispersion of continents, Permian to Present: *ibid.*, v. 75, p. 4939–4956.
DZULYNSKI, STANISLAW, AND WALTON, E. K., 1965, Sedimentary features of flysch and greywackes: Elsevier, Amsterdam, 274 p.
FELLOWS, L. D., 1964, Geology of eastern Winding Stair Range, Latimer and LeFlore Counties: Oklahoma Geol. Survey Circ. 65, 102 p.
FLAWN, P. T., GOLDSTEIN, AUGUST, JR., KING, P. B., AND WEAVER, C. E., 1961, The Ouachita system: Univ. Texas Pub. 6120, 401 p.
FREZON, S. E., AND GLICK, E. E., 1959, Pre-Atoka rocks of Northern Arkansas: U.S. Geol. Survey Prof. Paper 314-H, p. 171–187.
GILLULY, JAMES, REED, J. C., JR., AND CADY, W. M., 1970, Sedimentary volumes and their significance: Geol. Soc. America Bull., v. 81, p. 353–376.
GOLDSTEIN, AUGUST, JR., 1959, Petrography of Paleozoic sandstones from the Ouachita Mountains of Oklahoma and Arkansas *in* Cline, L. M., Hilseweck W. J., and Feray, D. E. (eds.), The geology of the Ouachita Mountains, a symposium: Dallas and Ardmore Geol. Soc., p. 97–116.
———, AND HENDRICKS, T. A., 1953, Siliceous sediments of Ouachita facies in Oklahoma: Geol. Soc. America Bull., v. 64, p. 421–442.
———, 1962, Late Mississippian and Pennsylvania sediments of Ouachita facies, Oklahoma, Texas, and Arkansas *in* Branson, C. C. (ed.), Pennsylvanian System in the United States, a symposium: Am. Assoc. Petroleum Geologists, Tulsa, p. 385–430.
———, AND RENO, D. H., 1952, Petrography and metamorphism of sediments of Ouachita facies: Am. Assoc. Petroleum Geologists Bull., v. 36, p. 2275–2290.
GORDON, MACKENZIE, JR., AND STONE, C. G., 1973, Correlation of Carboniferous rocks of the Ouachita geosyncline with those of the adjacent shelf: Geol. Soc. America Abstracts with Programs, v. 5, p. 259.
GUZMAN, E. J., AND DE CSERNA, Z., 1963, Tectonic history of Mexico *in* Childs, O. E. and Beebe, B. W. (eds.), Backbone of the Americas: Am. Assoc. Petroleum Geologists Mem. 2, p. 113–129.
HAM, W. E., 1959, Correlation of Pre-Stanley strata in the Arbuckle-Ouachita Mountain region *in* Cline, L. M., Hilseweck, W. J., and Feray, D. E. (eds.), The geology of the Ouachita Mountains, a symposium: Dallas and Ardmore Geol. Soc., p. 71–86.
HAMILTON, C. T., 1973, A stratigraphic investigation of the Hot Springs sandstone and Tenmile Creek Formation in the southern Ouachita Mountains of Arkansas (M. S. thesis): Northern Illinois Univ., DeKalb, 93 p.
HAMMES, R. R., 1965, Ouachita overthrusting—stratigraphic appraisal: Am. Assoc. Petroleum Geologists Bull., v. 49, p. 1666–1679.
HARLTON, B. H., 1938, Stratigraphy of the Bendian of the Oklahoma salient of the Ouachita Mountains: *ibid.*, v. 22, p. 852–914.
———, 1959, Age classification of the upper Pushmataha series in the Ouachita Mountains *in* Cline, L. M., Hilseweck, W. J., and Feray, D. E. (eds.), The geology of the Ouachita Mountains, a symposium: Dallas and Ardmore Geol. Soc., p. 130–139.
HART, O. D., 1963. Geology of the Winding Stair Range, LeFlore County, Oklahoma: Oklahoma Geol. Survey Bull. 103, 87 p.
HASS, W. H., 1950, Age of lower part of Stanley shale: Am. Assoc. Petroleum Geologists Bull., v. 34, p. 1578–1588.
———, 1951, Age of the Arkansas Novaculite: *ibid.*, v. 35, p. 2526–2541.
HATCHER, R. D., JR., 1972, Developmental model for the southern Appalachians: Geol. Soc. America Bull., v. 83, p. 2735–2760.
HILL, J. G., 1967, Sandstone petrology and stratigraphy of the Stanley Group (Mississippian), southern Ouachita Mountains, Oklahoma (Ph.D. dissertation): Univ. Wisconsin, Madison, 121 p.

HOWARD, C. E., 1963, Petrology of the Jackfork Sandstone at the DeGray Dam, Clark County, Arkansas (Ph.D. dissertation): Louisiana State Univ., Baton Rouge, 204 p.

JOHNSON, K. E., 1966, A depositional interpretation of the Stanley Group of the Ouachita Mountains, Oklahoma *in* Flysch facies and structure of the Ouachita Mountains: Kansas Geol. Soc. 29th Field Conf. Guidebook, p. 140–163.

KELLER, G. R., AND CEBULL, S. E., 1973, Plate tectonics and the Ouachita system in Texas, Oklahoma, and Arkansas: Geol. Soc. America Bull., v. 83, p. 1659–1666.

KING, P. B., 1973, Sedimentary and tectonic evolution of Ouachita belt in Marathon region, Texas: *ibid.,* Abstracts with Programs, v. 5, p. 267–268.

KLEIN, G. DEV., 1966, Dispersal and petrology of sandstones of the Stanley-Jackfork boundary, Ouachita fold belt, Arkansas and Oklahoma: Am. Assoc. Petroleum Geologists Bull., v. 50, p. 308–326.

LAUDON, R. B., 1959, Stratigraphy and zonation of the Stanley shale of the Ouachita Mountains, Oklahoma (Ph.D. dissertation): Univ. Wisconsin, Madison, 117 p.

LAJOIE, J. (ed.), 1970, Flysch sedimentology in North America: Geol. Assoc. Canada Special Paper No. 7, 272 p.

LOPICCOLO, R. D., 1973, The origin and implications of some sedimentary structures in the Jackfork Group of the Ouachita Mountains, Oklahoma: Geol. Soc. America Abstracts with Programs, v. 5, p. 269–270.

LOWE, D. J., 1973a, Arkansas Novaculite—some physical aspects of its sedimentation: *ibid.,* p. 270.

———, 1973b, Regional controls of silica sedimentation in the Ouachita system: *ibid.,* p. 270–271.

LUTTRELL, E. M., 1965, Sedimentary analysis of some Jackfork sandstones, Big Cedar, Oklahoma (M.S. thesis): Univ. Wisconsin, Madison, 122 p.

MAHER, J. C., AND LANTZ, R. J., 1953, Correlation of Pre-Atoka rocks in the Arkansas Valley, Arkansas: U.S. Geol. Survey Oil and Gas Inv. Chart OC-51.

MCBRIDE, E. F., 1970, Flysch sedimentation in the Marathon region, Texas *in* Lajoie, J. (ed.), Flysch sedimentology in North America: Geol. Assoc. Canada Special Paper No. 7, p. 67–83.

MEYERHOFF, A. A., 1967, Future hydrocarbon provinces of Gulf of Mexico-Caribbean region: Gulf Coast Assoc. Geol. Soc. Trans., v. 17, p. 217–260.

MISER, H. D., 1921, Llanoria, the Paleozoic land area in Louisiana and eastern Texas: Am. Jour. Sci., 5th Ser., v. 2, p. 61–89.

———, 1959, Structure and vein quartz of the Ouachita Mountains of Oklahoma and Arkansas *in* Cline, L. M., Hilseweck, W. J., and Feray, D. E. (eds.), The geology of the Ouachita Mountains, a symposium: Dallas and Ardmore Geol. Soc., p. 30–49.

———, AND HENDRICKS, T. A., 1960, Age of the Johns Valley Shale, Jackfork Sandstone, and Stanley Shale: Am. Assoc. Petroleum Geologists Bull., v. 44, p. 1829–1834.

———, AND PURDUE, A. H., 1929, Geology of the DeQueen and Caddo Gap quadrangles, Arkansas: U.S. Geol. Survey Bull. 808, 195 p.

MORETTI, F. J., 1958, Petrographic study of the sandstones of the Jackfork Group, Ouachita Mountains, southeastern Oklahoma (Ph.D. dissertation): Univ. Wisconsin, Madison, 135 p.

MORRIS, R. C., 1964, Geologic investigation of Jackfork Group of Arkansas (Ph.D. dissertation): Univ. Wisconsin, Madison, 179 p.

———, 1971a, Stratigraphy and sedimentology of the Jackfork Group, Arkansas: Am. Assoc. Petroleum Geologists Bull., v. 55, p. 387–402.

———, 1971b, Classification and interpretation of disturbed bedding types in Jackfork flysch rocks (upper Mississippian), Ouachita Mountains, Arkansas: Jour. Sed. Petrology, v. 41, p. 410–424.

———, (in press), Carboniferous rocks of the Ouachita Mountains, Arkansas—A study of facies patterns along the unstable slope and axis of a flysch trough *in* Briggs, Garrett (ed.), Symposium on the Carboniferous rocks of the southeastern United States: Geol. Soc. America Special Paper 148.

PARK, D. E., JR., AND CRONEIS, CARY, 1969, Origin of Caballos and Arkansas Novaculite Formations: Am. Assoc. Petroleum Geologists Bull., v. 53, p. 94–111.

PITT, W. D., CAHOON, R. R., LEE, H. C., ROBB, M. G., AND WATSON, J., 1961, Ouachita Mountain core area, Montgomery County, Arkansas: *ibid.,* v. 45, p. 72–94.

RAINWATER, E. H., 1971, Possible future petroleum potential of peninsular Florida and adjacent continental shelves *in* Cram, I. H. (ed.), Future petroleum provinces of the United States—their geology and potential: *ibid.,* Mem. 15, v. 2, p. 1311–1341.

REINEMUND, J. H., AND DANILCHIK, WALTER, 1957, Preliminary geologic map of the Waldron quadrangle and adjacent areas, Scott County, Arkansas: U.S. Geol. Survey Oil and Gas Inv. Map OM-192.

REPETSKI, J. E., AND ETHINGTON, R. L., 1973, Conodonts from graptolite facies in the Ouachita Mountains, Arkansas and Oklahoma: Geol. Soc. America Abstracts with Programs, v. 5, p. 277.

RUSSELL, R. H., 1969, Geology of the Blue Ouachita Mountain area, frontal Ouachita Mountains, Arkansas (M.S. thesis): Northern Illinois Univ., DeKalb, 105 p.

SEELY, D. R., 1963, Structure and stratigraphy of the Rich Mountain area, Oklahoma and Arkansas: Oklahoma Geol. Survey Bull. 101, 173 p.

SELLARS, R. T., 1967, The Siluro-Devonian rocks of the Ouachita Mountains *in* Toomey, D. F. (ed.), Silurian-Devonian rocks of Oklahoma and environs: Tulsa Geol. Soc. Digest, v. 35, p. 231–241.

SHELBURNE, O. B., JR., 1960, Geology of the Boktukola syncline, southeastern Oklahoma: Oklahoma Geol. Survey Bull. 88, 84 p.

SHIDELER, G. L., 1970, Provenance of Johns Valley boulders in late Paleozoic Ouachita facies, southeastern Oklahoma and southwestern Arkansas: Am. Assoc. Petroleum Geologists Bull., v. 54, p. 789–806.

SIMONIS, E. K., 1967, Geology of part of the Oden quadrangle, Ouachita Mountains, Arkansas (M.S. thesis): Northern Illinois Univ., DeKalb, 116 p.

STERLING, P. T., STONE, C. G., AND HOLBROOK, D. F., 1966, General geology of eastern Ouachita Mountains, Arkansas *in* Field conference on flysch facies and structure of the Ouachita Mountains: Kansas Geol. Soc. 29th Field Conf. Guidebook, p. 177–194.

Stone, C. G., Haley, B. R., and Viele, G. W., 1973, A guidebook to the geology of the Ouachita Mountains, Arkansas: Arkansas Geol. Comm., 114 p.

Sullivan, D. A., Jr., 1966, A paleocurrent study of upper Mississippian and lower Pennsylvanian rocks of the frontal Ouachita Mountains and Arkansas Valley (Ph.D. dissertation): Washington Univ., St. Louis, 134 p.

Thomas, W. A., 1972, Regional Paleozoic stratigraphy in Mississippi between Ouachita and Appalachian Mountains: Am. Assoc. Petroleum Geologists Bull., v. 56, p. 81–106.

Tomlinson, C. W., and McBee, William, Jr., 1962, Pennsylvanian sediments and orogenies of Ardmore district, Oklahoma *in* Branson, C. C. (ed.), Pennsylvanian system in the United States, a symposium: Am. Assoc. Petroleum Geologists, Tulsa, p. 461–500.

van der Gracht, W. A., J. M. van Waterschoot, 1931, Permo-Carboniferous orogeny in south-central United States: Am. Assoc. Petroleum Geologists Bull., v. 15, p. 991–1057.

Vernon, R. C., 1971, Possible future petroleum potential in Pre-Jurassic, western Gulf basin *in* Cram, I. H. (ed.), Future petroleum provinces of the United States—their geology and potential: *ibid.*, Mem. 15, v. 2, p. 954–979.

Viele, G. W., 1973, Tectonic history of eastern Ouachita Mountains, Arkansas: Geol. Soc. America Abstracts with Programs, v. 5, p. 284–285.

Walker, R. G., 1967, Turbidite sedimentary structures and their relationship to proximal and distal depositional environments: Jour. Sediment. Petrology, v. 37, p. 25–43.

Walthall, B. H., 1967, Stratigraphy and structure, part of Athens Plateau, southern Ouachitas, Arkansas: Am. Assoc. Petroleum Geologists Bull., v. 51, p. 504–528.

Weaver, C. E., 1958, Geologic interpretation of argillaceous sediments; part 2, Clay petrology of upper Mississippian-lower Pennsylvania sediments in central United States: *ibid.*, v. 42, p. 272–309.

White, David, 1934, The age of the Jackfork and Stanley formations of the Ouachita geosyncline as indicated by plants: *ibid.*, v. 18, p. 1010–1017.

———, 1937, Fossil plants from the Stanley Shale and Jackfork Sandstones in southeastern Oklahoma and western Arkansas: U.S. Geol. Survey Prof. Paper 186-C, p. 43–68.

Wilhelm, Oscar, and Ewing, Maurice, 1972, Geology and history of the Gulf of Mexico: Geol. Soc. America Bull., v. 83, p. 575–600.

Williams, H., Turner, F. J., and Gilbert, C. M., 1954, Petrography: Freeman, San Francisco, 406 p.

PALEODRAINAGE PATTERN AND LATE-OROGENIC BASINS OF THE CANADIAN CORDILLERA

G. H. EISBACHER, M. A. CARRIGY, AND R. B. CAMPBELL
Geological Survey of Canada, Vancouver, British Columbia; Research Council of Alberta, Edmonton, Alberta; and Geological Survey of Canada, Vancouver, British Columbia

ABSTRACT

Recent models for the tectonic evolution of the Canadian Cordillera can be tested by relating uplift of the orogenic core zones to the depositional record of the foreland and successor basins, but this requires a comprehensive understanding of paleodrainage.

The Canadian Cordillera is made up of two orogens: the Columbian Orogen on the east and the Pacific Orogen on the west. Most of the late-orogenic molasse deposits are related to the Columbian Orogen (Omineca Crystalline Belt and Rocky Mountain Belt). In the Columbian Orogen three structural elements placed controlling restraints on late-orogenic drainage: uplifts, re-entrants and salients, and longitudinal intramontane fault zones. The uplift boundary can be defined as a zone of great vertical displacement of pre-orogenic sedimentary and volcanic rocks and probable involvement of terrane intruded or metamorphosed in Precambrian time. The uplift boundary separated the aggradational molasse basins from the erosional domain of the drainage system. Most of the clastic sediment derived from the uplifted core zone was transported by rivers flowing in longitudinal, intramontane valleys. These rivers merged near re-entrants of the uplift boundary and discharged their load into elongate molasse basins. During growth of folds and thrust faults within the orogen, valleys near the structural re-entrants constituted the shortest dispersal paths between the rising core zones and the subsiding foreland and successor basins. The area in front of the regional re-entrants (Crowsnest, Peace, Liard, and Peel on the east side; Chukachida and Thompson on the west side) therefore display the best developed molasse deposits in foreland basins to the east and successor basins to the west. From the structural salients only short, though locally vigorous, streams issued directly into the late-orogenic basins.

The molasse of the Columbian foreland basin displays two upward coarsening megacycles: an uppermost Jurassic through Lower Cretaceous cycle (Kootenay-Blairmore Assemblage), and an Upper Cretaceous through Oligocene cycle (Belly River-Paskapoo Assemblage). Regional drainage during deposition of the first cycle was directed to the north, whereas during the second cycle streams flowed predominantly to the southeast. Straight drainage lines connecting the re-entrants with the depositional basins are probably valid concepts for the earliest stages of uplift only. Progressive growth of folds and thrust faults in the Rocky Mountain Belt, and faulting near intramontane valleys, produced curved and even U-shaped river systems, which merged near re-entrants and effected thorough mixing of compositionally diverse sediment loads. The molasse of the successor basins reflects progressive unroofing of the crystalline core zone. The two cycles of the Columbian foreland are similar to molasse sequences in the Alps and seem to reflect two phases of isostatic uplift related to repeated intervals of tectonic crustal thickening.

INTRODUCTION

During the past five years the authors have been involved in the study of tectonics and sedimentation in parts of the Canadian Cordillera in British Columbia and Alberta. One of us has recently synthesized the evolution of eugeosynclinal successor basins of the Intermontane Belt (Eisbacher, 1974). The discussion resulting from this compilation, and the new tectonic subdivision of the Canadian Cordillera by Wheeler and Gabrielse (1972) in the aftermath of plate-tectonic models, has led us to consider the total effect of tectonic uplift on late-orogenic sedimentation during the growth of both the Columbian and Pacific Orogens.

Major parts of the Canadian Cordillera are still known in little more than rudimentary fashion, but with delineation of the major tectonic belts shown in figure 1 it is now possible to study the interaction of different segments of the two orogens that previously were seen only in terms of more or less characteristic cross-sections (Douglas *et al.*, 1970). Much of the information included in this study has been recently acquired. Acknowledgement for help from various sources can therefore be considered only a poor substitute for the debt we owe to many fellow Canadian geologists whose stimulating discussions or unpublished contributions have encouraged our work. The following individuals have greatly aided our effort: F. G. Young, J. F. Lerbekmo, R. A. Rahmani, C. J. Yorath, H. Gabrielse, H. Tipper, and D. F. Stott contributed unpublished information; D. J. Tempelman-Kluit, H. U. Bielenstein, G. C. Taylor, D. K. Norris, R. A. Price, J. D. Aitken, J. Monger, and A. Okulitch discussed tectonic problems with us; H. Meixner and R. Dumas helped with compilation and drafting. Discussion by one of us (G.H.E.) with H. Fuchtbauer and F. B. Van Houten convinced us of the usefulness of the term "molasse" in de-

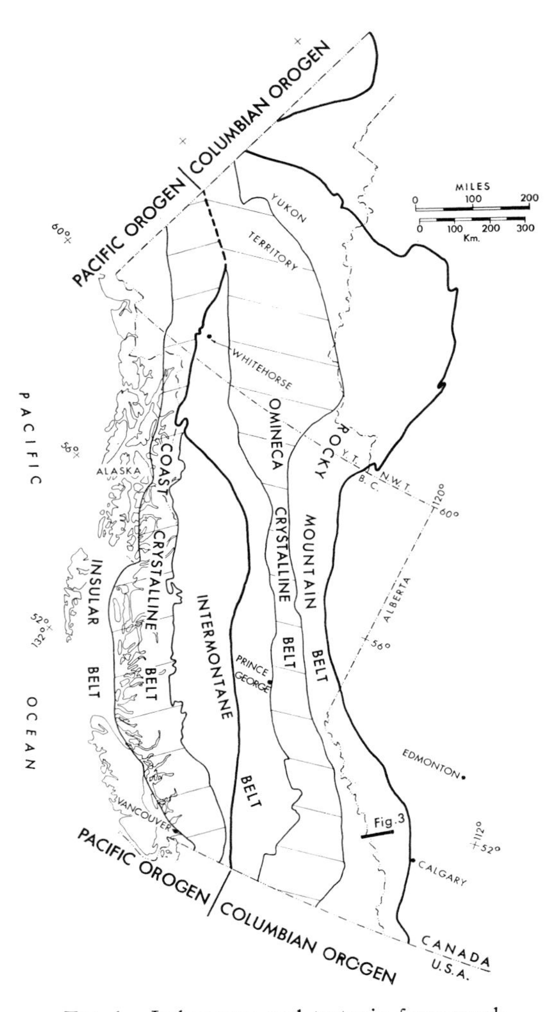

FIG. 1.—Index map and tectonic framework of the Canadian Cordillera.

scribing the late-orogenic depositional basins of the Columbian Orogen. W. R. Dickinson critically read the manuscript and suggested several improvements.

CONCEPT OF LATE-OROGENIC MOLASSE BASINS

Deformation and uplift are known to influence the depositional environment, composition, and thickness of clastic sedimentary rocks in basins adjacent to growing mountain chains. The tectono-stratigraphic significance of clastic wedges has been well illustrated by work in the Alpine-Mediterranean region where at an early stage of investigation the clastic mantle around the mountain chains was thought to represent a precise record of tectonic activity within the mountains. In the Alps the monotonous clastic "flysch" facies, made up of marine shale-sandstone sequences, is generally followed in time by varied shallow-water marine or non-marine "molasse" deposits, made up chiefly of coarse conglomerate, sandstone, marl, and minor coal. It is the molasse facies that reflects in many ways the gradual emergence and progressive erosional unroofing of rising orogenic belts. The search for oil and gas in the basins near the Alpine mountain chains has stimulated intensive recent research on the facies of the classic Flysch and Molasse, and has resulted in a number of excellent reviews of both in their type regions (see Hsu, 1970; Schmidt-Thome, 1963; Fuchtbauer, 1967; Van Houten, 1974). Mazarovich (1972) recently discussed the use of the molasse concept in the USSR and came to the conclusion that although molasse may be found in different settings it represents a distinct facies of upward and outward coarsening clastic basin-fill resulting from erosional disintegration of tectonically active mountain chains. In this paper the term molasse follows closely the definition recently proposed by Van Houten (1973) and refers to *late-orogenic clastic wedges laid down in shallow-marine and nonmarine environments adjacent to rising mountain chains.*

Most studies of molasse in the past have yielded a wealth of data on mineral composition, paleocurrents, provenance and depositional environments. Conclusions about the patterns of river systems within growing orogens have been approximate only (e.g. Schiemenz, 1960). The potential of this approach, however, has been realized in a recent study of the Devonian Wood Bay Formation in Spitzbergen (Friend and Moody-Stuart, 1972). It must be stressed that toward a proper understanding of lateral facies changes and the distribution of heavy minerals in molasse, some knowledge of the size and geometry of the river systems *outside* the depositional basins is essential. The geometry of a river course prior to entering the depositional basin determines which of the clastic components will ultimately form a sedimentary association, and to what extent alluvial sediments will be deposited in orogenic foreland basins, intramontane valleys, and eugeosynclinal successor basins. Towards such an understanding it is critical to separate tectonic elements that strongly influence the developing topography within a rising mountain chain from those that exert only a minor influence on the course of rivers from their highland sources to their lowland basins of deposition. Present-day drainage in areas of previous oro-

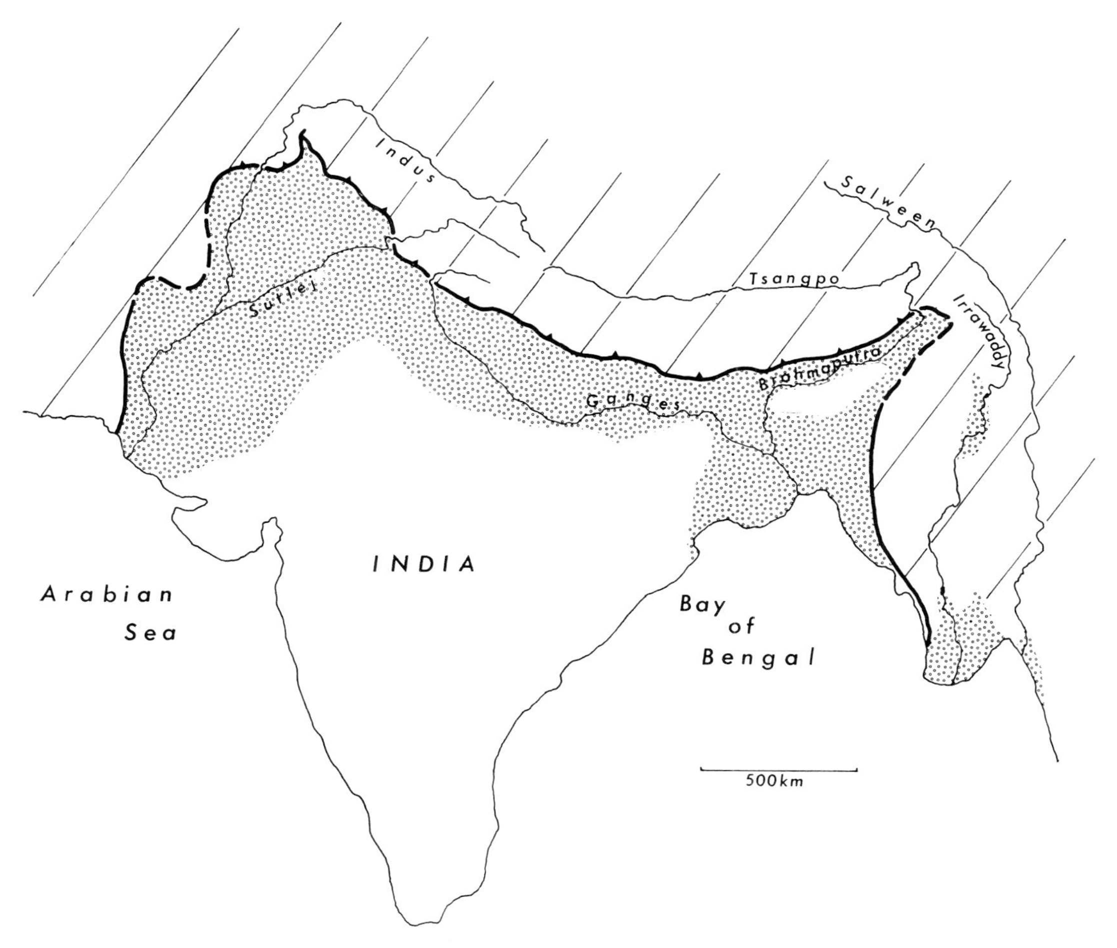

FIG. 2.—Tectonic elements influencing late-orogenic drainage in the Himalayan Orogen. Ruled pattern indicates uplifted part, circles denote active molasse deposition. For explanation see text.

genic movements may, to a limited degree, indicate drainage during the molasse phase, but in most cases such an inference from the mature drainage to the immature orogenic drainage must be hazardous, as will be shown in this paper.

It is therefore useful to examine cases of fold belts that, at present, seem to display deposition of molasse into basins adjacent to actively rising mountain chains. The most instructive example, but not the only one available, is the Himalayan mountain chain which will be used to illustrate the relevant points. The lofty ranges of the Himalayas supply debris for a large aggrading molasse basin in front of active boundary faults along the northern edge of the Indian subcontinent (fig. 2). This basin contains the predominantly nonmarine Siwalik beds (upper Miocene to Pleistocene) which are presently being reworked by or buried under sediments carried by the major rivers and their tributaries (Gansser, 1964). As exemplified by the actively aggrading Himalayan molasse, three principal structural features seem to control drainage evolution more than others:

(1) The *erosional domain* of drainage is distinctly separated from the *aggradational basins* by a zone of reverse faults roughly defining the *uplift boundary* of the mountain core.

(2) Drainage within the mountains is mainly *longitudinal* and follows pre-existing longitudinal fault zones. Most of the drainage in the aggradational molasse basin is also longitudinal with respect to active structures, except near the mountain front where vigorous but short transverse streams supply sediment to the main longitudinal rivers.

(3) The major longitudinal rivers enter the

molasse basins through *structural re-entrants,* where they commonly change their direction of flow before depositing their sediment load. Structural re-entrants constitute the shortest connection between molasse basins and the longitudinal intramontane valleys and therefore offer the easiest route for headward stream capture during uplift of mountainous terrain. *Structural salients* between re-entrants are obstacles to headward capture, unless they are ruptured by major transverse fault systems.

In the following we attempt to identify (a) the fundamental uplift boundary, (b) the longitudinal intramontane fault zones, and (c) the structural re-entrants of the Columbian Orogen. The evolution of the molasse basins will be discussed in the light of these tectonic elements within the framework of the Canadian Cordillera.

TECTONIC FRAMEWORK OF THE COLUMBIAN OROGEN

The Canadian Cordillera consists of two orogenic belts (Wheeler and Gabrielse, 1972): the Pacific Orogen on the west and the Columbian Orogen on the east (fig. 1). The two orogenic belts are separated by the Intermontane Belt that is characterized by volcanic and granitoid terrane overlain unconformably by non-volcanic successor basins of Middle Jurassic through early Tertiary age (Douglas and others, 1970; Eisbacher, 1974). The Columbian Orogen developed at the junction between miogeoclinal shelf sediments (Proterozoic to Lower Jurassic) and oceanic or island arc assemblages (Paleozoic to Middle Jurassic). Miogeoclinal sediments, oceanic volcanics and sediments, and parts of their crustal substratum are incorporated in the metamorphic complexes of the Omineca Crystalline Belt which represents the core zone of the Columbian Orogen. The geologic history leading up to the tectonic juxtaposition of oceanic and shallow shelf deposits has recently been synthesized in terms of a plate-tectonic model by Monger and others (1972). The Pacific Orogen consists essentially of the Coast Crystalline Belt, the Insular Belt, and the western part of the Intermontane Belt. The Middle Jurassic to Late Jurassic sedimentary history of the Intermontane Belt reflects uplift of the Columbian Orogen, and between latest Cretaceous to Eocene time uplift and deformation in the Pacific Orogen. The overwhelming part of the molasse deposits in the Canadian Cordillera is related to the Columbian Orogen and was laid down in foreland basins, intramontane valleys and eugeosynclinal successor basins (fig. 3). Only locally do deposits related to uplift in the Pacific Orogen overlap molasse deposits of the Columbian Orogen. We shall therefore discuss tectonic elements and paleodrainage of the Columbian Orogen, and demonstrate the influence of uplift in the Pacific Orogen on Eocene drainage in the successor basins.

Uplift Boundary

Uplift is a significant factor in the evolution of all orogenic belts involving major sialic core zones of metamorphic and/or large granitic intrusions (Zwart, 1969). The extent to which the core zones interact in a "brittle" or "mobile" fashion with the overlying or adjacent strata during orogenic displacements varies significantly from one mountain chain to the other (Wegmann, 1961; Rodgers, 1964; Julivert, 1970), and constitutes a fundamental tectonic problem. In the Columbian Orogen the nature of basement tectonics has recently become a subject of great interest and controversy because of its critical role in the interpretation of the deep subsurface geology (Bally and others, 1966; Roeder, 1967; Price and Mountjoy, 1970; Taylor, 1971; Campbell, in press). Basement, in the context of this paper, includes sedimentary rocks intruded, penetratively deformed, or metamorphosed in Precambrian time. Intrusion or metamorphism, wherever it affected sedimentary units, probably changed the mechanical properties to such an extent that the term basement is justified for these units with respect to later Mesozoic and Cenozoic displacements.

The Foothills of the Rocky Mountain Belt seem to be underlain, at least in the southern part, by an undeformed surface of crystalline basement, the presence of which has been documented by borehole and seismic data. Low-angle back-limb thrust faults and concentric folds dominate the structure in the sedimentary formations of the Foothills (Douglas, 1950; Dahlstrom, 1970). The Foothills structural model, combined with seismic evidence for an apparently undisturbed basement surface west of the Foothills near the International Boundary, was subsequently extrapolated along strike to interpret the deeper subsurface structure of the western Rocky Mountains (Bally and others, 1966; Price and Mountjoy, 1970). This structural model of the Rocky Mountains implies a planar basement surface as far west as the exposed crystalline core zone of the Columbian Orogen, and assumes crustal shortening of about 50 percent within the sedimentary sequence of the Rocky Mountains. As a consequence, the juxtaposition of metamorphosed and unmetamorphosed strata of the same age demands large-scale horizontal movements of

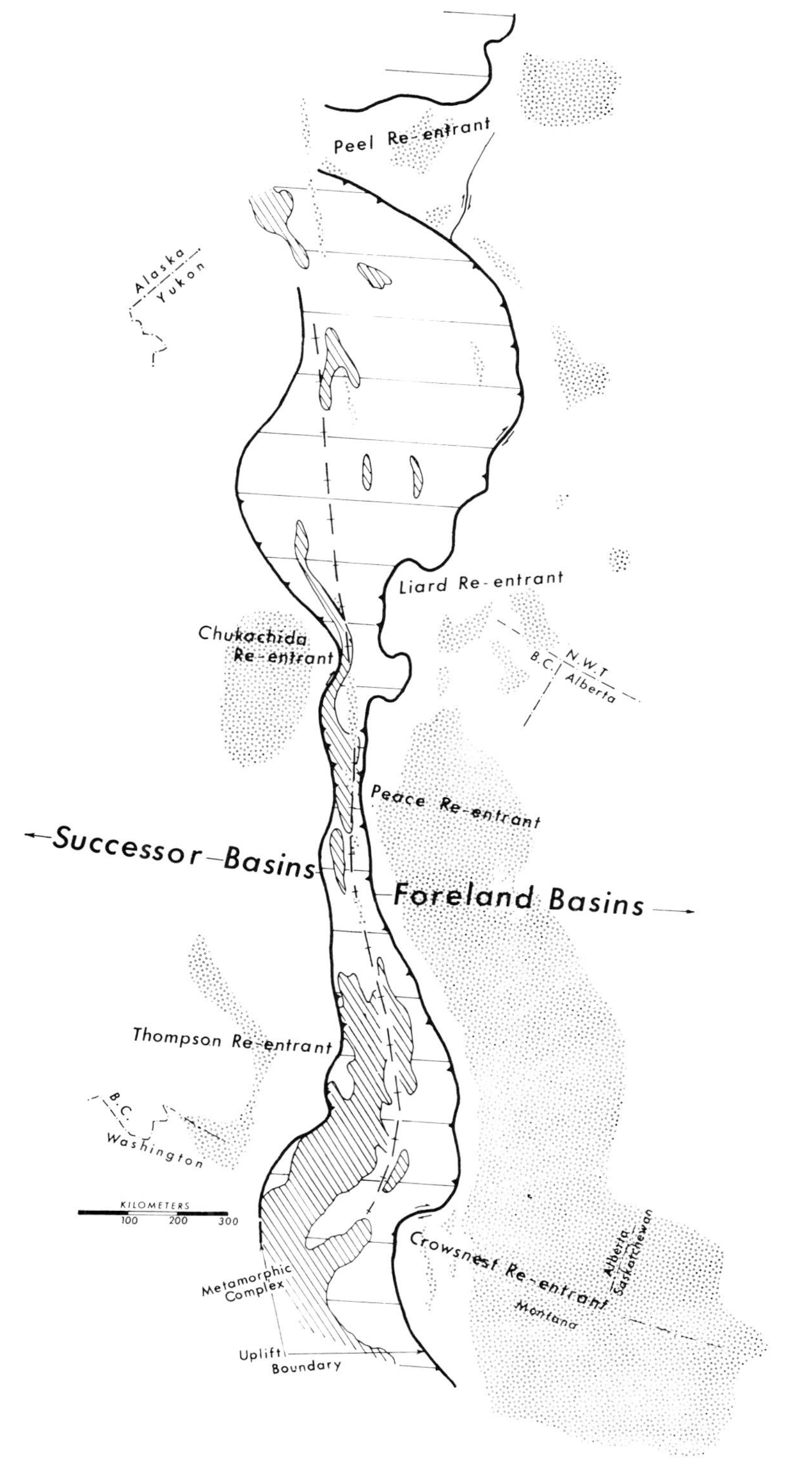

FIG. 3.—Distribution of molasse facies in successor basins, intramontane valleys, and foreland basin of the Columbian Orogen. Closely ruled pattern indicates high-grade metamorphic complexes.

mobile sialic material in the Omineca Crystalline Belt over *undisturbed* crystalline basement in the Rocky Mountain Belt (see Price and Mountjoy, 1970, fig. 2-1). Such a model also assumes low-angle thrust faults as a dominant structural pattern for the Paleozoic carbonates and the Proterozoic shale-sandstone sequence of the western Rocky Mountains. However, recent work near the thermal and structural roof of the southern Omineca Crystalline Belt suggests involvement of basement in the folds of the western Rocky Mountains. This basement is probably represented in extensive exposures of Malton Gneiss near the Rocky Mountain Trench (Campbell, 1973). Detailed mapping in the western Rocky Mountains generally reveals large faulted anticlines and broad synclines in the Paleozoic carbonates, and closely spaced folds, cleavage and decollement within the Upper Proterozoic sandstone-shale sequence (Charlesworth and others, 1967). Overall shortening of the Rocky Mountain Belt near the structural roof of the southern Omineca Belt appears to be less than 25 percent (Campbell, 1973). Even less shortening is reported for the northern part of the Rocky Mountain Belt (Taylor, 1972 personal communication; Norris, 1972a), where estimates of total shortening range between 10 and 20 percent. The possible involvement of basement and the structural style of the upper Proterozoic sequence permit an alternative to the estimates of shortening and the deep subsurface structure of the western Rocky Mountains as proposed by Price and Mountjoy (1970).

A magnificent set of maps for the Banff region by Price and Mountjoy (1969–1972) represents the basis for our reinterpretation of subsurface structure in the western Rocky Mountains of southern Canada. For the cross sections shown in figure 4 we have taken surface geology as mapped by Price and Mountjoy (1969–1972), used known thicknesses of stratigraphic sections, and extrapolated the known structural style of the Proterozoic sandstone-shale sequence as mapped by Charlesworth and others (1967), to derive the outline of deep subsurface structure. As reinterpreted, the structural style of the western Rockies is dominated by large synclines and faulted anti-

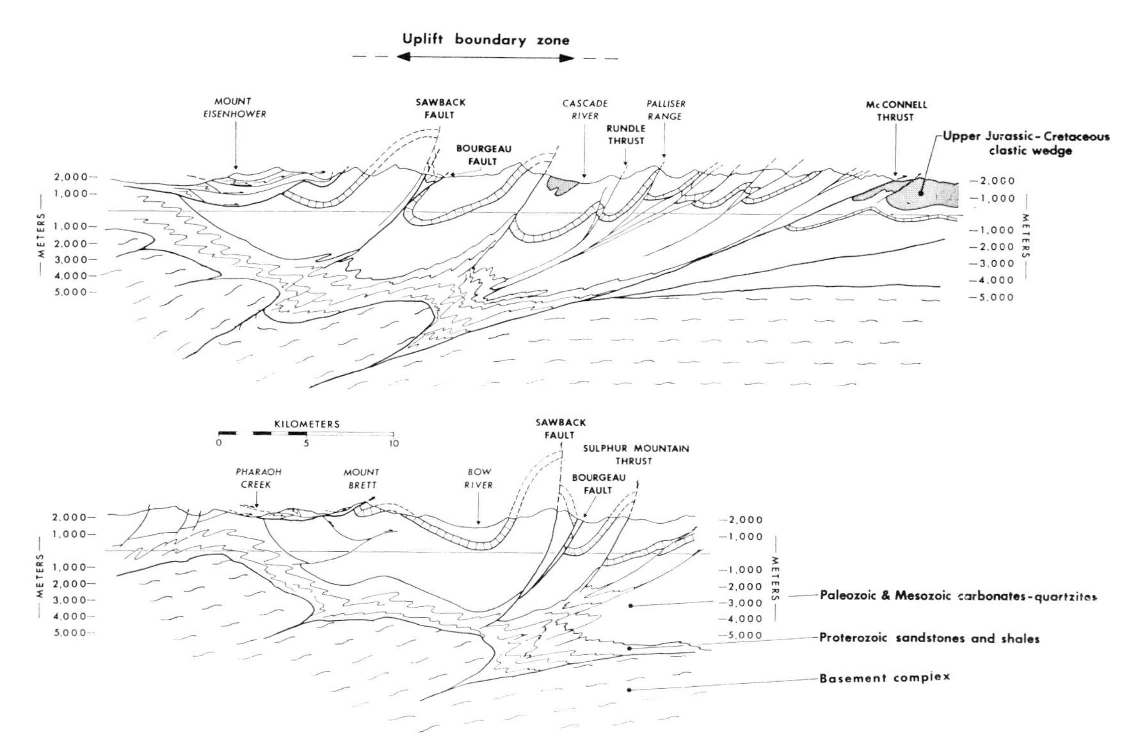

FIG. 4.—Hypothetical structural cross-sections through the uplift boundary zone between the eastern and western Rocky Mountain Belt near Banff; from surface geological mapping by Price and Mountjoy (1969–1972). The Mississippian Exshaw-Banff Formation is shown as a marker within the Paleozoic-Mesozoic carbonate-quartzite packet. For location see figure 1.

clines within the bedded Paleozoic carbonate sequence and tight folds in the cleaved Proterozoic sandstone-shale sequence (fig. 4). It seems to us that the "closely spaced array of discrete, discontinuous, interleaved, and overlapping thrust faults" that is considered typical for the whole southern Canadian Rocky Mountains by Price and Mountjoy (1970) is documented only in the eastern Rockies and Foothills, where flat thrust faults and apparently undeformed basement have been intersected by boreholes. The large synclines and faulted anticlines of the western Rockies can probably better be explained by uplift and decollement within the Proterozoic sandstone-shale sequence. Along the margins of major, basement-cored uplifts in the western Rockies gravity slides may have produced numerous tectonic slides and klippen. If the subsurface structural concept illustrated by the cross-sections in figure 4 is applied to a cross-section completely across the southern Canadian Rocky Mountains the total amount of shortening is about 30 percent and more in agreement with estimates reported from south (Mudge, 1970) and north (Campbell, 1973) of the area. Regardless of this modification, the southern Rocky Mountains still display the greatest crustal shortening in the Canadian part of the Columbian Orogen.

Schematic cross-sections through the eastern half of the Columbian Orogen in three different segments illustrate the zone along which vertical displacement becomes an important component of deformation along the eastern margin of the Columbian Orogen (fig. 5). This zone, along which lower Paleozoic units show great vertical separation, is here described as the eastern uplift boundary zone and is indicated in the cross-sections of figure 5. It is apparent from preceding discussion that the uplift boundary can be related only hypothetically to basement tectonics along most of the Rocky Mountain Belt. Vertical separation of stratigraphic units along this zone, however, is up to several kilometers and probably marks it as the fundamental boundary between the region of orogenic uplift and the molasse basins of the foreland. A similar fundamental zone of differential vertical movement also separates the western metamorphic-igneous core zone from the Mesozoic volcanics of the Intermontane Belt. In places this zone is a strike-slip fault and in others is a reverse fault or a large anticlinorium overturned westerly. Structures related to the two uplift boundaries reflect relative tectonic transport to the east along the eastern edge, and relative tectonic transport to the west along the western edge of the Columbian Orogen. Consequently, a broadly defined zone of structural divergence within the central uplift separates an area of relative tectonic transport to the east from an area of relative tectonic transport to the west (fig. 5). East of this axis the rocks of the Columbian Orogen consist mainly of deformed miogeoclinal carbonate-quartzite successions, whereas west of the axis high-grade metamorphic complexes are dominant (fig. 3). Because of the different rock types involved and the different metamorphic grades the structural style is markedly dissimilar east and west of the zone of structural divergence.

Longitudinal Intramontane Fault Zones

The longitudinal fault zones of the Columbian Orogen have been a focus of attention since McConnell's (1896) classic description of the Northern Rocky Mountain Trench. Some of the major faults follow closely the central zone of structural divergence as defined above. Towards the west stringers of ultramafic bodies are closely related to some of the lineaments. A recent compilation of this central fault system of the Cordillera shows that most of its strands are long-lived, repeatedly reactivated deep-crustal fractures which at times may have reached to the upper mantle (Eisbacher and Tempelman-Kluit, 1972; Monger and others, 1972).

Across the Northern Rocky Mountain Trench and Tintina Fault a major offset of miogeoclinal facies belts suggests possible early Mesozoic right-lateral transcurrent movement (Roddick, 1967; Tempelman-Kluit, 1972; Gabrielse, 1972). Other strands of the central fault system probably also participated in regional right-lateral movements. In the southern Canadian Cordillera the early Mesozoic lineaments have been obliterated by later widespread regional metamorphism, and major lineaments are generally related to late Mesozoic reverse faults. Late Mesozoic to early Tertiary displacements along the Rocky Mountain Trench and related fault zones are mainly vertical, and strike-slip components are probably small (Leech, 1966; Eisbacher, 1972). Numerous linear inliers of Upper Cretaceous and lower Tertiary continental clastic deposits have been described from near the major lineaments, and their existence suggests that the course of depositing rivers probably followed the fault zones. Longitudinal transport of sediment within the intramontane basins must therefore be seriously considered during late orogenic evolution of drainage and will be examined in more detail for the case of the Northern Rocky Mountain Trench.

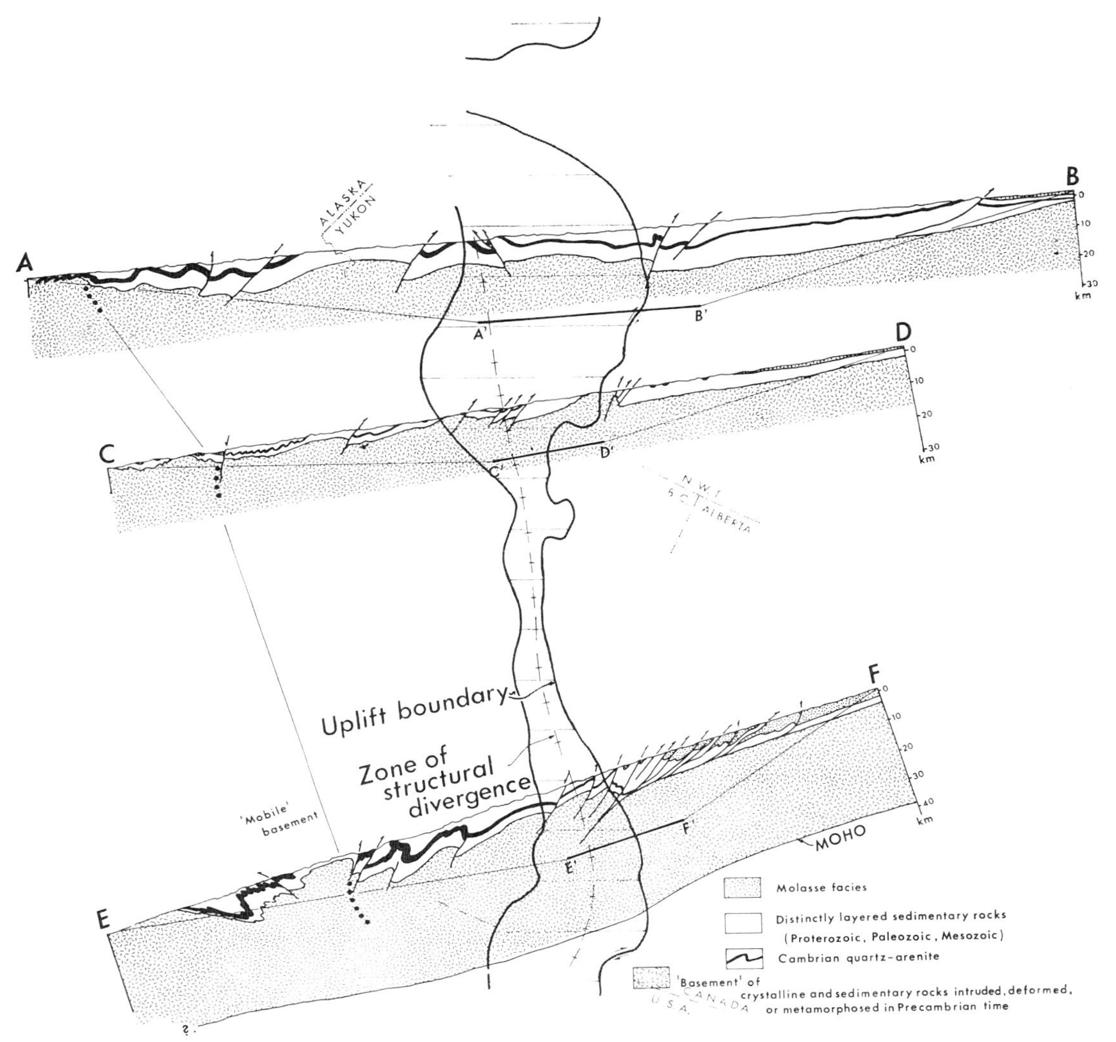

FIG. 5.—Schematic cross-sections through the eastern part of the Columbian Orogen indicating probable basement involvement near the uplift boundary and increased basement mobility near the zone of structural divergence. Modified from Douglas and others (1970).

Salients and Re-entrants

Orogenic belts are commonly segmented along trend into a series of structural salients which are linked to each other by structural re-entrants. Orogenic belts are usually wide across salients and narrow across re-entrants. The arcuate structure (virgation) of mountain chains is illustrated in an extreme form by the young Mediterranean-Himalayan orogens, whose curvature has been the subject of much speculation. In the North American Cordillera recent work by Crosby (1969), Norris (1972a), and Yates (1973) has shown clearly that much of the curvature along the eastern margin of the orogen is related to the configuration of the pre-orogenic continental margin and to the shape of the cratonic basement surface as reflected in the trend of isopach lines for miogeoclinal sedimentary units. The general significance of the inherited nature of curvature along orogenic belts was first clearly recognized by Bucher (1933, p. 259). Beutner (1972) emphasized that contemporaneous basement uplifts in front of orogenic belts may retard the advance of thrust sheets and thus also create structural re-entrants. During progressive crustal shortening the development of the primary curvature of salients is further accentuated by radial movements along transverse faults compatible with the direction of thrusting (Crosby, 1969; Bielenstein, 1969). The resulting convergence of folding near re-entrants is generally expressed by superposition of differently oriented structural trends (Price, 1967; Gabrielse and Blusson, 1969). In the eugeosynclinal and crystalline parts of the orogen curvature is probably mainly due to relative displacement of crustal segments along deep transcurrent faults and their interaction with pre-existing structures along continental margins (Carey, 1958).

Along the Columbian Orogen six major well-documented re-entrants with intervening salients define the shape of the mountain front along the uplift boundaries as defined above (fig. 3). The re-entrants on the east side, related to the foreland basin, are the Crowsnest, Peace, Liard and Peel re-entrants; those on the west side are the Chukachida and Thompson re-entrants near which the successor basins are localized.

TRANSITION FROM SUBSIDENCE TO UPLIFT

The youngest clearly miogeoclinal beds of the Canadian Cordillera, described by Pelletier (1965), Gibson (1971) and others, are the westerly prograded Triassic carbonate-quartz arenite units of the Rocky Mountain Belt. These miogeoclinal shallow-water deposits are overlain disconformably, and in places unconformably, by black phosphatic shale, dark argillaceous limestone, calcareous siltstone and chert of the Lower to Middle Jurassic part of the Fernie Formation (Frebold, 1957; Stott, 1967a). The Lower to Middle Jurassic part of the Fernie Formation is about 100 to 200 m thick; its lithology suggests that it was deposited in a starved basin along the continental slope. Thin layers of tuff indicate distant volcanic activity (Deere and Bayliss, 1969).

Far to the west of the miogeocline, the Lower to Middle Jurassic nonvolcanic Laberge Basin received flysch deposits many times as thick as the contemporaneous Fernie beds (Souther, 1971; Wheeler, 1961). The close similarity of invertebrate faunas in the Laberge flysch and the Fernie shales, and the isopach trend of the Lower to Middle Jurassic Fernie shales suggests strongly that open sea connected the two clastic basins. It is here proposed that the Laberge-Fernie Basin may represent an Early to Middle Jurassic flysch basin occupying an oceanic trough between the craton on the east and mobile volcanic belts on the west (Eisbacher, 1974). Figure 6 illustrates the present location of the Laberge and Fernie clastics and the contemporaneous Hazelton volcanics of Early to Middle Jurassic age. A possible proximal-distal relationship between the Laberge Assemblage and the Fernie Formation seems to be indicated by the mainly westerly source of the Laberge flysch.

Middle Jurassic deformation of the Laberge Assemblage into southwesterly overturned structures followed by uplift of the Columbian core zone disrupted the sedimentary basin. During this deformation the presently missing sedimentary facies between the Laberge flysch and the Fernie shales may have disappeared underneath thrust faults along the western uplift boundary (fig. 6). Contemporaneous with uplift along the western part of the Columbian Orogen extensive delta complexes began to prograde over fine-grained marine clastics near the Thompson, Chukachida, Crowsnest and Peace re-entrants. Shoaling of the sea initiated the first of two major molasse cycles in the southern foreland basin of the Columbian Orogen. In the northern foreland basin only the second molasse cycle is developed and marine conditions prevailed in the Early Cretaceous.

The gradual inversion of the subsiding miogeocline into an uplifted area and the related deposition of clastic wedges represents a most important interval of orogenic history. It reflects the beginning of accelerated intracrustal mass transfer that had to be balanced by erosion and transfer of sediment at the surface (Beloussov, 1962). In the Cordillera this transition coincided with emplacement of large epizonal plutons into low-grade metamorphic terrane of the Omineca Crystalline Belt (Eisbacher, 1974).

FORELAND BASINS

The southern foreland basin of the Columbian Orogen displays two major molasse sequences (or "megacycles") representing a twice repeated progradation of coarse-grained nonmarine clastics over marine shales and sandstones (fig. 7). The lower sequence consists of the marine Upper Fernie Assemblage and the nonmarine Kootenay-Blairmore Assemblage. The upper sequence is made up of the marine Alberta Assemblage and the nonmarine Belly River-Paskapoo Assemblage. The age brackets given for these four sedimentological assemblages in figure 7 are approximate and do not indicate isochronous boundaries. In the northern foreland basin only the upper of the two molasse sequences is developed and consists of an Upper Cretaceous to lower Tertiary continental clastic wedge deposited over Lower Cretaceous marine sandstone-shale assemblages.

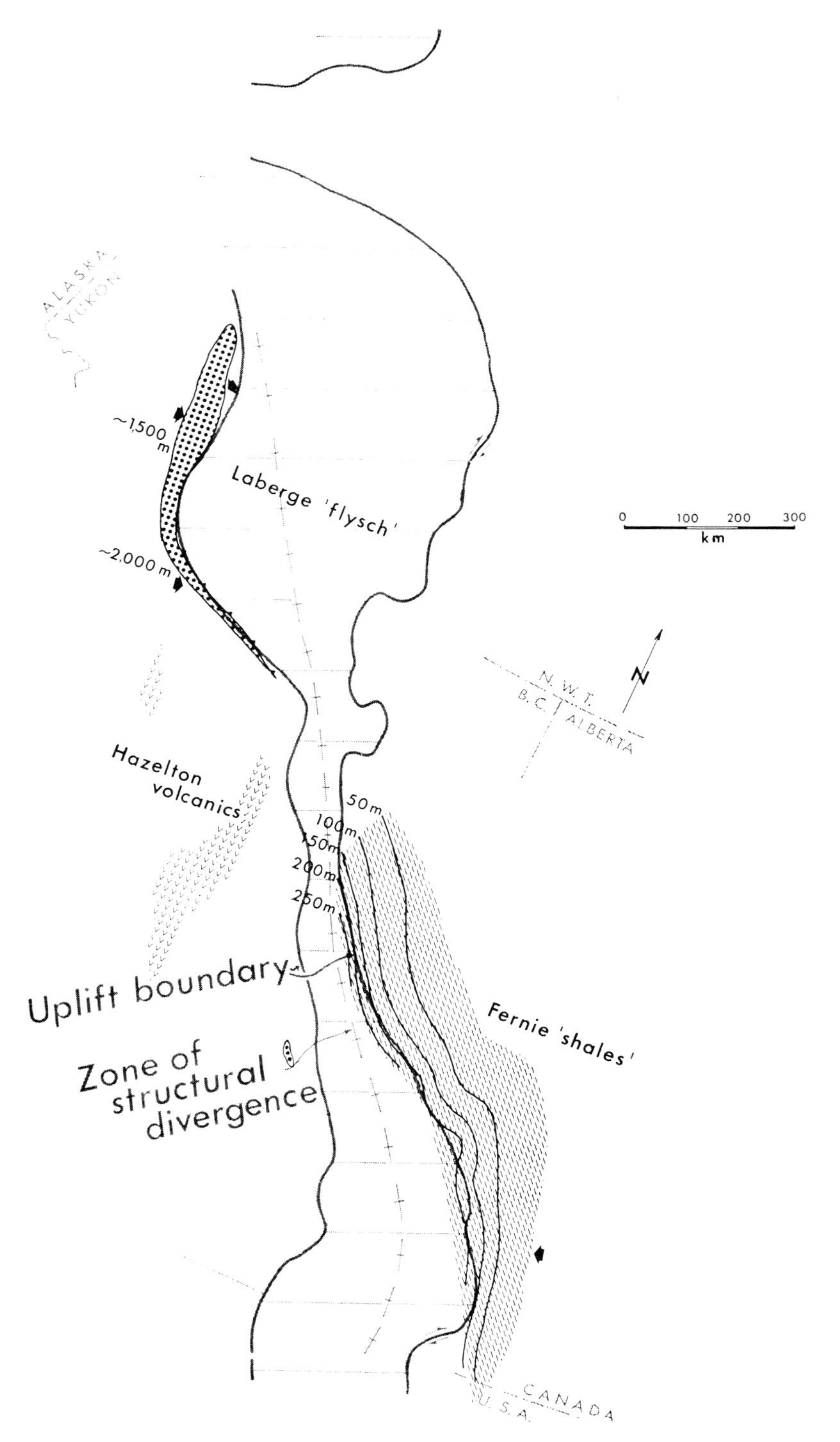

FIG. 6.—Distribution and thickness of Lower to Middle Jurassic Laberge flysch and Fernie shales, and occurrence of contemporaneous Hazelton volcanics. Arrows indicate sediment transport.

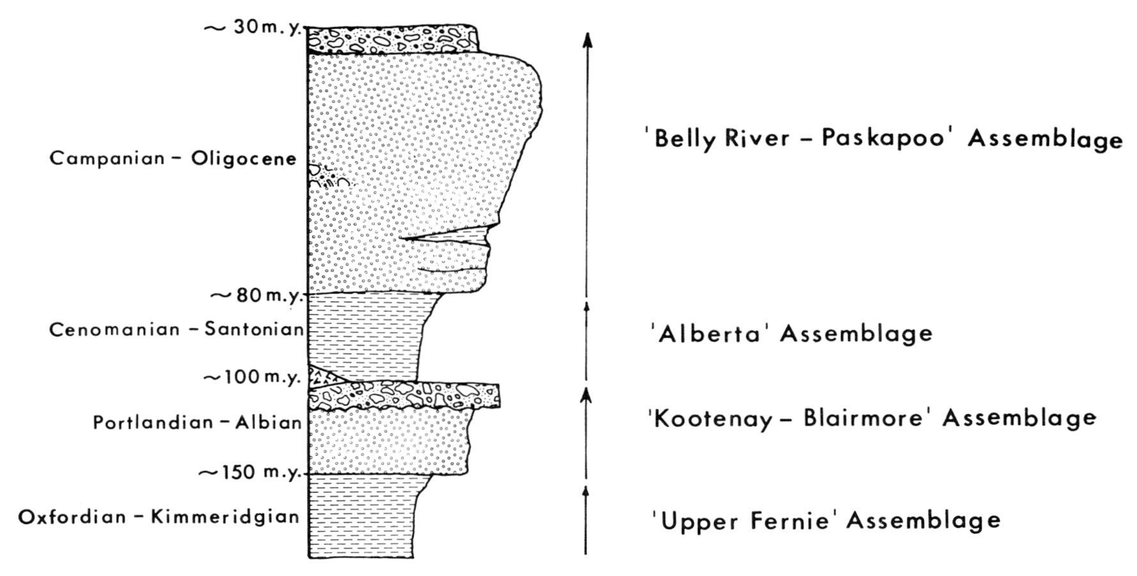

FIG. 7.—The four lithological assemblages of the southern foreland basin of the Columbian Orogen (see text).

Many of the stratigraphic complexities of the Columbian foreland basin have recently been illustrated by D. F. Stott in a series of highly instructive diagrams (*in* Douglas and others, 1970 and Stott, 1970b, c). The following generalizations are based largely on these compilations and recently obtained field data.

Lower Molasse (Kootenay-Blairmore Assemblage)

Crowsnest re-entrant.—Continental deposition of the lower molasse sequence in the Columbian foreland basin was preceded by the appearance of numerous sandstone beds in the upper part of the Fernie Formation. In the Crowsnest re-entrant these Upper Jurassic "Passage Beds" have recently been interpreted as prodelta deposits of an easterly prograding clastic wedge, and are overlain by the predominantly deltaic coal-swamp deposits of the Kootenay Formation (Jansa, 1972). Following an erosional interval, alluvial sediments of the Blairmore Group covered extensive areas of the southern foreland basin (Norris, 1964). Redbeds associated with these alluvial plain sediments are confined to the immediate vicinity of the Crowsnest re-entrant (Mellon, 1967; Holter and Mellon, 1972). The Kootenay-Blairmore Assemblage is there about 1500 m thick and thins to the east and north. A phase of volcanism (Crowsnest Volcanics) in the westernmost parts of the foreland basin terminated the lower molasse cycle in the Crowsnest re-entrant.

The modal composition of the Kootenay-Blairmore Assemblage has been studied by Mellon (1967), Rapson (1965), and Jansa (1972). In the lower part, sandstones and conglomerates contain mainly quartz, quartzite and chert. Schultheis and Mountjoy (1971) suggested outcrops of Cambrian quartzite and Carboniferous chert of the western Rocky Mountains as the principal source areas for the siliceous conglomerate clasts. Jansa (1972) and Rapson (1965) suggested Proterozoic terrane as another possible source for quartzose components. In the highest parts of the Kootenay-Blairmore Assemblage feldspar and volcanic rock fragments are abundant and make up as much as 50 percent of the detrital framework of sandstones (Mellon, 1967). Also within higher levels of the section the first influx of abundant granitic clasts (10 to 30 percent) indicates unroofing of plutons in the source area. Granitic pebbles in these conglomerates yielded K-Ar whole rock dates between 113 and 174 my (Norris and others, 1965). Similar granite-pebble conglomerates have also been reported by Mudge and Sheppard (1968) from equivalent foreland basin deposits in Montana. Heavy mineral suites in the lower part of the Kootenay-Blairmore Assemblage contain only ubiquitous zircon, tourmaline, and rutile, whereas the upper part contains epidote and apatite as significant constituents as well (Mellon, 1967).

Peace re-entrant.—The gradual upward coarsening of the Fernie Formation near the Peace re-entrant has been documented by Stott (1967a,b; 1972a,b). The transition between the predominantly marine Fernie shales and the Kootenay-Blairmore molasse takes place within the shallow-water marine deposits of the Nikanassin Formation or Minnes Group. As near the Crowsnest re-entrant, coarse alluvial plain deposits overly the transitional beds with a regional surface of erosion at the base (Stott, 1968, 1972c). The total sequence consists of about 2000 m of sandstone, coal and intercalated shale that grades into a marine shale succession towards the north.

Sandstones near the Peace re-entrant are composed mainly of quartz (30 to 50 percent) and chert (20 to 30 percent). Quartz, quartzite, chert, and minor volcanic pebbles are the principal components of the conglomerates (Stott, 1968). No redbeds are known to occur near the Peace re-entrant. From the maximum clast size, isoliths and isopachs, Stott (1968, p. 110–111) inferred two contributing river systems that coalesced on an alluvial plain sloping gently to the northeast.

Eastern basin margin.—Contemporaneous with the deposition of the Kootenay-Blairmore molasse along the mountain front, general emergence of the foreland basin is also reflected by the northwesterly progradation of fluviatile-deltaic quartz sandstone bodies along the eastern margin of the basin, and by the erosion of Jurassic and older sedimentary sequences from emergent intrabasinal topographic highs. These quartzose sandstones are known as the McMurray Formation (or Athabasca Oil Sands), and appear to have been derived from the crystalline Precambrian shield terrane to the east and southeast of the basin (Carrigy, 1966, 1967).

Paleodrainage and tectonic setting.—A hypothetical paleodrainage pattern can be inferred for the Kootenay-Blairmore Assemblage based to a great extent on paleogeographical reconstructions by Holter and Mellon (1972) and Stott (1972c). The development of redbeds in the vicinity of the Crowsnest re-entrant, the distribution of boulder conglomerates, the northerly directed paleocurrents in the McMurray Formation, and northerly trending Early Cretaceous paleocurrents in Montana (McGookey and others, 1972), suggest that sediment transport was northerly along the axis of the central part of the southern foreland basin. Lateral clastic input on an open alluvial plain was fed by streams issuing from near the Peace and Crowsnest re-entrants (fig. 8).

The source areas included quartz arenites and chert of the Western Rocky Mountains and granitic intrusives of the Omineca Crystalline Belt. Detailed paleontological evidence presented by Jeletzky (1971) suggests that during late Early Cretaceous time great dissimilarities began to develop between marine invertebrates of the foreland basin and the eugeosynclinal successor basins. This faunal diversification indicates the emergence of a continuous land area near the axis of the Omineca Crystalline Belt. In spite of this well documented uplift no folded structures are known to unconformably underlie the Kootenay-Blairmore Assemblage. The erosional disconformity at the base of the Blairmore Group truncates miogeoclinal sedimentary rocks dipping gently to the west. Progressively older strata were exposed towards the east (Norris, 1964). Parts of the northern Rocky Mountains were covered by a sea which received fine-grained marine sediments from the south (fig. 8). In contrast to the relationships in the eastern foreland basin, intense deformation must have preceded the deposition of Upper Jurassic and Lower Cretaceous molasse in the successor basins along the western margin of the Columbian Orogen, where underthrusting of Laberge flysch was complete by Early Cretaceous time (Eisbacher, 1974).

Comparison of figures 6 and 8 shows that the distribution of the major lower molasse fans mimiced the trend of the Laberge-Fernie Basin although the source region was now near the axis of the Omineca Crystalline Belt. The present outline of the Rocky Mountain Belt must have originated later.

Upper Molasse (Belly River-Paskapoo Assemblage)

After deposition of the lower molasse in the southern foreland basin a widespread marine transgression proceeded from north to south during Albian time. This transgression was followed by the deposition of about 1000 m of predominantly fine-grained shallow-water marine shale and sandstone studied by Stott (1963, 1967c) and Jeletzky (1971). This unit is here described as the Alberta Assemblage. It consists of numerous shoaling-upward cycles consisting of black shale, argillaceous carbonate, sideritic concretionary shale and siltstone, and mature pebbly sandstone tongues. These cycles are particularly well developed in the Cardium Formation where they represent shifting and prograding barrier bars and nearshore environments along a subsiding basin margin (Sinha, 1970; Michaelis and Dixon, 1969). Near its base

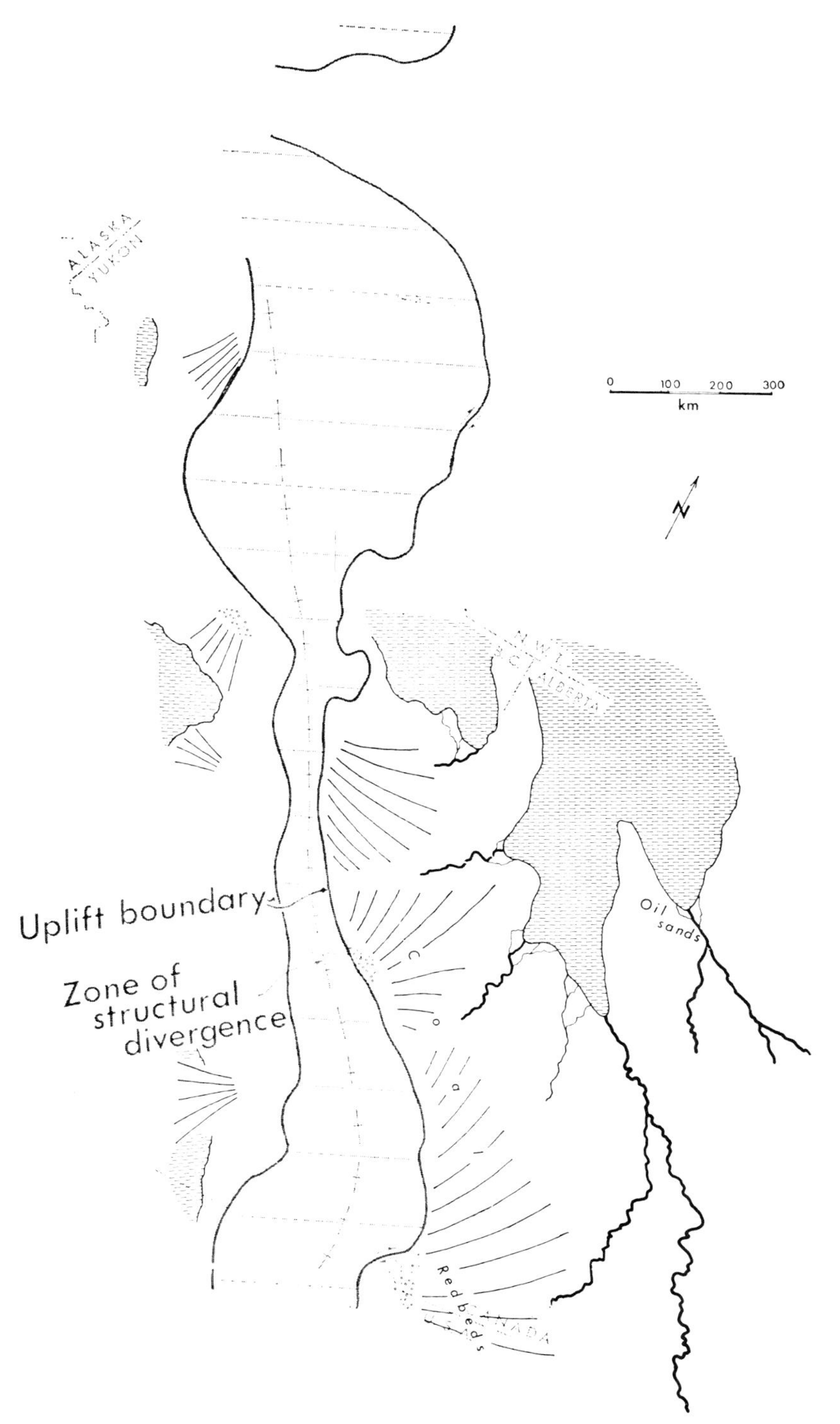

FIG. 8.—Inferred paleodrainage during deposition of the Kootenay-Blairmore molasse.

the Alberta Assemblage includes a river-dominated deltaic sequence, the Cenomanian Dunvegan Sandstone, which was apparently derived from the rising northern Rocky Mountain Belt. It is coarsest and thickest near the Liard re-entrant where Stott (1960) reported about 200 m of conglomeratic sandstone. A paleocurrent reconnaissance by one of us (G. H. E.) east of the Peace re-entrant indicated that in this area the Dunvegan sands were deposited by southeasterly flowing streams. The Dunvegan sandstones are composed of about 60 percent quartz and 20 percent chert; the presence of muscovite suggests that low-grade metamorphic rocks were already exposed in the source area. In general, the chemical stability, textural maturity and inferred environment of deposition of the clastics of the Alberta Assemblage indicates that most of the sands were reworked from underlying older clastic rocks.

The transition between the marine Alberta Assemblage and the easterly prograding Belly River-Paskapoo molasse is gradual and occurred mainly during Campanian time (Jeletzky, 1971; Stott, 1963). At approximately the same time some of the fault-controlled intramontane valleys of the Columbian Orogen received clastic sediments from surrounding highlands. Contemporaneous explosive volcanism in the western Cordillera delivered a significant airborne ash phase to the Belly River-Paskapoo Assemblage. Ashfall tuffs in the foreland basin generally yield ages between 77 and 55 m.y. (Folinsbee and others, 1965; J. F. Lerbekmo, 1973 personal communication).

Crowsnest re-entrant.—The transition between the shales of the Alberta Assemblage and the continental Belly River-Paskapoo Assemblage has been studied in detail by Lerbekmo (1963) and Campbell and Lerbekmo (1963) near the Crowsnest re-entrant. In the east numerous marine tongues interfinger with the basal fluvial-deltaic sandstone members, but towards the Rocky Mountains only a thin marine to non-marine transition zone marks the base of the continental molasse. The Belly River-Paskapoo Assemblage consists of about 3000 m of sandstone, mudstone and minor coal. Conglomerate is rare except in the youngest (Eocene-Oligocene) deposits which are related to the final morphogenic uplift of the Rocky Mountain Belt. The composition of sandstones has been studied in detail by Lerbekmo (1963), Nelson (1968), and Carrigy (1971). The most common framework constituents are quartz (30 to 40 percent), feldspar (10 to 20 percent), detrital carbonate (10 to 20 percent), volcanic rock fragments (5 to 10 percent), and chert (10 to 20 percent). No abrupt compositional changes occurred during deposition of the Belly River-Paskapoo Assemblage. In the higher parts of the sequence chert and detrital carbonate increase at the expense of volcanic rock fragments. This trend is also expressed laterally, with chert and detrital carbonate increasing and volcanic rock fragments decreasing, towards the Rocky Mountain Belt. Along the western margin of the foreland basin Gwinn and Mutch (1965) have reported intertonguing of Upper Cretaceous volcanics and contemporaneous clastic wedges. Erosion of these volcanics could be directly responsible for some of the feldspar and volcanic rock fragments in the Belly River-Paskapoo sandstones. Heavy mineral suites are rich in apatite, biotite, garnet, and epidote indicating a substantial contribution of clastic material from metamorphic terrane (Carrigy, 1971).

Paleocurrents near the Crowsnest re-entrant have been measured by Carrigy (1971 and unpublished data) and are incorporated in figure 9. They indicate a roughly triangular alluvial plain with an elevated apex near the Crowsnest re-entrant. Deposition must have occurred mainly within low-gradient meandering alluvial systems and overbank coal swamps near sea level. Towards the east river-dominated deltas caused slow eastward progradation of the shoreline (Shepheard and Hills, 1970).

During the final depositional phase in Eocene-Oligocene time clastic input was probably also supplied by northeasterly flowing rivers that had their source in western Montana (Axelrod, 1963; Vonhof, 1965). At this stage, considerable changes in gradient led to the formation of coarse alluvial fans in fault-controlled intramontane basins (Price, 1965; Jones, 1969) and to the deposition of coarse gravel as far as 200 km east of the mountain front. These coarse conglomerates of the foreland basin are preserved only as small remnants resting with erosional unconformity on Belly River-Paskapoo clastics of eastern Alberta and western Saskatchewan (Vonhof, 1965; Carrigy, 1971). They indicate the transition from longitudinal drainage to a superimposed transverse river system as a reflection of the final morphogenic uplift of the Columbian Orogen (Bally and others, 1966).

Peace re-entrant.—The Belly River-Paskapoo Assemblage southeast of the Peace re-entrant consists of about 1000 m of fluvial sandstone, mudstone, and coal. The composition of both the framework components and heavy minerals

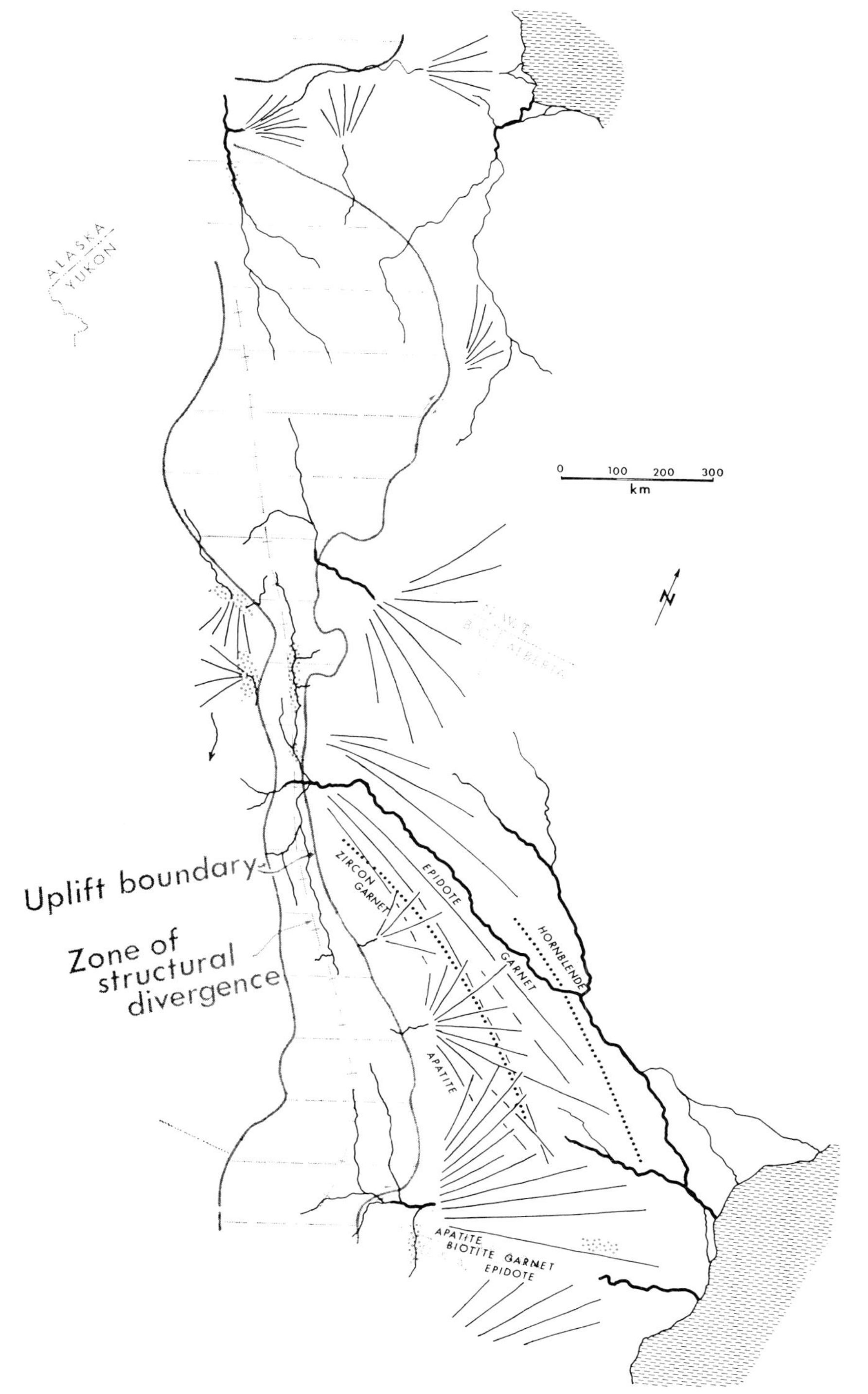

FIG. 9.—Inferred paleodrainage during deposition of the Belly River-Paskapoo molasse. Note heavy mineral facies indicated.

in sandstones displays a distinct zonation parallel to the trend of the Rocky Mountains (Carrigy, 1971; Rahmani, 1973). Among the major framework grains volcanic rock fragments and feldspars are more common in an outer eastern zone but clastic carbonate dominates a western zone close to the mountain front. Similarly, zircon and apatite dominate the westernmost heavy mineral assemblages, epidote and garnet are most common in a central zone, and hornblende is common in an eastern zone. Rahmani (1973) obtained K-Ar ages between 114 and 155 my from seven hornblende concentrates sampled in the eastern belt. The hornblende of the eastern heavy mineral zone was possibly derived from the northern core zone of the Columbian Orogen from where it may have been transported via the Liard re-entrant (fig. 5) to southeasterly flowing rivers of the foreland basin (Rahmani, 1973). The regional longitudinal mineral zonation, the hornblende ages, and limited paleocurrent evidence suggests that the major drainage in the foreland basin southeast of the Peace re-entrant was longitudinal to the southeast. The volcanic rock fragments (5 to 15 percent) were probably derived from Takla-Hazelton volcanics outcropping about 200 km west of the Peace re-entrant.

Southerly drainage documented in the intramontane basin near the northern Rocky Mountain Trench carried micaceous, quartz-chert sands toward the Peace re-entrant. A mica concentrate from these intramontane sandstones yielded an age of 117 my (Eisbacher, in press).

It can therefore be inferred that much of the *southern* foreland basin received sediment from the *northern* part of the Columbian Orogen. A schematic longitudinal section through the sedimentary deposits of the foreland basin between the Liard and Peace re-entrants (fig. 10 after Stott *in* Douglas and others, 1970) suggests that faulting along the Northern Rocky Mountain Trench and uplift of the northern Rockies may have rerouted drainage, terminated foreland deposition, and initiated erosional downcutting in the northern foothills. Numerous intramontane clastic deposits ranging in age from Late Cretaceous to Oligocene are known along the zone of structural divergence of the Columbian Orogen (Eisbacher, in press; Hopkins and others, 1972). These sedimentary inliers indicate that a highly diverse intramontane sediment load was carried by the longitudinal river systems before entering the depositional regime of the foreland. Most of the rivers probably merged near the Peace and Liard re-entrants where mixing of the sediment load took place. The two major drainage systems entering the foreland basin through the Peace and Liard re-entrant may be represented by the central and eastern mineral zones, respectively (fig. 9). Smaller streams flowing perpendicular to the mountain front discharged their locally derived or reworked sediment load directly into the foreland basin. These streams deposited the sediments of the western, carbonate-rich mineral zone and eventually captured more internal rivers during the morphogenic uplift of the Rocky Mountains during late Eocene, Oligocene and Miocene; these streams were the ancestors of the present rivers along the east side of the Rocky Mountains.

Peel Re-entrant.—The molasse near the Peel re-entrant generally overlies Lower to mid-Cretaceous flysch deposits or shallow-water marine barrier-bar and deltaic deposits (Tempelman-Kluit, 1970; Norris, 1972b; Yorath, 1970). The principal phase of tectonic shortening in this segment of the Columbian Orogen occurred soon after deposition of the Lower Cretaceous marine sedimentary rocks, because they themselves have been locally converted to low-grade metamorphic rocks yielding K-Ar mineral ages of between 81 and 93 my (D. J. Tempelman-Kluit, personal communication). In other parts of the deformed belt granitoid plutons yielding ages mainly between 80 and 100 my cut across regional tectonic structures (Douglas and others, 1970).

In the foreland basin the molasse facies con-

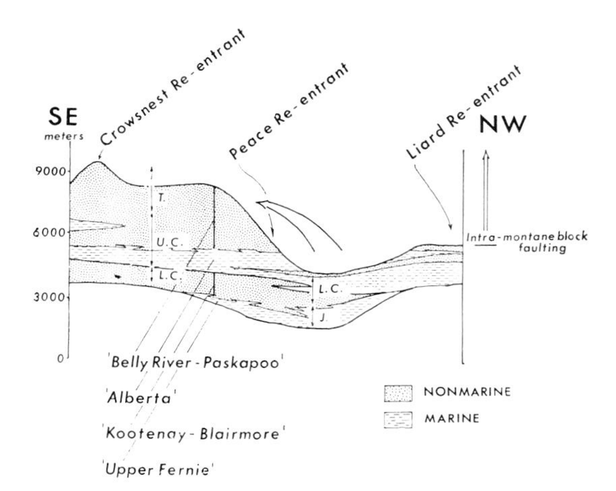

FIG. 10.—Schematic longitudinal cross-section through the molasse of the foreland basin between the Crowsnest and Liard re-entrants indicating the beginning of intramontane faulting and reworking of early molasse by later southeasterly flowing rivers. Ages of strata shown are Jurassic (J.), Lower Cretaceous (L.C.), Upper Cretaceous (U.C.), and Tertiary (T.).

sists of Upper Cretaceous (Campanian, Maestrichtian) to Tertiary clastic wedges equivalent to the Belly River-Paskapoo Assemblage to the south. This continental facies has been described by Mountjoy (1967), Green (1972), Young (1971), and Yorath (1970), who reported sedimentary sections up to 2000 m thick consisting of alluvial deposits containing sandstones made up of chert (20 to 50 percent), quartz (20 to 40 percent) and siliceous rock fragments (30 to 50 percent). The source areas for these clastic constituents are found mainly in the western part of the deformed belt within successions of Proterozoic and Paleozoic sedimentary rocks. Preliminary paleocurrent analyses by Young (1971, and personal communication) have been incorporated in the hypothetical paleodrainage model of the Peel re-entrant during the second molasse cycle (fig. 9). Intramontane continental clastics similar to those in the foreland basin are known from the Tintina Trench (Green, 1972) and from a fault-controlled basin within the Mackenzie Mountains (Blusson, 1973), but no information is available on paleocurrent trends.

Most of the molasse deposits in the Peel re-entrant were probably supplied by rivers draining the longitudinal Tintina Trench. Other rivers may have supplied debris from within the deformed belt, and probably flowed towards an area southeast of the present Mackenzie Delta.

SUCCESSOR BASINS

The sedimentary and tectonic evolution of the successor basins in the Canadian Cordillera has been reviewed recently (Eisbacher, 1974). In this paper, discussion is concentrated on drainage evolution near the Chukachida re-entrant where relationships have been studied in some detail. The molasse facies near this re-entrant is represented by the continental phases of the Upper Jurassic and Lower Cretaceous Bowser Assemblage and by the unconformably overlying and entirely continental Upper Cretaceous and lower Tertiary Sustut Assemblage (fig. 11). The Bowser and Sustut Basins developed on deformed and intruded volcanics, cherts and volcaniclastics of the upper Paleozoic Cache Creek Assemblage and the lower Mesozoic Takla-Hazelton Assemblage.

The first reflection of uplift in the Columbian core zone is seen in thick Upper Jurassic chert-pebble conglomerates of a southwesterly prograding delta complex along the northern margin of the Bowser Basin (Eisbacher, 1973). In latest Jurassic and Early Cretaceous time, alluvial fans and coal swamps formed concomitant with the regression of the sea to the southwest. This early phase of deposition by high-gradient rivers was followed by backcutting of the drainage system, pedimentation, and finally the unroofing of quartzose crystalline terrane in Albian time (Richards and Dodds, 1973; Eisbacher, in press). Paleocurrent indicators in the overlying Upper Cretaceous fluvial deposits of the Sustut Assemblage still reflect southwesterly flowing streams but suggest a lithologically more diverse source area (Eisbacher, in press). A dramatic rearrangement of drainage occurred within the Sustut Basin during the early Tertiary (fig. 12), and paleocurrents changed from a transverse pattern to a longitudinal pattern reflecting the growth of northwesterly trending structures in the Bowser Basin related to contemporaneous uplift of the Coast Crystalline Belt of the Pacific Orogen.

The composition of sandstones and conglomerates of the Chukachida molasse illustrates the progressive downcutting and backcutting of the drainage system. The compositional evolution of sandstones shown in figure 13 indicates an increase of quartz from Upper Jurassic to Upper Cretaceous and subsequent mixing of clastics derived from Bowser and lower Sustut successions during deposition of the upper Sustut. In Eocene time considerable amounts of potassic feldspar were added to the basin through ashfall tuffs related to volcanism that was active 200 km west of the Sustut Basin (Eisbacher, 1971). A similar increase of quartz in progressively younger sandstones has also been documented in other Mesozoic successor basins of the western Cordillera (Sutherland-Brown, 1968; Cole and others, 1973), suggesting widespread emergence of sialic crystalline terrane through a cover of mafic volcanics during mid-Cretaceous time.

Even stronger evidence of unroofing is obtained from the analysis of conglomerates. The example chosen for figure 14 represents a composite section in the northern part of the Sustut Basin. Clasts in Bowser conglomerates are almost entirely chert (or silicified tuffs), derived mainly from areas underlain by the Cache Creek Assemblage (see figure 11). Unconformably overlying basal Sustut conglomerate beds contain clasts of granitoids derived from Mesozoic plutons, andesite derived from the Takla volcanics, and chert derived from the Cache Creek Assemblage or reworked chert-pebble conglomerate of the Bowser Assemblage. A slight change in paleocurrents towards the southwest and the increase in quartzite fragments indicate that outcrops of Cambrian miogeoclinal quartz arenites formed part of a

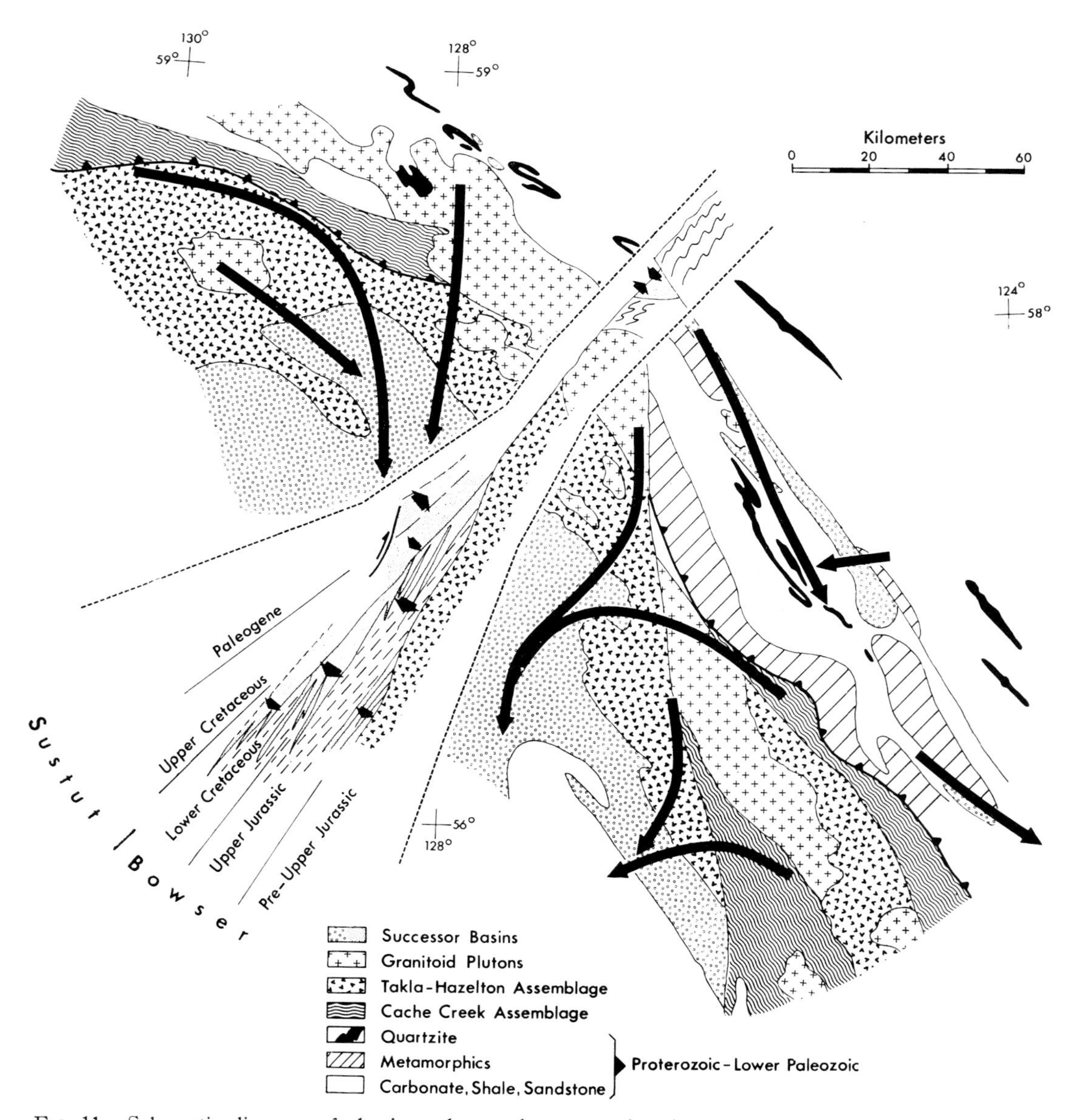

FIG. 11.—Schematic diagram of clastic wedges and sources of molasse near the Chukachida re-entrant.

major new source area about 100 km northeast of the Sustut Basin situated close to the drainage divide between the Sustut Basin and the intramontane Sifton Basin. A complete reversal of paleocurrents during deposition of the upper Sustut is reflected by a great influx of chert clasts derived from Bowser chert-pebble conglomerate. The final establishment of longitudinal drainage to the southeast continued the earlier phase of down-cutting into Mesozoic terrane and also initiated reworking of chert-bearing lower Sustut conglomerates.

TECTONIC INTERPRETATION OF CORDILLERAN MOLASSE FACIES

In comparing figures 6 and 8 it becomes clear that the evolution of the continental molasse facies began with the emergence of the marine Fernie-Laberge Basin above sea level. Folding and crustal shortening *prior* to the deposition of the Upper Jurassic and Lower Cretaceous molasse can be documented only west of the zone of structural divergence (Eisbacher, 1974). Most of the eastern Rocky Mountain Belt must have been deformed *after* deposition

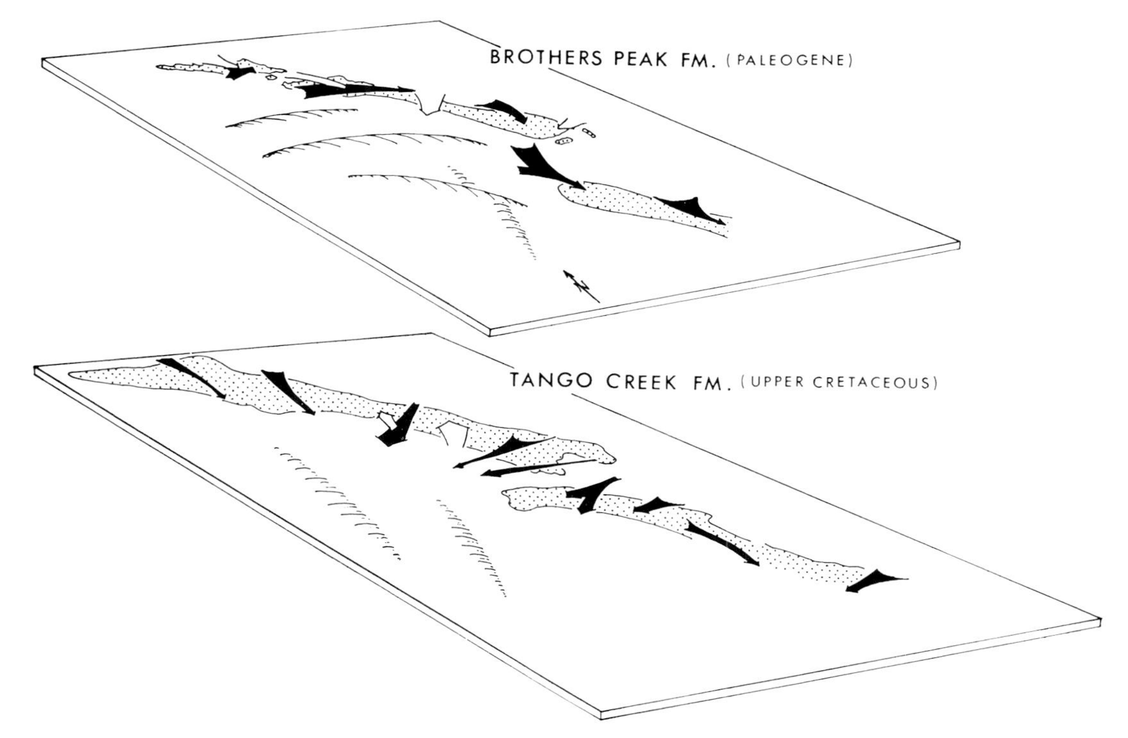

FIG. 12.—Paleocurrent changes and concomitant growth of structures during deposition of the continental Sustut Group near the Chukachida re-entrant.

of the first molasse sequence because no folded structures are known to underlie the Kootenay-Blairmore Assemblage in the Rocky Mountain Belt. During deposition of the second molasse phase in Late Cretaceous and early Tertiary time evidence for extensive shortening west of

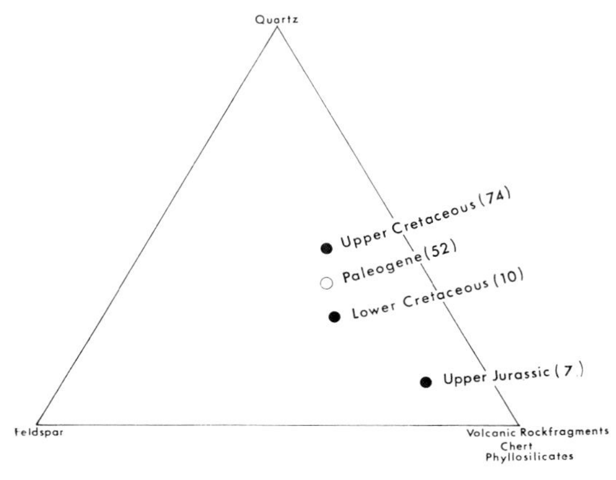

FIG. 13.—Evolution of sandstone composition in molasse of the Chukachida re-entrant. Numbers in brackets refer to the number of modal analyses averaged to arrive at the compositional trend.

the zone of structural divergence is not common. Reesor and Moore (1971) inferred from K-Ar age determinations within the Shuswap Metamorphic Complex of the southern Columbian Orogen that penetrative deformation had ceased by about 60–80 my. In the northern regions widespread Late Cretaceous pedimentation concomitant with progressive unroofing of crystalline complexes can be documented (Eisbacher, in press). The eastward displacement of the clastic facies in the foreland basin, however, indicates that crustal shortening must have affected the whole Rocky Mountain Belt. During deposition of the Belly River-Paskapoo Assemblage intramontane uplift near the zone of structural divergence created fault-bounded longitudinal river basins which are characterized by complex tectonic overprinting (Eisbacher, 1972). As noted above, tectonic shortening in the Rocky Mountain Belt increases from north (about 10 percent) to south (about 30 percent). Price (1973) suggested that loading by advancing thrust sheets must have been a significant factor in the creation of the southern foreland basin. Greater shortening and tectonic loading in the southern Rocky Mountain Belt may well have been the cause of the drain-

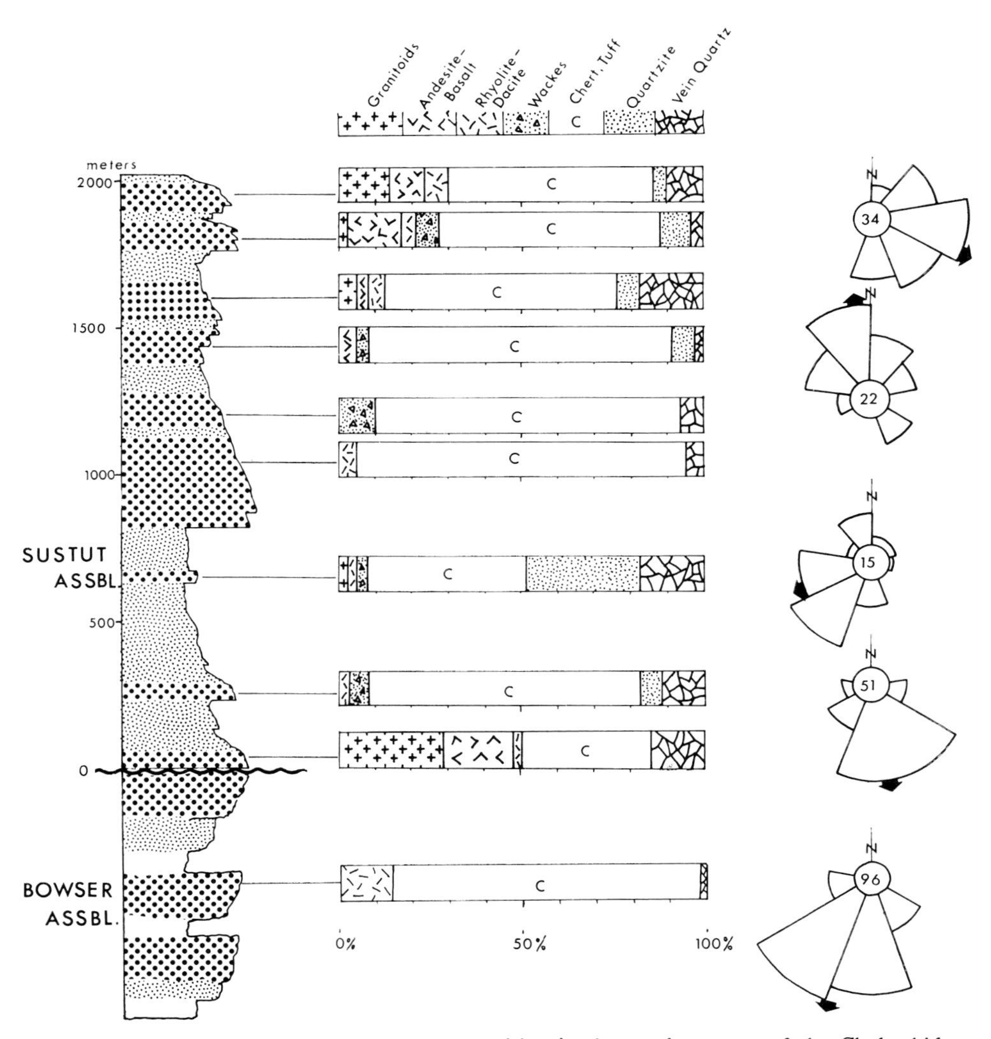

FIG. 14.—Paleocurrents and conglomerate-clast composition in the northern part of the Chukachida molasse.

age reversal towards the southeast during the second molasse cycle which resulted in a thick clastic wedge along the front of the southern Rocky Mountains (see fig. 10).

Figure 15 illustrates our concept of the evolution of the two molasse "megacycles" of the Columbian Orogen in terms of crustal shortening. Crustal shortening, deformation, and uplift west of the zone of structural divergence led to the deposition of the Kootenay-Blairmore Assemblage. Crustal shortening and thrust faulting in the Rocky Mountain Belt east of the zone of structural divergence led to deposition of the Belly River-Paskapoo Assemblage. The interval between these two phases of uplift was characterized by continued subsidence but only limited influx of clastic material indicating extensive peneplanation in the Columbian Orogen. During the second megacycle, underthrusting of sialic crust probably supplied additional sialic material to the deeper portions of the western part of the orogen (Charlesworth, 1959), where it may have, through cratonization, prevented further extensive shortening. The zone of structural divergence roughly separates the older from the younger belt of deformation and is characterized by complex cleavage fans, backfolding, and tectonic overprinting related to both structural domains. It also displays normal faults which are generally younger than all the other structures (Bally and others, 1966).

Great vertical uplift of the whole Columbian Orogen in Eocene and Oligocene time termi-

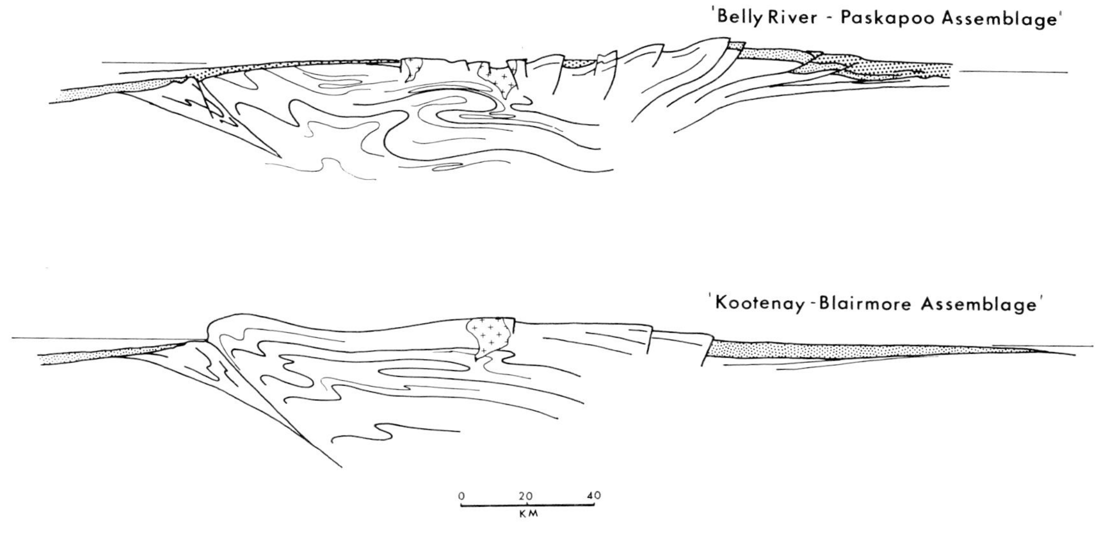

FIG. 15.—Schematic cross-section through the Cordilleran Orogen during deposition of the Kootenay-Blairmore and Belly River-Paskapoo Assemblages.

nated deposition and initiated erosion of all molasse assemblages. This final uplift must have been triggered by isostatic anomalies created by shortening of sialic crust underlying the easternmost part of the Cordilleran miogeocline. Complete regression of the sea from the successor basins and the foreland basins after the deformation of the Columbian Orogen suggests that the isostatic anomaly created by crustal shortening must have outweighed the effect of sedimentary and tectonic loading far beyond the uplift boundary. Tectonic and sedimentary loading only retarded the eventual uplift and erosion of the molasse deposits.

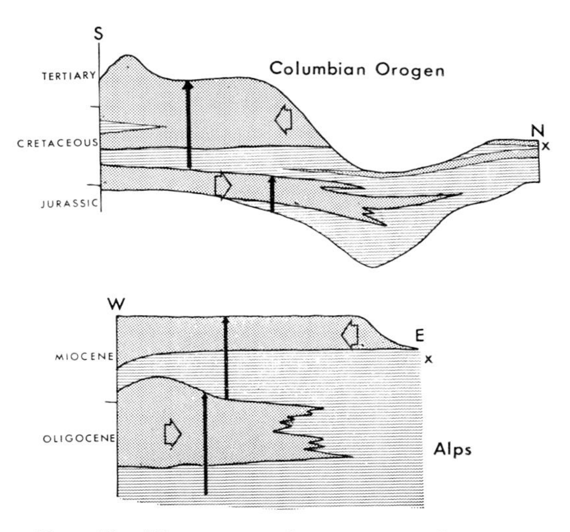

FIG. 16.—The two molasse megacycles of the southern Columbian Orogen and the northern Alps. "X" denotes the onset of intramontane deposition. Arrows indicate sediment transport direction along the basin axis (diagrams after Stott, 1972b; Schmidt-Thome, 1963).

The general geophysical significance of orogenic uplift has been recognized in other mountain belts and was formulated concisely by Winkler-Hermaden for the Eastern Alps (1957, p. 659):

"For the interpretation of the youthful uplift of the Alps I consider a motion that attempted to re-instate step-by-step the disturbed balance in the axial zone, brought about by the tectonic subduction ('Einsaugung') of the crust in the form of a sialic welt . . ." (free translation G.H.E.).

In the Alps, the molasse has long been subdivided into "Lower Marine-Lower Freshwater" and "Upper Marine-Upper Freshwater" megacycles (Schmidt-Thome, 1963; Fuchtbauer, 1967). Similar to the Columbian Orogen, deposition of the upper molasse phase coincided with extensive intramontane faulting and a reversal of the drainage direction along the axis of the foreland basin (fig. 16). It is probably worthwhile to consider the possibility of two major phases of crustal shortening along opposite orogenic flanks for the origin of the two distinct molasse cycles of the Alps, and analogous megacycles related to crustal shortening could exist in other orogenic belts as well. This similarity of isostatic uplift in otherwise quite dissimilar mountain belts may help to clarify the poorly understood causes and mechanisms of intracrustal mass transfer during the final stages of mountain building and also shed light on the timing of hydrocarbon migration in late-orogenic sedimentary basins.

REFERENCES CITED

Axelrod, D. I., 1968, Tertiary flora and topographic history of the Snake River basin, Idaho: Geol. Soc. America Bull., v. 79, p. 713–734.

Bally, A. W., Gordy, P. L., and Stewart, G. A., 1966, Structure, seismic data, and orogenic evolution of southern Canadian Rocky Mountains: Bull. Can. Petroleum Geology, v. 14, p. 337–381.

Beloussov, V. V., 1962, Basic problems in geotectonics (2nd ed.): McGraw-Hill, N.Y., 815 p.

Beutner, E. C., 1972, Curvature of the frontal Sevier Orogenic Belt: Geol. Soc. America Abstracts with Programs, v. 4, p. 451.

Bielenstein, H. U., 1969, The Rundle thrust sheet, Banff, Alberta (Ph.D. dissertation): Queen's Univ., Kingston, 149 p.

Blusson, S. L., 1973, Operation Stewart, Yukon Territory and District of Mackenzie: Geol. Surv. Canada Paper 73-1A, p. 17–18.

Bucher, W. H., 1933, The deformation of the earth's crust: Princeton Univ. Press, Princeton, 518 p.

Campbell, F. A., and Lerbekmo, J. F., 1963, Mineralogic and chemical variations between Upper Cretaceous continental Belly River shales and marine Wapiabi shales in western Alberta: Sedimentology, v. 2, p. 215–226.

Campbell, R. B., 1973, Structural cross-section and tectonic model of southeastern Canadian Cordillera: Can. Jour. Earth Sci., v. 10, p. 1607–1620.

Carey, S. W., 1958, A tectonic approach to continental drift, *in* Carey, S. W. (ed.), Continental drift: Univ. Tasmania Geology Dept. Symposium No. 2, p. 177–355.

Carrigy, M. A., 1966, Lithology of the Athabasca oil sands: Research Council Alberta Geol. Div. Bull. 18, 48 p.

———, 1967, Some sedimentary features of the Athabasca oil sands: Sed. Geology, v. 1, p. 327–352.

———, 1971, Lithostratigraphy of the uppermost Cretaceous (Lance) and Paleocene strata of the Alberta Plains: Research Council Alberta Geol. Div. Bull. 27, 161 p.

Charlesworth, H. A. K., 1959, Some suggestions on the structural development of the Rocky Mountains in Canada: Alberta Soc. Petroleum Geologists Jour., v. 7, p. 249–256.

———, Weiner, J. L., Akehurst, A. J., Bielenstein, H. U., Evans, C. R., Griffith, R. E., Remington, D. B., Stauffer, M. R., and Steiner, J., 1967, Precambrian geology of the Jasper region, Alberta: Research Council Alberta Geol. Div. Bull. 23, 74 p.

Cole, M. R., Pierson, T. C., and Tennyson, M. E., 1973, Petrologic intervals in the Jurassic-Cretaceous sandstones of the North Cascades fold belt, Washington—a west coast trend?: Geol. Soc. America Abstracts with Programs, v. 5, p. 24.

Crosby, G. W., 1969, Radial movements in the western Wyoming salient of the Cordilleran overthrust belt: *ibid.,* Bull., v. 80, p. 1061–1078.

Dahlstrom, C. D. A., 1970, Structural geology in the eastern margin of the Canadian Rocky Mountains: Bull. Can. Petroleum Geology, v. 18, p. 332–406.

Deere, R. E., and Bayliss, P., 1969, Mineralogy of the Lower Jurassic in west-central Alberta: *ibid.,* v. 17, p. 133–153.

Douglas, R. J. W., 1950, Callum Creek, Langford Creek, and Gap map-areas, Alberta: Geol. Survey Canada Mem. 255, 124 p.

———, Gabrielse, H., Wheeler, J. O., Stott, D. F., and Belyea, H. R., 1970, Geology of western Canada.

Eisbacher, G. H., 1971, A subdivision of the Upper Cretaceous-Lower Tertiary Sustut Group, Toodoggone map-area, British Columbia: *ibid.,* Paper 70-68, 16 p.

———, 1972, Tectonic overprinting near Ware, Northern Rocky Mountain Trench: Can. Jour. Earth Sci., v. 9, p. 903–913.

———, 1973, Tectonic Framework of Sustut and Sifton Basins, British Columbia: Geol. Survey Canada Paper 73-1A, p. 24–26.

———, 1974, Evolution of successor basins in the Canadian Cordillera, *in* Dott, R. H., Jr., and Shaver, R. H. (eds.), Modern and ancient geosynclinal sedimentation: Soc. Econ. Paleontologists Mineralogists Special Pub. 19, p. 274–291.

———, in press, Sedimentary history and tectonic evolution of the Sustut and Sifton Basins, north-central British Columbia: Geol. Survey Canada Paper.

———, and Tempelman-Kluit, D. J., 1972, Map of major faults in the Canadian Cordillera and southeast Alaska—Map and Abstract: Geol. Assoc. Canada Cordilleran Sec. Abs., p. 13–14, p. 22–25.

Folinsbee, R. E., Baadsgaard, H., Cumming, G. L., Nascimbene, J., and Shafiqullah, M., 1965, Late Cretaceous radiometric dates from the Cypress Hills of western Canada: Alberta Soc. Petroleum Geology 15th Ann. Field Conf. Guidebook, Pt. 1, p. 162–174.

Frebold, Hans, 1957, The Jurassic Fernie Group in the Canadian Rocky Mountains and Foothills: Geol. Survey Canada Mem. 287, 197 p.

Friend, P. F., and Moody-Stuart, M., 1972, Sedimentation of the Wood Bay Formation (Devonian) of Spitzbergen: Regional analysis of a late orogenic basin: Norsk Polarinstitutt, Skrifter 157.

Fuchtbauer, H., 1967, Die Sandsteine in der Molasse nordlich der Alpen: Geol. Rundschau, v. 56, p. 266–300.

Gabrielse, H., 1972, Sedimentary facies and Northern Rocky Mountain Trench: Geol. Assoc. Canada Cordilleran Section Abs., p. 15.

———, AND BLUSSON, S. L., 1969, Geology of Coal River map-area, Yukon Territory and District of Mackenzie: Geol. Survey Canada Paper 69-39, 22 p.

GANSSER, A., 1964, Geology of the Himalayas: Interscience, London, 289 p.

GIBSON, D. W., 1971, Triassic stratigraphy of the Sikanni Chief River-Pine Pass region, Rocky Mountains Foothills, northeastern British Columbia: Geol. Survey Canada Paper 70-31, 105 p.

GREEN, L. H., 1972, Geology of Nash Creek, Larsen Creek, and Dawson map-area, Yukon Territory: *ibid.*, Mem. 364, 157 p.

GWINN, V. E., AND MUTCH, T. A., 1965, Intertongued Upper Cretaceous volcanic and non-volcanic rocks, central-western Montana: Geol. Soc. America Bull., v. 76, p. 1125–1144.

HOLTER, M. E., AND MELLON, G. B., 1972, Geology of the Luscar (Blairmore) coal beds, central Alberta Foothills: Proc. First Geol. Conf. on Western Can. Coals, Research Council Alberta Inf. Ser. No. 60, p. 125–135.

HOPKINS, W. S., RUTTER, N. W., AND ROUSE, G. E., 1972, Geology, paleoecology, and palynology of some Oligocene rocks in the Rocky Mountain Trench of British Columbia: Can. Jour. Earth Sci., v. 9, p. 460–470.

HSU, K. J., 1970, The meaning of the word Flysch—a short historical search, *in* Lajoie, Jean (ed.), Flysch sedimentology in North America: Geol. Assoc. Canada Special Paper 7, p. 1–11.

JANSA, L., 1972, Depositional history of the coal-bearing Upper Jurassic-Lower Cretaceous Kootenay Formation, southern Rocky Mountains, Canada: Geol. Soc. America Bull., v. 83, p. 3199–3222.

JELETZKY, J. A., 1971, Marine Cretaceous biotic provinces and paleogeography of western and Arctic Canada illustrated by a detailed study of ammonites: Geol. Survey Canada Paper 70-22, 92 p.

JONES, P. B., 1969, The Tertiary Kishenehn Formation, British Columbia: Bull. Can. Petroleum Geology, v. 17, p. 234–246.

JULIVERT, M., 1970, Cover and basement tectonics in the Cordillera Oriental of Colombia, South America, and a comparison with some other folded chains: Geol. Soc. America Bull., v. 81, p. 3623–3646.

LEECH, G. B., 1966, The Rocky Mountain Trench: Geol. Survey Canada Paper 66-14, p. 307–329.

LERBEKMO, J. F., 1963, Petrology of the Belly River Formation, southern Alberta Foothills: Sedimentology, v. 2, p. 54–86.

MAZAROVICH, O. A., 1972, Geotectonic conditions for the formation of molasse: Geotectonics No. 1, p. 14–21.

MCCONNELL, R. G., 1896, Report on an exploration of the Finlay and Omineca Rivers: Geol. Survey Canada Ann. Rept., v. 7, p. 6C–40C.

MCGOOKEY, D. P., HAUN, J. D., HALE, L. A., GOODELL, H. G., MCCUBBIN, D. G., WULF, G. R., AND WEIMER, R. J., 1972, Cretaceous System, *in* Geologic Atlas of the Rocky Mountain Region, U.S.A.: Rocky Mountain Geol. Assoc., Denver, p. 190–228.

MELLON, G. B., 1967, Stratigraphy and Petrology of the Lower Cretaceous Blairmore and Mannville Groups, Alberta Foothills and Plains: Research Council Alberta Geol. Div. Bull. 21, 270 p.

MICHAELIS, E. R., AND DIXON, G., 1969, Interpretation of depositional processes from sedimentary structures in the Cardium sand: Bull. Can. Petroleum Geology, v. 17, p. 410–443.

MONGER, J. W. H., SOUTHER, J. G., AND GABRIELSE, H., 1972, Evolution of the Canadian Cordillera: a plate-tectonic model: Am. Jour. Sci., v. 277, p. 577–602.

MOUNTJOY, E. W., 1967, Upper Cretaceous and Tertiary stratigraphy, northern Yukon Territory and northwestern Distict of Mackenzie: Geol. Survey Canada Paper 66–16, 70 p.

MUDGE, M. R., 1970, Origin of the disturbed belt in northwestern Montana: Geol. Soc. America Bull., v. 81, p. 377–392.

———, AND SHEPPARD, R. A., 1968, Provenance of igneous rocks in Cretaceous conglomerates in northwestern Montana: U.S. Geol. Survey, Prof. Paper 600-D, p. D137–D146.

NELSON, H. W., 1968, Petrography of sandstone interbeds in the Upper Cretaceous-Paleocene of Gulf Spring Point: Bull. Can. Petroleum Geology, v. 16, p. 425–430.

NORRIS, D. K., 1964, The Lower Cretaceous of the southeastern Canadian Cordillera: *ibid.*, v. 12, p. 512–535.

———, 1972a, En echelon folding in the Northern Cordillera of Canada: *ibid.*, v. 20, p. 634–642.

———, 1972b, Structural and stratigraphic studies in the tectonic complex of northern Yukon Territory, north of Porcupine River: Geol. Survey Canada Paper 72-1B, p. 91–99.

———, STEVENS, R. D., AND WANLESS, R. K., 1965, K-Ar age of igneous pebbles in the McDougall-Segur Conglomerate, southeastern Canadian Cordillera: *ibid.*, Paper 65-2, 11 p.

PELLETIER, B. R., 1965, Paleocurrents in the Triassic of northeastern British Columbia, *in* Middleton, G. V. (ed.), Primary sedimentary structures and their hydrodynamic interpretation: Soc. Econ. Paleontologists and Mineralogists Special Pub. 12, p. 233–245.

PRICE, R. A., 1965, Flathead map-area, British Columbia and Alberta: Geol. Survey Canada Mem. 336, 221 p.

———, 1967, The tectonic significance of mesoscopic subfabrics in the Southern Rocky Mountains of Alberta and British Columbia: Can. Jour. Earth Sci., v. 4, p. 39–70.

———, 1973, Large-scale gravitational flow in supracrustal rocks, southern Canadian Cordillera, *in* Jong, K. A. de, and Scholton, Robert (eds.), Gravity and tectonics: Elsevier, Amsterdam, p. 491–502.

———, AND MOUNTJOY, E. W., 1970, Geologic Structure of the Canadian Rocky Mountains between Bow and Athabaska Rivers—a progress report: Geol. Assoc. Canada Special Paper 6, p. 7–25.

———, 1969–1972, Geol. Survey Canada Maps 1294A, 1297A, 1271A–1274A.

RAHMANI, R. A., 1973, Heavy mineral analysis of Upper Cretaceous and Paleocene sandstones in Alberta and adjacent areas of Saskatchewan (Ph.D. Thesis): Univ. Alberta, Edmonton.

RAPSON, J. E., 1965, Petrography and derivation of Jurassic-Cretaceous clastic rocks, southern Rocky Mountains, Canada: Am. Assoc. Petroleum Geologists Bull., v. 49, p. 1426–1452.

REESOR, J. E., AND MOORE, J. M., JR., 1971, Petrology and structure of Thor-Odin Gneiss Dome, Shuswap Metamorphic Complex, British Columbia: Geol. Survey Canada Bull. 195, 149 p.

Richards, T. A., and Dodds, C. J., 1973, Hazelton (East) map-area: *ibid.*, Paper 73-1, Pt. A, p. 38–42.
Roddick, J. A., 1967, Tintina Trench: Jour. Geology, v. 75, p. 23–33.
Rodgers, J., 1964, Basement and no-basement hypotheses in the Jura and the Appalachian Valley and Ridge: Virginia Polytechnic Inst. Dept. Geol. Sci. Mem. 1, p. 71–80.
Roeder, D. H., 1967, Rocky Mountains: Beitrage zur Regionalen Geologie der Erde (v. 5): Gebrueder Borntraeger, Berlin, 318 p.
Schiemenz, S., 1960, Fazies und Palaeogeographie der Subalpinen Molasse zwischen Bodensee und Isar: Beih. Geol. Jahrb., v. 38, 119 p.
Schmidt-Thome, P., 1963, Le bassin de la Molasse d'Allemagne du Sud, *in* Livre a la memoire du Professeur Paul Fallot: Soc. Geol. France, v. 2, p. 431–452.
Schultheis, N., and Mountjoy, E. W., 1971, Cadomin Conglomerate of Alberta derived from Main Range thrust sheets uplifted during Early Cretaceous time: Geol. Soc. America Abstracts with Programs, v. 3, p. 411–412.
Shepheard, W. W., and Hills, L. V., 1970, Depositional environments, Bearpaw-Horseshoe Canyon (Upper Cretaceous) transition zone, Drumheller "Badlands," Alberta: Bull. Can. Petroleum Geology, v. 18, p. 166–215.
Sinha, R. N., 1970, Cardium Formation, Edson area, Alberta: Geol. Survey Canada Paper 68-30, 65 p.
Souther, J. G., 1971, Geology and mineral deposits of Tulsequah map-area, British Columbia: *ibid.*, Mem. 362, 84 p.
Stott, D. F., 1960, Cretaceous rocks in the region of Liard and Mackenzie Rivers, Northwest Territories: *ibid.*, Bull. 63, 36 p.
———, 1963, The Cretaceous Alberta Group and equivalent rocks, Rocky Mountain Foothills, Alberta: *ibid.*, Mem. 317, 306 p.
———, 1967a, Fernie and Minnes strata north of Peace River, foothills of northeastern British Columbia: *ibid.* ,Paper 67-19A, 58 p.
———, 1967b, Jurassic and Cretaceous stratigraphy between Peace and Tetsa Rivers, northeastern British Columbia: *ibid.*, Paper 66-7, 73 p.
———, 1967c, The Cretaceous Smoky Group, Rocky Mountain Foothills, Alberta and British Columbia: *ibid.*, Bull. 132, 133 p.
———, 1968, Lower Cretaceous Bullhead and Fort St. John Groups between Smoky and Peace Rivers, Rocky Mountain Foothills, Alberta and British Columbia: *ibid.*, Bull. 152, 279 p.
———, 1972a, Jurassic and Cretaceous stratigraphy of Rocky Mountain Foothills, northeastern British Columbia and Alberta: *ibid.*, Paper 72-1A, p. 227–228.
———, 1972b, Cretaceous stratigraphy, northeastern British Columbia: Proceed. First Conf. on Western Can. Coal, Research Council Alberta Inf. Ser. No. 60, p. 137–150.
———, 1972c, The Cretaceous Gething delta, northeastern British Columbia: *ibid.*, p. 151–163.
Sutherland-Brown, A., 1968, Geology of the Queen Charlotte Islands, British Columbia: B.C. Dept. Mines and Petroleum Resources Bull. 54, 226 p.
Taylor, G. C., 1971, Structural style of the northern Canadian Rockies: Geol. Soc. America Abstracts with Programs, v. 3, p. 416–417.
Tempelman-Kluit, D. J., 1970, Stratigraphy and structure of the Keno Hill Quartzite in Tombstone River-Upper Klondike River map-areas, Yukon Territory: Geol. Survey Canada Bull. 180, 102 p.
———, 1972, Evidence for timing and magnitude of movement along the Tintina Trench: Geol. Assoc. Canada, Cordilleran Sec. Abs., p. 39.
van Houten, F. B., 1973, Meaning of molasse: Geol. Soc. America Bull., v. 84, p. 1973–1976.
———, 1974, Northern Alpine Molasse and similar Cenozoic sequences, southern Europe, *in* Dott, R. H., Jr., and Shaver, R. H. (eds.), Modern and ancient geosynclinal sedimentation: Soc. Econ. Paleontologists and Mineralogists Special Pub. 19, p. 260–273.
Von Hof, J. A., 1965, The Cypress Hills Formation and its reworked deposits in southwestern Saskatchewan: Alberta Soc. Petroleum Geology 15th Ann. Field Conf., Pt. 1, p. 142–161.
Wegmann, E., 1961, Anatomie comparee des hypotheses sur les plissement de couverture (le Jura plisse): Uppsala Univ. Geol. Inst. Bull. 40, p. 169–182.
Wheeler, J. O., 1961, Whitehorse map-area, Yukon Territory: Geol. Survey Canada Mem. 312, 156 p.
———, and Gabrielse, H., 1972, Cordilleran Structural Province, *in* Price, R.A., and Douglas, R. J. (eds.), Variations in tectonic styles in Canada: Geol. Assoc. Canada Special Paper 11, p. 1–81.
Winkler-Hermaden, A., 1957, Geologisches Kraeftespiel und Landformung: Springer Verlag, Wien, 822 p.
Yates, R. G., 1973, Significance of the Kootenay Arc in Cordilleran tectonics: Geol. Soc. America Abstracts with Programs, v. 5, p. 125.
Yorath, C. J., 1970, Cretaceous and Tertiary stratigraphy, District of Mackenzie: Geol. Survey Canada Paper 70-1A, p. 243–245.
Young, F. G., 1971, Mesozoic stratigraphic studies, northern Yukon Territory and northwestern District of Mackenzie: *ibid.*, Paper 71-1A, p. 245–247.
Zwart, H. J., 1969, Metamorphic facies series in the European orogenic belts and their bearing on the causes of orogeny: Geol. Assoc. Canada Special Paper 5, p. 7–16.

EVOLUTION OF SYNOROGENIC CLASTIC DEPOSITS IN THE INTERMONTANE UINTA BASIN OF UTAH[1]

DAVID W. ANDERSEN[2] AND M. DANE PICARD
University of Utah, Salt Lake City, Utah

ABSTRACT

In the Uinta Basin of northeastern Utah, the Uinta and Duchesne River Formations are composed of diverse fluvial sedimentary rocks. The rock units overlie extensive lacustrine deposits of Lake Uinta and provide a complete record of late Laramide (latest Eocene) tectonic events in this part of the Rocky Mountains. The fluvial deposits consist of heterogeneous, laterally discontinuous sandstone lenses with varying amounts of conglomerate and poorly stratified fine-grained rocks. Sedimentary structures and facies relationships indicate that the Duchesne River Formation was deposited mainly by relatively small, rapidly aggrading, southward-flowing streams. Most stream channels probably were braided, with high gradients and high velocities of flow.

Uplift of the Uinta Mountains changed geographic conditions and drainage patterns in Uinta Basin and strongly influenced the characteristics of contemporaneous sedimentary deposits. Important features of the stratigraphic sequence are: 1) the oldest major body of sediment (early Duchesnean) produced during uplift of the Uinta Mountains is considerably younger than the youngest preserved deposits of Lake Uinta (middle Uintan); 2) lower (early Duchesnean) and upper (late Duchesnean) conglomeratic rock units record two major episodes of uplift, each composed of several smaller events; and 3) volcanic ash deposits (now altered) accumulated during the time (middle Duchesnean) between major uplifts. Composition of the clastic rocks reflects the abundance of sedimentary source rocks in the Uinta Mountains. Recycled sedimentary material includes fragments of chert, carbonate and clastic sedimentary rocks. Much of the red pigment in the deposit is also derived from sedimentary source rocks. Altered volcaniclastic rocks probably originated in eruptive centers on the west, in the area of the present Wasatch Mountains or Basin and Range province. Differential movement of the Uinta Mountains and Uinta Basin during Duchesnean time represents a very late stage of Laramide tectonism. Volcanism in the Basin and Range province on the west may have begun during the Duchesnean. The latest Eocene and earliest Oligocene(?), $\sim$ 36–40 my BP, was thus a time of active tectonism in northeastern Utah.

INTRODUCTION

A significant aspect of Cenozoic tectonic activity in western North America has been relative movement of small, isolated mountain ranges and adjacent sedimentary basins. Although broad patterns of activity apparently exist, individual tectonic elements may have acted essentially independently. Each mountain range and adjacent basin must therefore be studied, and regional generalizations are possible only after many different records of tectonic style and chronology are established.

New concepts of plate tectonics require a reexamination of records of tectonic activity. Many tectonic events can be satisfactorily modeled, but plate tectonic theories are least successful in explaining major tectonic elements not clearly associated with plate margins. The central Rocky Mountain region of the United States is such an area. Studies of structural relationships in central Rocky Mountain ranges provide considerable insight into their histories, but diverse structural interpretations have not produced a generally accepted model of their origin. The Tertiary sedimentary record in Rocky Mountain intermontane basins is useful in refining tectonic chronologies, and one approach to the study of a Rocky Mountain range is to relate its history to other tectonic histories within the Rocky Mountains and at active plate margins.

In northeastern Utah, the Duchesne River Formation contains an excellent synorogenic sedimentary record of the latest Eocene and probably earliest Oligocene tectonic development of part of the Uinta Mountains and Uinta Basin. Stratigraphy, paleocurrent directions, depositional environments, petrography and history of the formation are related to tectonics of the Uinta Mountains area. This record of tectonism can be related to contemporaneous events in other places.

[1] Major financial support was received from the National Science Foundation (Grant GA-12570 to Picard) and the Uniform School Fund (Grant to Picard). Acknowledgment is also made to the donors of the Geological Research Fund, administered by the Department of Geological and Geophysical Sciences, University of Utah, for partial support of this research (Grant to Andersen). R. D. Cole, T. A. Maxwell, W. P. Nash and W. L. Stokes read the manuscript and contributed many suggestions for its improvement.

[2] Present address: California State University, San Jose, California.

Tectonic setting.—The Uinta Basin is a structural and topographic basin in northeastern Utah and northwestern Colorado (fig. 1). It lies at the northern limit of the Colorado Plateau province, near the boundaries of the Basin and Range province and the Rocky Mountains province. The Tertiary history of the Uinta Basin is closely related to that of the adjacent Uinta Mountains.

Apparently, tectonic activity in the western Rocky Mountain region began rather abruptly about 80 my ago (Eardley, 1962; Gilluly, 1963; Coney, 1972; Monger and others, 1972), and important synorogenic sedimentary deposits accumulated during the Late Cretaceous and early Tertiary (Spieker, 1949; van Houten, 1969). Especially in the middle Rocky Mountains, including the Uinta Mountains, tectonism involved considerable vertical uplift of roughly elliptical ranges and downwarping of adjacent basins (Eardley, 1962, 1963; Blackstone, 1963). During the late Eocene (about 40 my BP) major tectonism in the central and southern Rocky Mountains declined (Eardley, 1962; van Houten, 1969; Coney, 1972; Tweto, 1973). Shortly thereafter, widespread volcanic activity began in the area of the present-day Basin and Range province (Armstrong, 1970; McKee, 1971; Coney, 1972; Lipman and others, 1972). In Canada, deformation of the Rocky Mountains continued through the Eocene (Russell, 1954; Monger and others, 1972), but was largely completed by the early Oligocene (Roeder, 1967; Hopkins and others, 1972). Deposition of the Duchesne River Formation occurred during the time of transition from deformation in the Rocky Mountains to volcanism in the Basin and Range province (fig. 2).

The degree of continuity of tectonic activity within structural provinces and throughout the world has long been a matter of interest. The idea of short, worldwide orogenic periods separated by long intervals of quiescence was popular for many years (Chamberlin, 1909; Blackwelder, 1914; Stille, 1936). However, Gilluly (1949, 1963) challenged this concept and presented evidence that, at one place or another, tectonic activity in the western United States has been more or less continuous throughout the Cenozoic. Largely on the basis of new radiometric dates, Damon and Mauger (1966), Damon (1970) and Kistler and others (1971) revived the idea of periodic magmatic events separated by quiescence. Indeed, Damon

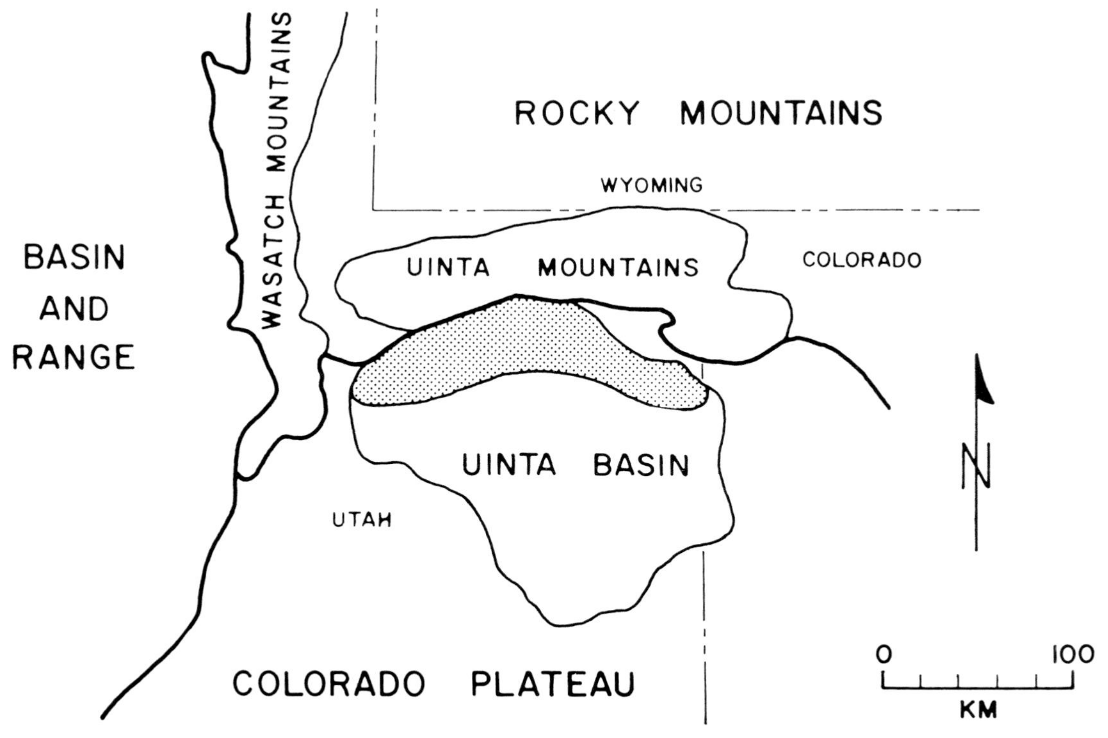

FIG. 1.—Index map of northern Uinta Basin, showing outcrop area (shaded) of synorogenic Duchesne River Formation.

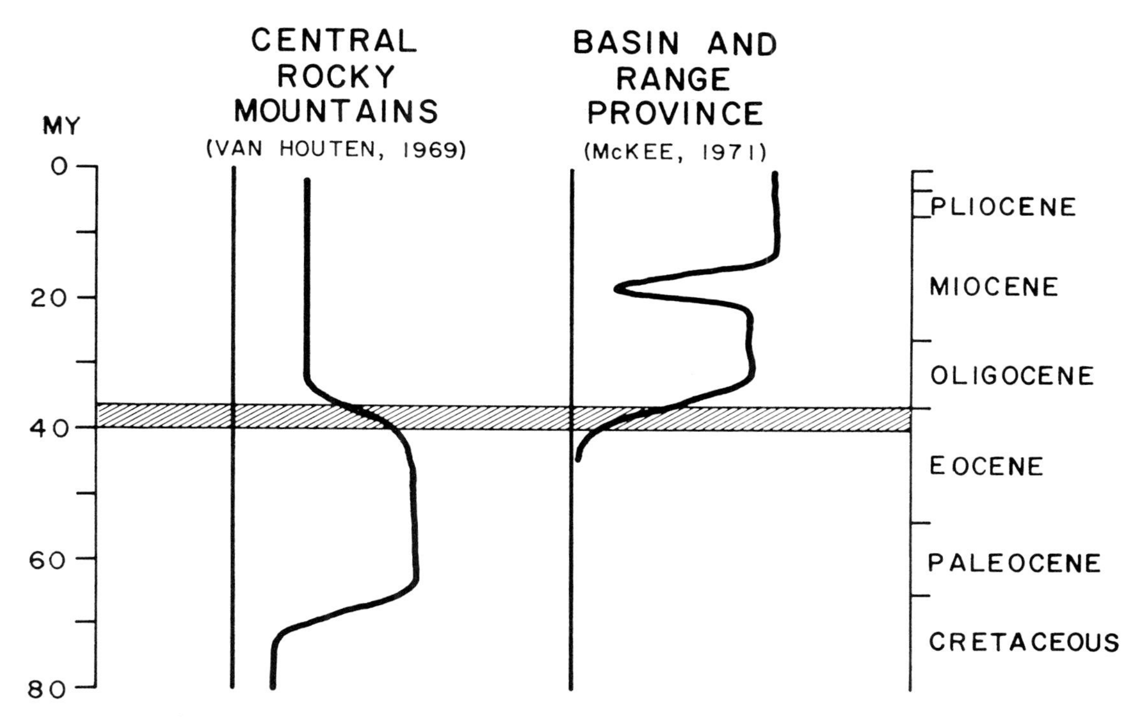

FIG. 2.—Relationship of Duchesnean Age to periods of molasse deposition in Rocky Mountains and volcanism in Basin and Range province. Curves indicate volume of molasse deposits (after van Houten, 1969) and frequency of radiometric dates of igneous rock units (after McKee, 1971).

and Mauger (1966) suggested that the entire late Eocene was a time of quiescence in western North America between sharply defined Laramide and Basin and Range orogenies. Such hypotheses are difficult to reconcile with continuous motion of an immense lithospheric plate (Gilluly, 1973) and have generally not been supported by additional radiometric dates (Crowder and others, 1973; Lanphere and Reed, 1973). However, the debate continues. The Duchesne River Formation constitutes a sedimentary record of the time of transition from tectonic activity in the Rocky Mountains to tectonic activity in the Basin and Range (fig. 2), and the formation provides an opportunity to evaluate the abruptness of the transition and the duration of possible quiescence at that time.

Paleogeographic setting.—The Uinta Basin was the site of accumulation of continental deposits throughout the Eocene. Early and middle Eocene fluvial and lacustrine deposits (Wasatch and Green River Formations) are economically important and have received considerable attention. Bradley (1929, 1931, 1948), Picard (1955), Cashion (1967) and Picard and High (1972) have described the paleogeographic conditions during the existence of Lake Uinta. The Uinta Mountains area was the site of a low barrier between Lake Uinta and Lake Gosiute to the north, but there is no evidence of high relief between the Uinta Mountains and the Uinta Basin. During the early late Eocene (Uintan), Lake Uinta slowly diminished in size. It first retreated from the eastern Uinta Basin and receded progressively westward and southward with time (Dane, 1954; Picard, 1955).

The fluvial deposits of the overlying Uinta Formation have received less attention than contemporaneous lacustrine deposits. Vertebrate paleontologists have described the general lithologic and stratigraphic relationships (Osborne, 1895; Kay, 1934). The paleogeographic significance of the eastern Uinta Formation was deduced by Stagner (1941), who demonstrated an eastern granitic source of the sediment on the basis of facies relationships, paleocurrent directions, grain-size distributions and petrography. The drab-colored claystone and minor sandstone of the eastern Uinta Formation (fig. 3) represent channel and floodplain deposits of a low-gradient stream system flowing westward across a nearly flat alluvial plain into a receding remnant of Lake Uinta (Stagner, 1941). The saline facies and the sandstone and limestone facies of the Uinta Formation represent

FIG. 3.—Fine-grained floodplain deposits of Uinta Formation in central Uinta Basin. Sediment was derived from low-lying granitic rocks on the east.

FIG. 5.—Fluvial conglomerate at base of Duchesne River Formation, western Uinta Basin. Stratification is distinct and framework clasts are closely packed.

the youngest deposits of the lake remnant in the west (Dane, 1954; Picard, 1955).

Overlying the youngest preserved lacustrine deposits in the western Uinta Basin, Bissell (1952) and Dane (1954) mapped red fluvial deposits which they assigned to the Uinta Formation. These, too, are predominantly floodplain and minor channel deposits of relatively low-gradient streams, but the red pigment of claystone and abundant sedimentary rock fragments in the sandstone indicate that the source probably was in low-lying flanks of the Uinta and Wasatch Mountains to the north and northwest.

After deposition of the youngest preserved lacustrine deposits, uplift of the Uinta Mountains affected drainage patterns and sediment types in the Uinta Basin. In the east, reddish brown sand and gravel, now the Duchesne River Formation, was derived from the north and deposited on greenish gray clay of the Uinta Formation (fig. 4). In the west, coarse gravel was deposited on older fine-grained red sediment of the Uinta Formation (fig. 5). In the Uinta Mountains, the Duchesne River Formation unconformably overlies pre-Tertiary rocks (fig. 6). Although exact synchroneity of these boundaries is unlikely, latest Eocene uplift of the Uinta Mountains was an event that strongly influenced paleogeography of the Uinta Basin and led to accumulation of the Duchesne River Formation.

STRATIGRAPHY

Age.—The age of the Duchesne River Formation was debated for many years. Vertebrate paleontologists working directly with the formation originally regarded it as Oligocene (Peterson, 1932; Burke, 1934). Other workers (Simpson, 1933; Wood and others, 1941), who compared the Duchesnean fauna with other assemblages, regarded it as Eocene. The Duchesnean Stage, equivalent in age to the Duchesne River Formation, was established as the young-

FIG. 4.—Reddish brown sandstone of Duchesne River Formation, derived from Uinta Mountains on the north, overlying varicolored claystone of Uinta Formation, eastern Uinta Basin.

FIG. 6.—Duchesne River Formation overlying inclined Triassic rocks, Uinta Mountains.

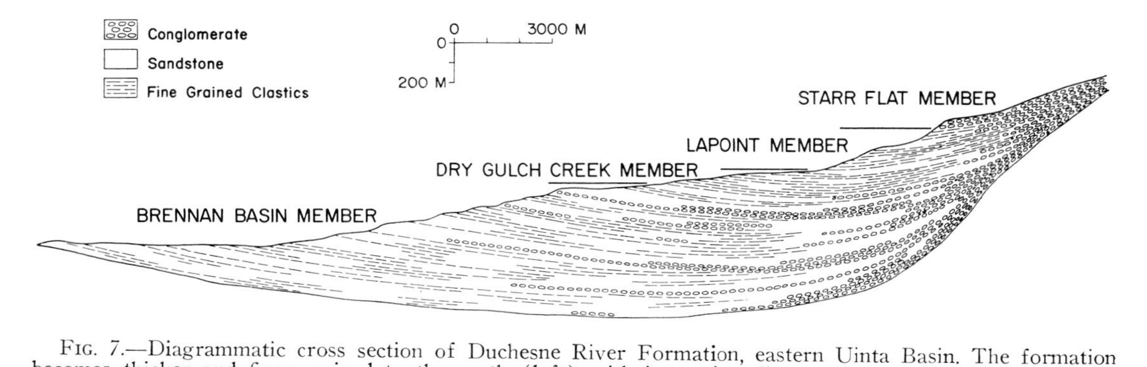

FIG. 7.—Diagrammatic cross section of Duchesne River Formation, eastern Uinta Basin. The formation becomes thicker and finer-grained to the south (left) with increasing distance from the Uinta Mountains. Members are illustrated in following photographs (figs. 8–11).

est subdivision of the Eocene (Wood and others, 1941). Since the convincing review by Simpson (1946), most workers have considered the Duchesnean to be latest Eocene (see, for example, Stokes and Madsen, 1961; Evernden and others, 1964; Black and Dawson, 1966). Thus, the fossiliferous portions of the Duchesne River Formation are, by current definition, of latest Eocene age. The uppermost part of the formation (Starr Flat Member) is undated and may be Oligocene (Andersen and Picard, 1972).

The absolute age of the Duchesnean has received less attention. Eardley (1952) and Gazin (1959) suggested a Duchesnean age for the Norwood Tuff of Utah. Evernden and others (1964) obtained three K-Ar dates of about 37.5 my from the Norwood Tuff and established limits of 36 and 40 my on the Duchesnean Age. More recent paleontological information (Nelson, 1971) indicates that most of the Norwood Tuff is Chadronian (earliest Oligocene). Also, the Keetley Volcanics, assumed to be closely related to the Norwood Tuff (Eardley, 1952, 1969; Nelson, 1971), are dominantly Oligocene in age (Bromfield and Crittenden, 1971). Thus, the dates obtained by Evernden and others (1964) may not represent the Duchesnean. Further calibration of subdivisions of the Eocene and additional dating of the Eocene-Oligocene boundary would be of considerable interest. A single K-Ar date of 40 my was obtained by John Clark (M. R. Dawson, 1972, personal commun.) from an ashy siltstone at the base of the Lapoint Member of the Duchesne River Formation. More dates from the formation itself are needed.

Lithofacies.—Lithofacies distributions in the synorogenic Duchesne River Formation were established by Andersen and Picard (1972). Relationships are best known in the area of extensive exposure near Vernal. Figure 7 illustrates the general lithofacies pattern established from six correlated measured sections in the Vernal area. Grain size decreases dramatically southward with increasing distance from the mountain front. Simultaneously, the thickness of the formation increases markedly to the south. Measured at the surface, the thickness of the formation in the Vernal area is 878 m (Andersen and Picard, 1972), but because deformation continued during deposition the formation probably is not this thick in any single vertical section. Relatively poorly stratified conglomerate and conglomeratic sandstone near the mountain front thus appear to represent lag deposits that accumulated gradually as finer sediment was carried farther south to accumulate more rapidly in the Uinta Basin.

In addition to abrupt lateral changes, vertical changes in the Duchesne River Formation are also important. The formation has been formally divided into four members. From oldest to youngest, they are the Brennan Basin, Dry Gulch Creek, Lapoint and Starr Flat Members (Andersen and Picard, 1972). At the base of the formation, the Brennan Basin Member consists dominantly of coarse- to medium-grained sandstone and conglomerate with subordinate interbedded fine-grained rocks (fig. 8). The overlying Dry Gulch Creek Member contains a much greater proportion of fine-grained rocks, with some sandstone and little conglomerate (fig. 9). Fining upwards through the two members occurs through a section as thick as 748 m (fig. 7), and is similar to that observed by Belt (1968) and McCormick and Picard (1969) in other fluvial units. It probably represents decreasing relief as erosion exceeded uplift in the source area.

Overlying the fine-grained Dry Gulch Creek Member is the Lapoint Member, composed of thin, persistent beds of bentonitic claystone

FIG. 8.—Conglomeratic sandstone of Brennan Basin Member in western Uinta Basin.

FIG. 10.—Interbedded gray, bentonitic claystone and reddish brown, detrital claystone of Lapoint Member in eastern Uinta Basin.

interbedded with sandstone, detrital fine-grained rocks and minor conglomerate (fig. 10). The youngest unit in the Duchesne River Formation is the Starr Flat Member, composed of sandstone and poorly stratified conglomerate (fig. 11). In the north-central Uinta Basin the member is 234 m thick. Conglomerate is much coarser and more abundant in this member where it is thin (fig. 7). The Starr Flat Member is partly a coarse-grained marginal facies, but it clearly overlies the finer-grained Lapoint Member in many places and apparently represents renewed relative uplift of the Uinta Mountains near the close of Duchesnean time. The vertical sequence thus indicates two major episodes of differential uplift and subsidence, separated by an interval when relief was less and volcanic material accumulated.

SEDIMENTARY STRUCTURES

Stratification and sedimentary structures.—A great variety of stratification types and sedimentary structures is present in the Duchesne River Formation. Only a few are abundant, including trough and planar cross-stratification and cross-lamination (McKee and Weir, 1953) and horizontal stratification. Ripple-stratification, ripple marks, shrinkage cracks and parting lineation are less common.

Conglomerate beds display indistinct horizontal stratification or form broad, shallow troughs. Isolated sandstone lenses within conglomerate are common, and they usually are horizontally or cross-stratified (fig. 12). Commonly, conglomerate and sandstone beds disconformably overlie finer grained rocks, and erosional surfaces are marked by elongate, trough-shaped scour marks up to 1 m in depth.

In sandstone, trough and planar cross-stratification are abundant. Medium- and small-scale cross-stratification are both present, and troughs are filled with festoon-shaped foreset laminae (fig. 13). Thickness of troughs ranges from a few centimeters to about a meter, and most troughs are from 25 to 30 cm thick. Horizontal and inclined planar stratification are also common (fig. 14), and many apparently struc-

FIG. 9.—Dry Gulch Creek Member in north-central Uinta Basin.

FIG. 11.—Conglomeratic sandstone of Starr Flat Member in northern Uinta Basin.

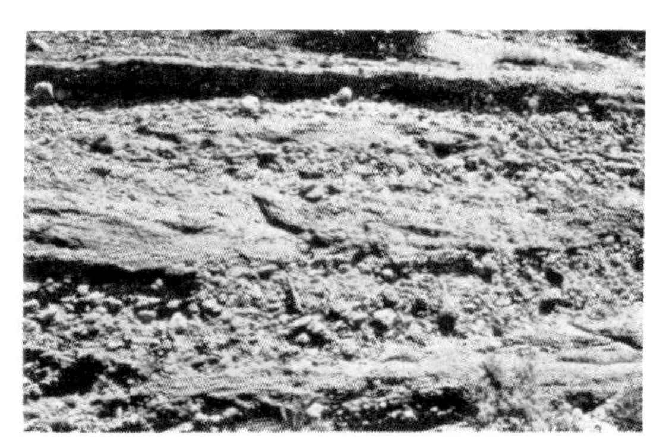

FIG. 12.—Indistinctly stratified fluvial conglomerate and sandstone in Brennan Basin Member.

FIG. 14.—Horizontal and planar cross-stratification in sandstone, Brennan Basin Member.

tureless sandstone beds are probably horizontally stratified.

Fine-grained rocks are horizontally stratified or laminated. Ripple-stratification is rare. Commonly, irregularly interstratified sandstone defines a wavy horizontal stratification (fig. 15). The internal lamination of the fine-grained rocks generally is indistinct. Extensive mottling or burrowed structure (fig. 16) and very poor sorting of the fine-grained material suggest organic disturbance.

Microfacies.—Conglomerate and sandstone occur in sharply bounded, broad lenses set in finer grained rocks (fig. 17). Basal erosional surfaces generally have low relief (<1 m), but locally attain rather high relief (10 m). Sandstone lenses are elongate, and some erosional surfaces are traceable for more than 1000 m in a north-south direction. In east-west cross-sections, most sandstone bodies are less than 300 m in width, and many are less than 50 m. Thicknesses range from 1 to about 15 m, and width-to-thickness ratios generally are between 50 and 300. Locally, especially in the Brennan Basin Member southwest of Vernal, sandstone lenses are present with width-to-thickness ratios less than 10.

Within sandstone bodies, sedimentary structures are typically associated in a consistent assemblage. Horizontally stratified coarse-grained sandstone or conglomeratic sandstone layers are commonly interstratified with trough or planar cross-stratified sandstone. Long, low-angle trough cross-strata typically are present at the top of such a two-part sequence (fig. 18), and these upper strata commonly are pebbly. Individual two-part sequences may represent discrete sedimentation units formed in one event during uniformly changing depositional conditions. Most sandstone bodies are composed of many lenses of these sequences, probably formed at different times under similar conditions. In many places, especially in the central and western parts of the Uinta Basin, individual two-part sandstone sequences defined by sedimentary structures are separated by thin partings of fine-grained rocks.

Sandstone passes abruptly into overlying fine-

FIG. 13.—Trough cross-stratification in sandstone of Brennan Basin Member.

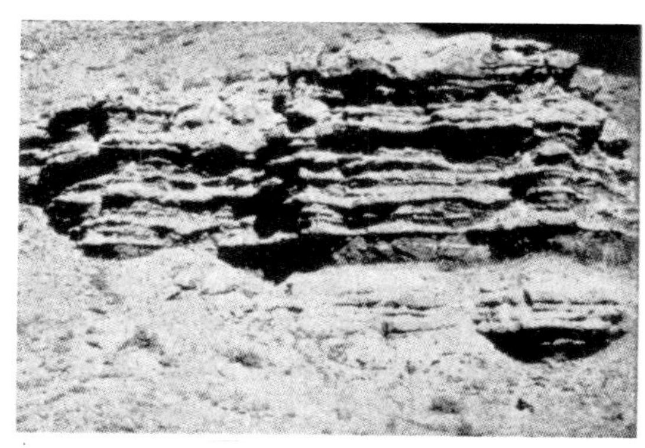

FIG. 15.—Interbedded thin, persistent sandstone and poorly stratified fine-grained rocks in floodplain deposits.

FIG. 16.—Mottled and burrowed claystone near base of Brennan Basin Member in eastern Uinta Basin. Length of rod is 2 m.

grained rocks (fig. 17). Typically, pebbly strata at the top of a sandstone body are overlain by mottled claystone. Sandstone bodies with deeply erosional bases and unusually low width-to-thickness ratios are finer grained than other sandstone bodies and generally are overlain by broader lenses of coarser grained sandstone.

Paleocurrents.—Of the wide variety of directional sedimentary features in the Duchesne River Formation, medium-scale trough cross-stratification (McKee and Weir, 1953) is the most abundant. Orientations of troughs record local flow directions and are believed to correlate well with ancient stream channel orientations. Seventy-eight outcrops were selected for paleocurrent study, and the orientations of the plunge axes of an average of 11 trough cross-strata were measured at each outcrop. Modal directions were established according to a method modified from Tanner (1959) in which 30 degree intervals containing more than one standard deviation above the average number of readings per interval are considered significant modes. The dominantly unimodal patterns of trough orientations at each outcrop and for the formation as a whole suggest that the data can be treated as a circular-normal distribution (Fisher, 1953) about a single preferred orientation.

To estimate the orientation of the preferred direction, the vector resultant ("vector mean") was calculated for each outcrop. The areal pattern of paleocurrent directions was established by moving vector averages (Pelletier, 1958). The six-mile twonship-and-range grid pattern was used, and grouped data from four adjacent townships were plotted at their common corner (fig. 19). The length of each vector is proportional to the vector strength (vector resultant divided by number of measurements) for that corner. For the total formation, all measurements are represented on a circular histogram, and the vector resultant direction (S 12.2 W) and 95 percent confidence limit ($\pm$ 4.2°) are indicated. The general southward orientation of paleocurrent flow and the remarkable uniformity of directions are evident.

During the late Uintan, streams flowed westward across the Uinta Basin (Stagner, 1941). At the beginning of Duchesnean time, uplift of the Uinta Mountains strongly influenced the drainage system, resulting in the flow directions indicated in figure 19. Once established, the southward flow direction persisted throughout Duchesnean time and is represented in the present drainage pattern of the Uinta Basin.

DEPOSITIONAL ENVIRONMENTS

The Duchesne River Formation is underlain and overlain by continental clastic rocks and is far removed from known late Eocene marine deposits. The fossils in the formation are nonmarine vertebrates (Andersen and Picard, 1972). Rocks of widely ranging grain size are

FIG. 17.—Discontinuous channel-sandstone lenses interbedded with fine-grained floodplain deposits, Brennan Basin Member.

FIG. 18.—Low-angle trough cross-stratification at top of sandstone lens in Brennan Basin Member.

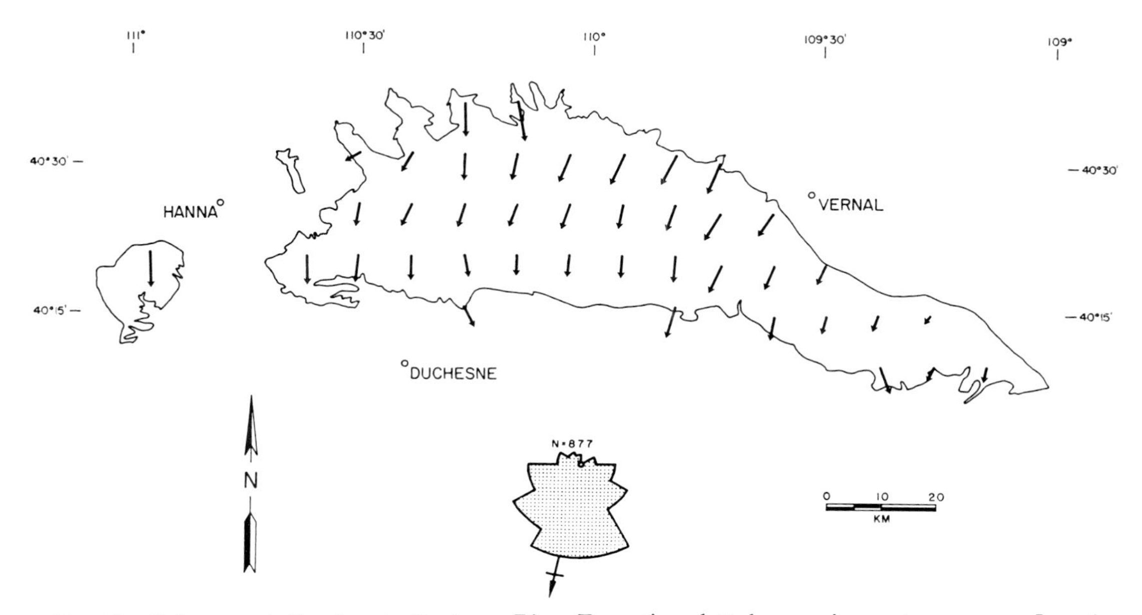

FIG. 19.—Paleocurrent directions in Duchesne River Formation plotted as moving vector averages. Lengths of vectors are proportional to vector strength. Histogram, vector resultant direction and 95 percent confidence limits for entire formation shown at bottom.

present as discontinuous, lenticular strata with rapid changes in facies (figs. 8, 17). A fluvial environment of deposition is indicated for the formation. The coarse grain size of some of the rocks and the abrupt north-south facies changes suggest that high-gradient streams were important.

Debris flows.—Clastic rocks very near to the Uinta Mountains generally are conglomeratic, with a maximum observed particle size of 2 m. In some conglomerates, large clasts are dispersed in a poorly sorted matrix of finer material (fig. 20), resembling the deposits of debris flows described by Bull (1964) and Fisher (1971). In general, however, debris-flow deposits are not abundant in the Duchesne River Formation, and much material originally moved in debris flows may have been reworked by streams.

Channel deposits.—Lenticular conglomerate and sandstone bodies are interpreted as stream channel deposits. Coarse clast size and high proportion of bed load material suggest high velocities and steep gradients. Abundant trough and planar cross-stratification and horizontal stratification indicate deposition in the upper part of the lower flow regime and the lower part of the upper flow regime (Harms and Fahnestock, 1965).

The inhomogeneous channel deposits near the Uinta Mountains resemble sediments deposited in modern braided stream systems described by Ore (1964, 1965), Williams and Rust (1969) and Smith (1970). Poorly sorted, horizontally stratified coarse-grained material (fig. 12) probably accumulated in longitudinal bars (Ore, 1964, 1965; Smith, 1970), and more uniformly sorted, cross-stratified sandstone (figs. 13, 14) resemble sediment of transverse bars (Smith, 1970, 1971). The wide range of grain sizes, numerous local scours and abrupt superposition of diverse sedimentary structures indicate extreme fluctuations in local flow conditions. Probably, rapid lateral shifting of channels was common, and seasonal variation of flow was possibly also important. Silt and clay might

FIG. 20.—Poorly sorted debris-flow deposit in Starr Flat Member. Framework clasts are dispersed in fine-grained matrix.

have filled small, abandoned segments of channels or been trapped by vegetation and coarse sediment on bars, but most of the silt and clay probably was carried downstream in suspension or deposited on the floodplain during periods of high flow.

Ephemeral stream deposits commonly share many characteristics with deposits of braided streams (Picard and High, 1973). However, some distinctive features result from waning floods in ephemeral streams. Ripple-stratified and ripple-marked sand and fine-grained sediment are extremely abundant (Williams, 1971; Karcz, 1972; Picard and High, 1973). Interference ripple marks, dendritic surge marks, fluted steps, abundant tracks and trails, shrinkage cracks, and mud curls characterize postflood fine-grained sediment in ephemeral streams (High and Picard, 1968; Picard and High, 1969, 1973). The scarcity of these features in the Duchesne River Formation suggests that the channel deposits in it were formed by perennial streams.

Farther from the Uinta Mountains, lenticular sandstone bodies lack extremely coarse-grained bed load material and contain a more limited assemblage of sedimentary structures. Upper flow regime structures indicate high-energy conditions, but limited lateral extent of these sandstone bodies suggests that channels were effectively confined within cohesive banks. These sandstone bodies are interpreted as deposits of meandering streams, most of which apparently were entrenched in older floodplain material. Laterally homogeneous, fining-upward cycles of channel deposits, typical of point bar deposits of large meandering rivers (Allen, 1970) are not common in the Duchesne River Formation. Sinuous streams of the formation probably were too small or too highly confined to form these types of deposits.

Floodplain deposits.—Interstratified with the channel deposits are abundant finer grained rocks probably deposited outside of major channels. Stratification in these floodplain deposits generally is indistinct, and contorted beds and burrowed structures are common. Most rocks are mottled and very poorly sorted, with sand grains dispersed in silty or clayey matrix, further indicating postdepositional mixing of sediments. Probably, finer grained floodplain materials supported abundant vegetation and numerous burrowing animals. In the Brennan Basin Member, fine-grained rocks are liberally interstratified with thin but laterally persistent sandstone beds (fig. 15). These generally are fine grained, never pebbly, and usually unevenly stratified horizontally or trough cross stratified. Beds a few centimeters in thickness are almost certainly results of overbank flooding from nearby channels, and vaguely stratified beds of similar appearance as much as 50 cm thick may have a similar origin.

In many modern fluvial settings, vertical accretion of extensive overbank deposits is minor, and most fine-grained materials accumulate laterally as local deposits associated with laterally migrating channels (Wolman and Leopold, 1957). In the Duchesne River Formation, evidence of extensive lateral migration of channels is rare. Furthermore, intertonguing relationships of channel and floodplain deposits (fig. 17) indicate that most floodplain deposits accumulated by vertical accretion. Abundant vegetation on the floodplain has been postulated from evidence of extensive biogenic sediment disturbance. Schumm and Lichty (1963) noted that vegetation can effectively trap and hold sediment from overbank floods and contribute substantially to vertical accretion on a floodplain. Possibly, vertical accretion of floodplain deposits in the Duchesne River Formation was similarly accelerated by vegetation.

Most fine-grained floodplain deposits in the formation are dark reddish brown in color, and recognizable organic material is rare. Oxidizing conditions prevailed. Organisms apparently were abundant enough to disturb the sediment but did not reduce its pigment.

Some light-colored thin beds are present in the floodplain deposits. X-ray diffraction analyses indicate that these beds have higher total carbonate content and much higher proportion of dolomite than associated redbeds. Evidence of vadose cementation (Dunham, 1969; James, 1972) is absent. High dolomite content is unusual in modern caliche layers (Goudie, 1972), and dolomitic carbonate beds are best explained as deposits of temporary ponds. In general, however, permanent standing water on the alluvial plain must have been relatively restricted.

Discussion.—On modern alluvial fans and alluvial plains, coarse sediment accumulates rapidly where confined streams emerge from mountains. Aggradation can be caused by a decrease in gradient on the valley floor, but it can also take place on a nearly uniform gradient where depth and velocity decrease as stream width increases at the mountain front (Bull, 1964). Near the Uinta Mountains, channel deposits of the Duchesne River Formation represent high-gradient braided streams. Shallow trough cross-strata suggest shallow water depths for depositing currents. Wide lateral extent of sandstone and conglomerate bodies and extreme lithologic inhomogeneity indicate widely fluctuating local flow conditions as unconfined streams shifted laterally across a broad alluvial plain of nearly uniform slope. Some debris-flow deposits are preserved near the Uinta Mountains, but most of them apparently were reworked by the rapidly shifting streams.

Farther from the Uinta Mountains, lenticular sandstone bodies represent entrenchment of streams in older floodplain material. These streams probably had lower gradients and higher sinuosities than the more proximal braided streams. Backfilling of entrenched channels with sandy sediment occurred as vertical accretion of floodplain material continued. During the early and late Duchesnean, the braided conditions extended southward, but in the middle Duchesnean the zone of braided stream conditions was limited to a narrow belt near the Uinta Mountain front.

PALEOCLIMATE

Paleoclimatic reconstructions for western North America during mid-Tertiary time are based largely on paleobotanical evidence. The early Tertiary apparently was warm and moist compared with the present, and gradual cooling and drying probably has occurred since the Cretaceous (Chaney, 1940; Axelrod and Bailey, 1969). Floras from Pacific coastal lowland sites indicate subtropical and humid, warm temperate conditions during the late Eocene and early Oligocene (Penney, 1969). Floras from the Rocky Mountain region are diverse, possibly representing a variety of habitats over a considerable range of elevations. Axelrod (1966) has reconstructed a general altitude zonation for the northwestern United States involving more than 1300 m of relief during the late Eocene and early Oligocene.

Fossils are extremely rare in the Duchesne River Formation. Large browsing mammals and carnivores are found in the formation (Andersen and Picard, 1972), but little is known of the paleoecological conditions indicated by them. Crocodile remains (Peterson, 1932) suggest ponds or slowly flowing streams. However, only a few individuals have been found, and so these conditions may not have been widespread. Tortoise (Clark, 1932) probably lived in dry interchannel areas. The reptiles indicate warm conditions throughout the year. A diverse community of smaller vertebrates, invertebrates, and plants undoubtedly existed on the alluvial plain, but the coarse-grained, oxidized deposits probably are not favorable for their preservation. Where preserved in such deposits, small fossils are extremely difficult to collect, and environmentally distinctive fossils have not been found in the formation.

Rocks of the Duchesne River Formation indicate depositional and post-depositional conditions, and some climatic implications are suggested. Channel deposits reflect turbulent, relatively high-gradient stream conditions. Small, perennial streams probably had large seasonal variations in discharge. Dolomitic carbonate beds probably record temporary ponds on the floodplain, but they are rare. Gypsum is related to modern groundwater; no persistent beds have been found to support the interpretation of Warner (1963, p. 148) that gypsum represents saline lake deposits in the formation. Vegetation may have been abundant, but post-depositional conditions were oxidizing. Most floodplain sediments probably were dry during most of the year, and accumulations of abundant organic matter in soils certainly were rare.

Because of the apparent high relief, conditions in the Uinta Mountains must have been quite different from those of the depositional site. At the present time, neither set of conditions can be reconstructed confidently. Climatic conditions probably were warm, temperate and relatively dry, with the mountains considerably colder and wetter than the depositional basin. The discovery of small mammals, invertebrates, and plant remains in the formation would contribute greatly to the reconstruction of paleoclimatic conditions.

PETROGRAPHY

Texture

Most of the Duchesne River Formation consists of clastic sedimentary rocks. Rock types range from coarse conglomerate to claystone, and there is a full gradation of intermediate sizes. Terminology used here for clastic rocks is from Wentworth (1922) and Picard (1971). Sandstone is the most abundant rock in the formation, comprising about 50 percent. Conglomerate and fine-grained rocks are less abundant and constitute about 10 and 40 percent of the formation, respectively. Although fine-grained rocks are abundant, fissility is generally poor and shale is rare.

Sorting is very poor in all Duchesne River rocks. In conglomerate, grain size distributions are generally bimodal, and matrix material is abundant. Some coarse-grained rocks are so poorly sorted that framework clasts are not in contact (fig. 20). Sandstone is also poorly sorted. Authigenic cements constitute between 10 and 40 percent of most sandstones, and apparently much interstitial matrix material has been replaced by carbonate.

Fine-grained rocks consist largely of common varieties, especially silty sandstone, silty claystone and claystone. More poorly sorted rocks, including sandy claystone, sandy mudstone and clayey mudstone, which are rare in other settings (Picard, 1971), are relatively abundant in the Duchesne River Formation. Burrowed structures or very poor stratification, mottled coloring, and very poor sorting are typical of most fine-grained rocks in the formation. Subangular to angular sand grains are scattered throughout fine-grained rocks, and organic mixing of sediment probably is responsible for the poor sorting of these rocks.

Composition

The composition of framework constituents is described according to the sandstone classification of Folk (1968). Compositional terms are modified and are applied to clastic rocks of all grain sizes. Figure 21 illustrates the compositions of 121 clastic rocks studied and shows the differences in compositions of rocks of different grain size. Nearly all rocks in the formation are lithic, sublithic or quartzose in composition, and only one sandstone studied is a subarkose. Conglomerate generally contains a higher proportion of rock fragments than other rocks, and the highest proportion of quartz grains is found in sandstone.

Quartz.—Quartz grains are grouped according to optical properties into monocrystalline grains with undulatory or nonundulatory extinction, and polycrystalline grains (Blatt and Christie, 1963). Metaquartzite fragments from the Precambrian Uinta Mountain Group are locally abundant in coarse-grained rocks and are counted with polycrystalline quartz. Chert is considered a rock fragment.

The relationship between quartz extinction types and grain size is illustrated in figure 22. Percent nonundulatory of monocrystalline quartz is less in coarse-grained sandstone than in finer grained rocks. Percent monocrystalline of total quartz is high and increases steadily with decreasing grain size. In conglomerate, not shown in figure 22, only a very small pro-

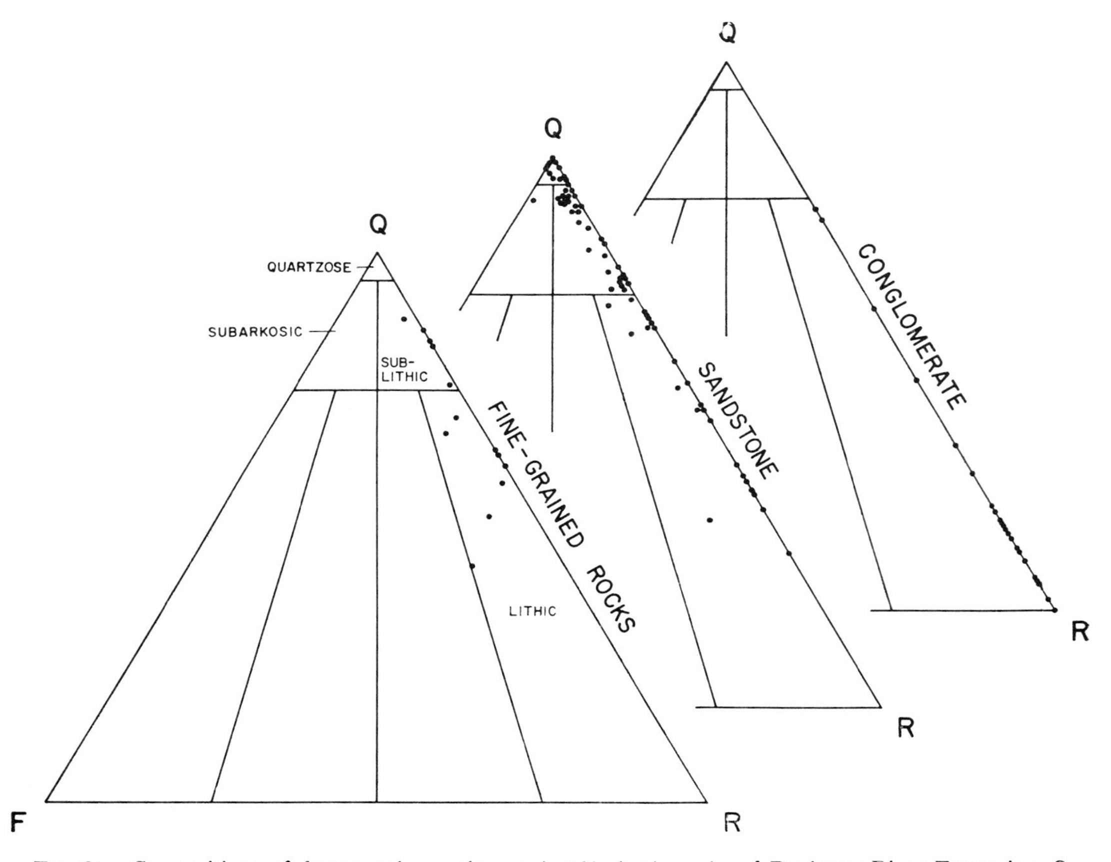

Fig. 21.—Compositions of framework constituents in 121 clastic rocks of Duchesne River Formation. Q—quartz; F—feldspars; R—rock fragments. Classification modified from Folk (1968). Conglomerate contains a significantly higher proportion of rock fragments than do finer-grained rocks.

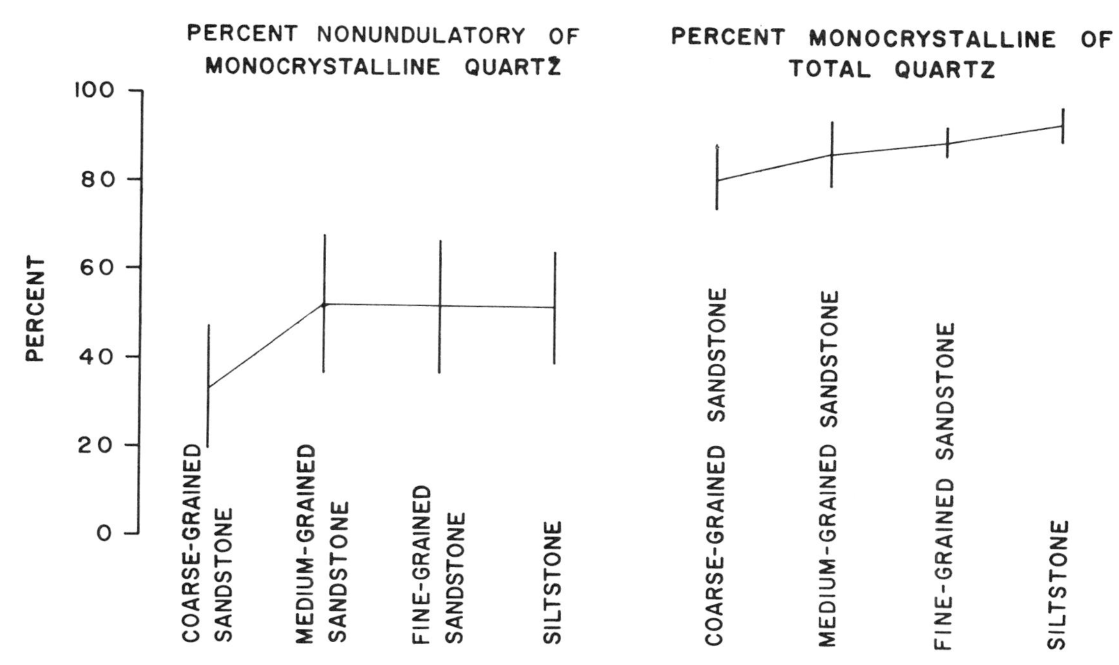

FIG. 22.—Relationship between quartz extinction types and median grain size in 56 rocks studied in thin section. Bars represent means ±1 standard deviation.

portion of quartz is monocrystalline. Decreases in the amounts of undulatory and polycrystalline quartz with decreasing grain size have been observed in fresh detritus weathered from massive plutonic and metamorphic rocks, and unstable quartz types apparently are selectively destroyed during grain size reduction (Blatt, 1967a).

Feldspar.—Feldspars are rare in the Duchesne River Formation, generally constituting less than 5 percent of the grains in a sample. In the samples studied, the order of abundance is: microline > orthoclase > plagioclase. Feldspars are most abundant in the bentonitic claystone of the Lapoint Member. These are considered to be altered volcaniclastic rocks and are discussed separately. The feldspar-poor Duchesne River sandstones contrast sharply with the arkosic and subarkosic sandstones of the underlying Uinta Formation (Stagner, 1941), and the difference provides strong evidence of a change in source area from Uintan to Duchesnean time.

Rock fragments.—Rock fragments consist of detrital material that was derived primarily from the Uinta Mountains on the north and include many types of polycrystalline rock fragments as well as monocrystalline grains other than quartz and feldspar, such as apatite, mica, zircon and opaque minerals. The Uinta Mountains consist of Mesozoic and Paleozoic sedimentary rocks overlying mildly metamorphosed Precambrian quartzite and argillite. Unlike many ranges of the central Rocky Mountains, The Uinta Mountains lack a granitic core. Fragments of carbonate and clastic sedimentary rocks are most abundant in the Duchesne River Formation, and chert fragments are less abundant. Other rock fragments include probable volcanic clasts, micas and minor amounts of ferromagnesian and heavy minerals. Examples of several rock fragment types are shown in figure 23. Relative amounts of sedimentary rock fragments in lithic and sublithic rocks are shown in figure 24, and a wide scatter of rock fragment compositions is evident. Fields outlined in figure 24 indicate concentrations of sandstone and conglomerate compositions. In general, conglomerate is rich in fragments of clastic rocks and carbonates and poor in chert grains. Sandstone is rich in carbonate and poor in clastic rock fragments, but the fields of conglomerate and sandstone compositions overlap considerably. Fine-grained rocks are also diverse in composition, but the small sample size (5) prevents generalization.

Heavy minerals in the formation were studied by Warner (1963), and apatite is the most abundant. As with other rock fragments, most of the heavy minerals originated in sedimentary and low-grade metamorphic rocks of the Uinta Mountains area. Biotite is abundant in

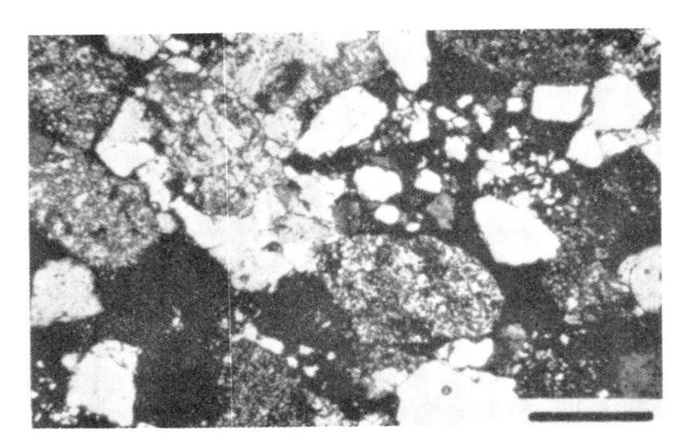

FIG. 23.—Photomicrograph of litharenite from Starr Flat Member, showing fragments of chert, carbonate and clastic sedimentary rocks. Crossed nicols; bar is 0.5 mm.

altered volcaniclastic rocks, but the diverse assemblage of heavy minerals derived from igneous and metamorphic rocks common in other late Eocene rocks of the Rocky Mountain region (Sato and Denson, 1967) has not been observed in the Duchesne River Formation.

Clay minerals.—Kaolinite and illite are the common clay minerals in fine-grained rocks and as matrix material in sandstone and conglomerate. Montmorillonite and mixed-layer clays are relatively rare in most of the formation. Distinctive montmorillonite-rich claystone is interpreted to be volcaniclastic and is considered separately.

Cements.—Authigenic silica and carbonate are cementing agents in rocks of the Duchesne River Formation. Cement generally comprises between 10 and 40 percent of sandstone and siltstone. Very porous rocks contain less abundant cement, probably because of weathering at the outcrop. In conglomerate, cement has not been quantitatively evaluated, but most conglomerate is loosely cemented. Cement in claystone and mudstone is difficult to resolve microscopically. Calcite is evident, however, in X-ray diffraction patterns of nearly all fine-grained rocks studied, and some of this probably is authigenic cement.

Silica cement is minor in Duchesne River sandstone, and most of the cementing material is carbonate. X-ray diffraction indicates that essentially all of the authigenic carbonate is calcite. Petrographic relationships indicate that silica cement was the first cement to form, followed by at least two generations of calcite cement. Calcite cement apparently replaced much of the original clay matrix of the sandstone and clearly replaced a small amount of quartz.

Some of the red pigment in sandstone predates even the earliest cement. Silica cement was deposited over and around stained grain boundaries and interstitial red matrix. During subsequent carbonate cementation, much red pigment from detrital rock fragments and grains of opaque oxides was mobilized and further oxidized. Cement of this stage generally is stained a pale reddish brown color. In many samples, a younger carbonate cement, representing the third stage of cementation, is free of red pigment.

Provenance

One of the fundamental factors controlling the nature of a sedimentary rock unit is the composition of the source material from which the sediment was derived. Sedimentary rock fragments are generally quite distinctive and easily recognized, and they provide the best evidence of recycling of sediment from preexisting sedimentary rocks. However, quantitative data by which to evaluate the representation of sedimentary source material in younger deposits are rare (Blatt, 1967b). Clastic rocks of the Duchesne River Formation were derived almost entirely from sedimentary and very low-grade metasedimentary source rocks of the Uinta Mountains. The formation therefore provides an excellent example of a rock unit composed mainly of recycled sediment.

Quartz.—Historically, types of quartz grains were among the very earliest features used to interpret the provenance of sandstone (Sorby, 1877). Krynine (1946) assumed great genetic significance of quartz types. "Common" (non-

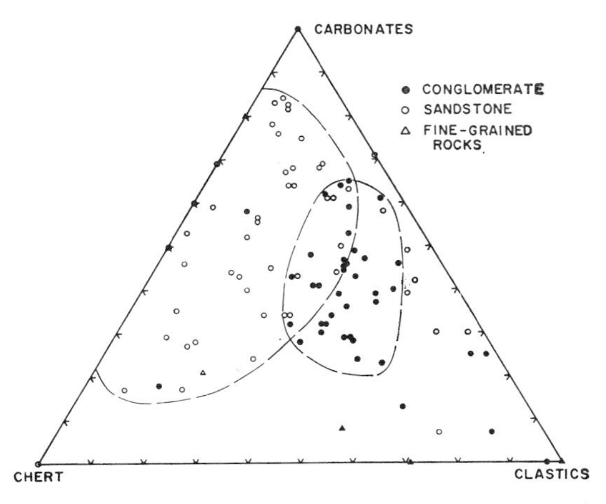

FIG. 24.—Relative amounts of sedimentary rock fragments in 98 lithic and sublithic rocks. Fields of concentration of conglomerate and sandstone compositions include more than 80 percent of samples in each group.

undulatory) quartz was attributed to plutonic igneous source rocks, and "strained" (undulatory) and "metamorphic" (polycrystalline) quartz were thought to represent metamorphic source rocks. Blatt (1967a) found that undulatory and polycrystalline quartz are abundant in plutonic igneous rocks as well as gneisses and schists. Nonundulatory quartz apparently is enriched in sedimentary rocks, especially in quartzarenites (Potter and Pryor, 1961; Blatt and Christie, 1963). Seemingly, undulatory and polycrystalline quartz grains are less stable than nonundulatory quartz and are selectively destroyed during recycling of sediment.

In the Duchesne River Formation, quartz grains are about 44 percent nonundulatory, 44 percent undulatory and 12 percent polycrystalline, and quartz type depends somewhat on grain size (fig. 22). All of this is recycled quartz. Although extensive information on source rocks of the Uinta Mountains area is not available, a small sample of rocks from the eastern Uinta Mountains suggests that undulatory and polycrystalline quartz are more abundant in the Paleozoic and Mesozoic sandstones than in the Precambrian quartzite. Thus, the unstable quartz types may survive recycling in sedimentary rocks to a greater extent than is generally believed.

Rock fragments.—Sedimentary rock fragments provide unambiguous evidence of polycyclic origin. Generally, chert is the only common sedimentary rock fragment even in sediment with known sedimentary source material (see, for example, Siever, 1949; Todd and Folk, 1957; Anderson, 1972). Only rarely have abundant extrabasinal carbonate grains been noted in sedimentary rocks. Gravels derived from the Sevier orogenic belt contain much reworked sedimentary material (Armstrong, 1968; Mullens, 1971; van de Graaff, 1972). The deposits may be a major source of information about the former distribution of some Paleozoic and Mesozoic rock types in the Basin and Range province. Unfortunately, lack of comparative information by which to judge the deposits has discouraged such an attempted reconstruction (Mullens, 1971).

The Paleozoic and Mesozoic rocks of the south flank of the Uinta Mountains range from about 3660 to 6400 m thick and consist of approximately 7 percent carbonates, 2 percent chert and 91 percent clastic sedimentary rocks (Huddle and McCann, 1947; Huddle and others, 1951; Kinney, 1955). Comparison of these data with figure 24 indicates that sedimentary rock fragments in the Duchesne River Formation contain a much higher proportion of carbonate and chert than does the source area. The explanation for this probably is that clastic sedimentary source rocks break down into quartz grains and clays, leaving the rock-fragment fraction relatively enriched in carbonate grains and chert.

Figure 25 shows the variation of sandstone composition with location in an east-west direction along the mountain front. Considerable scatter in points is evident, but sandstone rich in quartz and poor in rock fragments is much more common in the west end of the basin than in the east. Much of the Paleozoic section is faulted out in the west end of the Uinta Mountains, and Precambrian quartzite contributed more to Duchesne River sandstones in the west than in the east.

The sandstones rich in rock fragments in the eastern Duchesne River Formation provide an opportunity to study the effect of transport distance on sandstone composition. In figure 26 the percent rock fragments of grains in 40 sandstone samples is plotted against distance from the mountain front. Evidently, rock fragments have been selectively destroyed during transportation, but the effect is not strong. As far as 45 km south of the south flank (80 km from the present crest) of the mountains, sandstone still contains as much as 15 percent rock fragments. Carbonate rock fragments should be among the least stable in a fluvial environment. Figure 27 shows, however, that the percent carbonate grains of rock fragments in sandstone does not diminish substantially with transportation over these distances under the climatic conditions that existed here during the Duchesnean.

Pigment.—The origin of red pigment in sedimentary rocks has long been a matter of interest to geologists. Krynine (1949) established a system of genetic classification of redbeds. Based largely on previous work by himself (Krynine, 1935) and van Houten (1948) he concluded that "primary" redbeds derived from lateritic soils were the only important type of redbed in the geologic record (Krynine, 1949). "Post-depositional" redbeds, in which pigment is developed largely in place after deposition, and "secondary" redbeds, derived from erosion of previously existing red sedimentary rocks, were considered rare (Krynine, 1949). Picard (1965) and Walker (1967) subsequently demonstrated the importance of post-depositional mobilization of red pigment in sedimentary rocks. Clearly, some deposits are post-depositional redbeds, and diagenetic changes may affect all kinds of redbeds to some extent.

Characteristics of primary and post-depositional redbeds and the processes responsible for them have been the subjects of consider-

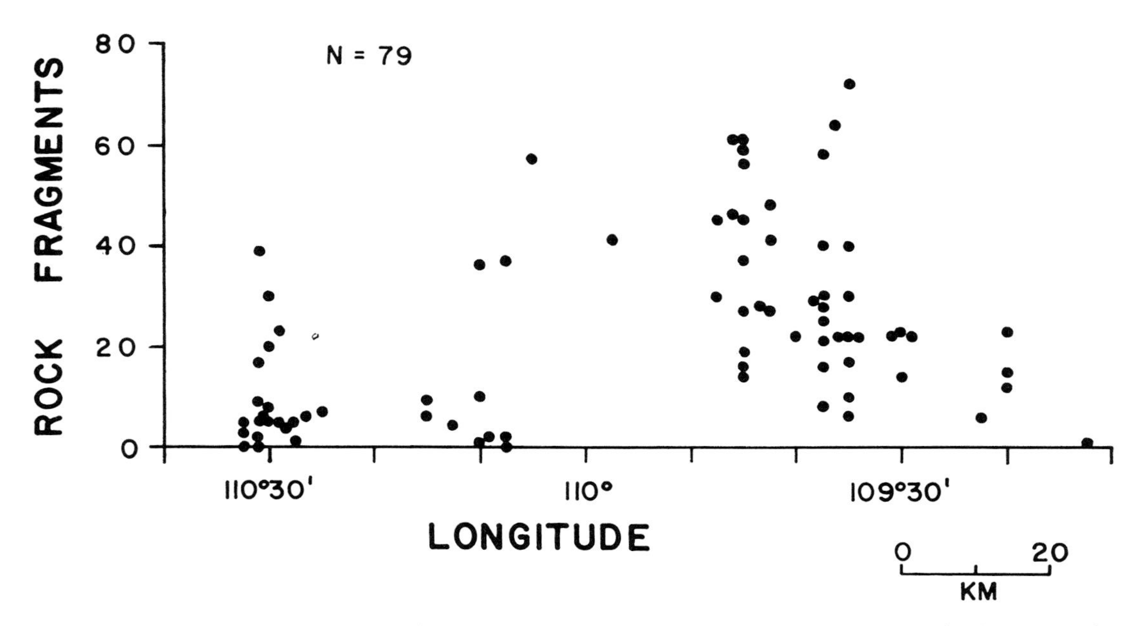

FIG. 25.—Percent rock fragments of grains in 79 Duchesne River sandstone and fine-grained rock samples plotted against longitude. Greater proportion of rock fragments in the east reflects greater exposure of sedimentary source rocks.

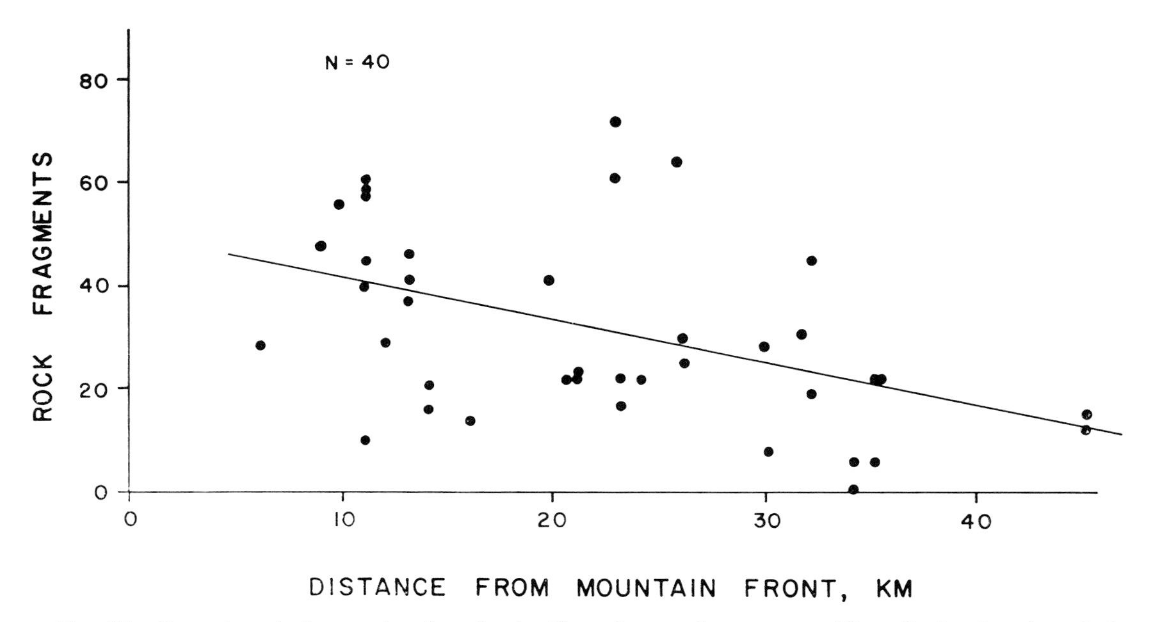

FIG. 26.—Percent rock fragments of grains in 40 sandstones from eastern Uinta Basin plotted against distance from mountain front. Regression line indicates tendency to decreasing proportion of rock fragments with increasing distance of transport. Correlation coefficient is only —0.46.

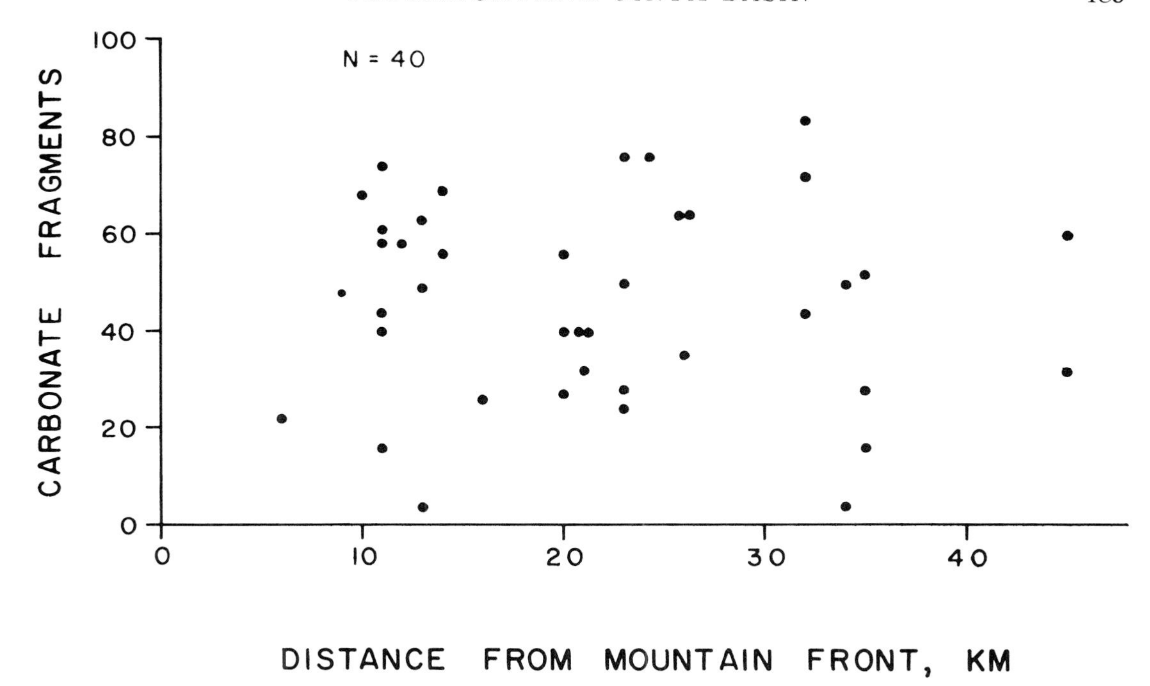

FIG. 27.—Percent carbonate grains of rock fragments in 40 sandstones from eastern Uinta Basin plotted against distance from mountain front.

able attention and debate. Secondary redbeds, however, have received far less attention. Krynine (1949) and Clark (1962) considered the Tertiary basins of the Rocky Mountain region a likely area for the accumulation of secondary redbeds, but the behavior of red sedimentary source material during transportation, deposition and diagenesis is poorly understood.

The Duchesne River Formation contains detritus derived from numerous red sedimentary source rocks (fig. 6). The deep reddish brown color of the formation contrasts sharply with most other Tertiary units in the Uinta Basin. Conglomerate of the Duchesne River is not strongly pigmented, and color that is present is in the interstitial matrix material. Most red or reddish brown beds are sandstone or fine-grained rocks. Sandstone contains red rock fragments (fig. 23) and quartz grains with rims of red stain (fig. 28). These grains apparently have survived transportation from red source rocks in the Uinta Mountains. Most of the pigment in cement is clearly related to red rock fragments (fig. 29) or to grains of iron oxide, but some of the pigment is not. Apparently, secondary red pigment was mobilized after deposition, and probably additional post-depositional pigment was formed as well. Fine-grained rocks are more difficult to study than sandstone. Hematite is the only important pigment in fine-grained rocks of the formation (van Houten, 1968), but its distribution is very uneven. The wide variety of colors in fine-grained rocks, ubiquitous mottling, and preservation of bleached animal burrows suggest that mobility of pigment in these rocks has not been as great as in sandstone. Possibly, much of the hematite was deposited along with other detrital minerals from the Uinta Mountains.

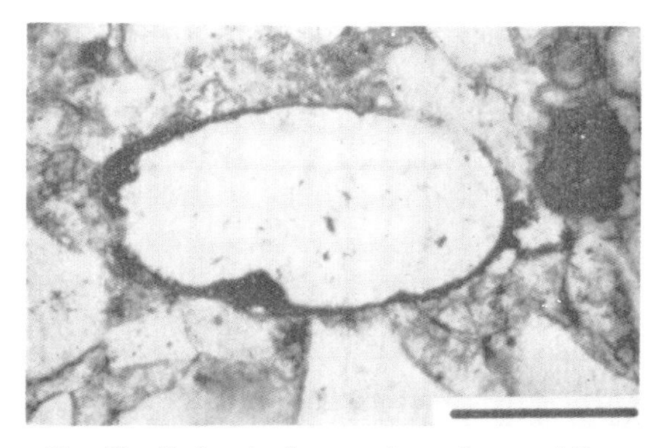

FIG. 28.—Red-stained quartz in sandstone of Brennan Basin Member. Well rounded, red-stained quartz grains are abundant in sedimentary rocks of the Uinta Mountains. Plane light; bar is 0.2 mm.

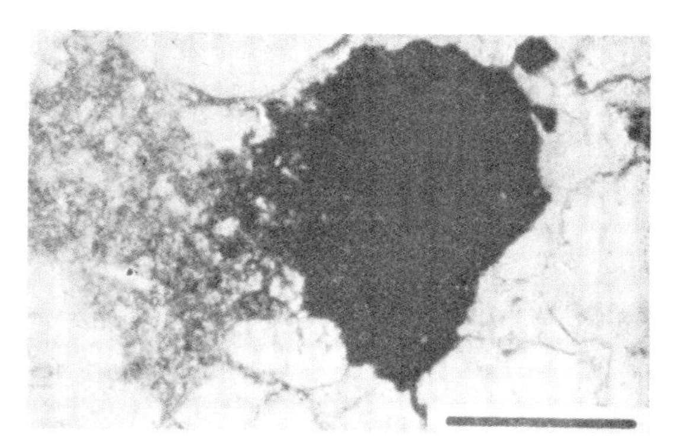

FIG. 29.—Reddish brown cement associated with detrital rock fragment in sandstone of Brennan Basin Member. Plane light; scale is 0.2 mm.

Some additional oxidation may have taken place, and at least locally some pigment was destroyed by reduction.

Volcaniclastic Rocks

Minor amounts of bentonitic claystone are present throughout the Duchesne River Formation, but bentonite is especially abundant in the Lapoint Member. Bentonites are mainly light gray (N 6, N 7) or greenish gray (5G 5/1, 5GY 7/1; Goddard and others, 1948), thin bedded and micaceous. Typically, individual bentonite beds are only a few centimeters thick, but they locally persist laterally for many kilometers. Warner (1963) demonstrated that individual bentonite beds correlate over large areas.

The mineralogy of the bentonitic claystone is unusual for the Duchesne River Formation. Sand-size grains constitute from 0 to about 40 percent of the rock, and these are as much as 25 percent feldspar and 30 percent biotite (fig. 30). Other common grains are quartz and opaque minerals. Detrital carbonate grains and chert occur rarely. Median grain size is generally about 200 μm, and biotite grains as large as 600 μm are present. Light-colored, relatively highly birefrigent clay matrix constitutes 60 percent or more of the rocks, and X-ray diffraction shows this to be almost pure montomorillonite. Because of the unusual lateral persistence of these strata and their unique composition, the bentonitic claystone is interpreted to be altered volcanic ash. Relict vitroclastic texture has not been observed, and no glass has been found. Sedimentary structures in the bentonite are obscure, probably because of interaction with groundwater, and the exact mechanism of deposition is unknown. Probably, the ash was at least partly reworked by streams before burial, but it can still be considered pyroclastic.

In the eastern Uinta Basin, bentonitic claystone is present mainly in thin beds interstratified with normal Duchesne River redbeds (fig. 10). Farther west, bentonitic material is more abundant and forms beds as much as 5 m thick. Eardley (1952, 1969) and Gazin (1959) suggested a Duchesnean age for some of the volcanism in the Park City area, near the western end of the Uinta Mountains. Andersen and Picard (1972) suggested that the Park City area supplied volcaniclastic material to the Duchesne River Formation.

Jack and Carmichael (1969) demonstrated that trace element composition can be used to distinguish between volcanic glasses of similar major element chemistry. Using the rapid-scan X-ray fluorescence technique described by Jack and Carmichael (1969), relative abundances of trace elements can be used to correlate volcanic materials, although absolute abundances are not measured. Figure 31 shows the relative abundances of Rb, Sr and Zr in whole rock samples of five Duchesne River bentonites. The similarity of these rocks indicates that either the original chemistry was retained or that changes in chemistry during alteration of the volcanic glass to montmorillonite were systematic. Shown for comparison are data from rapid scans of whole rock samples of the Absaroka Volcanics from the east entrance of Yellowstone National Park in Wyoming, the Keetley Volcanics from the Park City area, volcanics from the Challis Group near Carey, Idaho, and redbeds from the Duchesne River Formation. The similarity of Duchesne River bentonites to the Keetley Volcanics corroborates that the Park City area may have been the source of volcaniclastic material.

Armstrong and others (1969) and Armstrong

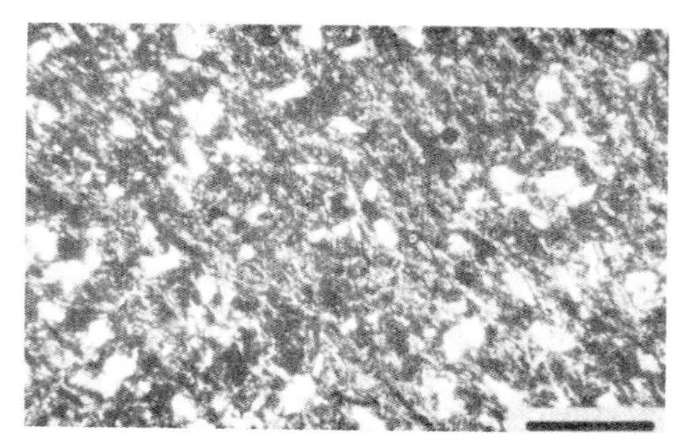

FIG. 30.—Photomicrograph of bentonitic claystone in thin section. Quartz, feldspars and biotite are set in matrix of montmorillonite. Crossed nicols; bar is 0.5 mm.

(1970) noted that volcanic activity in the Wasatch and Oquirrh Mountains and nearby areas of the eastern Basin and Range province represented some of the earliest Cenozoic volcanism in the area of the modern Great Basin. Probably this constituted an early phase of intermediate and silicic volcanism and ignimbrite eruption that became more widespread in the Basin and Range province during the Oligocene (Armstrong and others, 1969; McKee, 1971). If the correlation of bentonitic material in the Lapoint Member with this early volcanism is correct, deposition of the Duchesne River Formation was contemporaneous with the onset of silicic volcanism in and near the Basin and Range province in northern Utah.

SUMMARY

The late Eocene stratigraphic sequence in the Uinta Basin reflects significant tectonic events in the area. Uintan deposits accumulated in a broad, flat basin as lacustrine and fluvial deposits. Some sediment was derived from the north, but there is no evidence of high relief between the Uinta Mountains and the Uinta Basin. The youngest preserved lacustrine deposits in the basin are mid-Uintan in age, and they are overlain by pre-orogenic sediments of late Uintan age. Either the lake receded to the south and late Uintan lacustrine deposits are now lost to erosion, the lake dried up from lack of water, or exterior drainage was established and the lake disappeared. Although conclusive evidence is not yet available, the fluvial deposits of the Uinta Formation suggest relatively large streams and moist floodplain conditions. The third alternative, that exterior drainage was established during Uintan time, is presently favored.

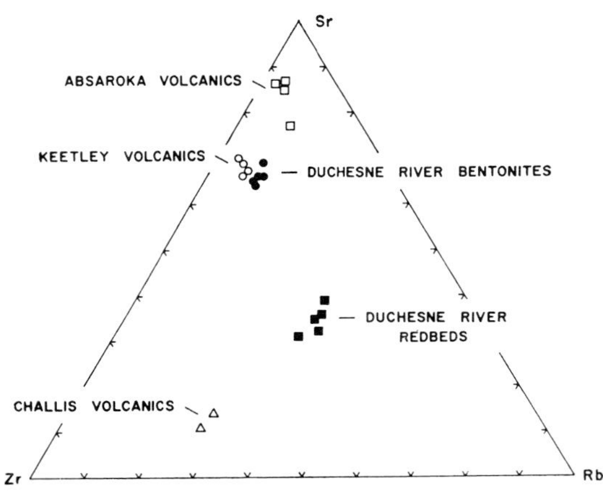

FIG. 31.—Relative amounts of Rb, Sr and **Zr** in Duchesne River Formation and three groups of volcanic rocks. See text for discussion.

At the beginning of Duchesnean time, somewhat later than deposition of the youngest preserved lacustrine deposits in the basin, uplift of the Uinta Mountains strongly affected drainage patterns and sediment types in the basin. Duchesnean paleocurrent directions indicate southward-flowing streams throughout the northern Uinta Basin. Coarse sediment accumulated as debris-flow deposits and as channel and floodplain deposits of relatively small, high-gradient, braided streams. Petrographic relationships clearly indicate that nearly all of the clastic material was derived from sedimentary and low-grade metamorphic source rocks of the Uinta Mountains.

Finer sediment and lower-gradient stream deposits of middle Duchesnean age suggest decreasing relief in the Uinta Mountains area. While relief was moderately low, abundant volcaniclastic rocks accumulated in the basin. Preliminary chemical data suggest that the Park City area of Utah was the source of most of the material. Increasing relief during late Duchesnean time probably resulted from renewed uplift of the Uinta Mountains and led to the accumulation of more conglomeratic material in the upper Duchesne River Formation.

TECTONIC IMPLICATIONS

Laramide chronology.—Deformation in the Uinta Mountains, as in many other ranges in the middle Rocky Mountains, involved mainly vertical uplift of the range relative to adjacent sedimentary basins (Eardley, 1962; Gilluly, 1963; Armstrong, 1968). Folding and basinward thickening of the Duchesne River Formation suggest that local arching uplift of typical Laramide style was important, rather than block-faulting of Basin and Range style. It therefore seems reasonable to relate uplift of the Uinta Mountains to deformation of other middle Rocky Mountain ranges.

The term "Laramide orogeny" has been applied to Late Cretaceous and Tertiary deformation within the Rocky Mountains, but its use has also been questioned. The early concept of the "Laramide revolution" was that it represented a sudden, worldwide diastrophic event marking the Mesozoic-Cenozoic boundary (Dana, 1895; Blackwelder, 1914; Stille, 1936). Additional work in the Rocky Mountain region led to the recognition of several orogenic phases, and the Laramide orogeny came to be regarded as a period of mountain building and erosion in the Rocky Mountain region that began in the Late Cretaceous and ended in the early Tertiary (Wilmarth, 1938, p. 1149). As evidence of deformation throughout the Mesozoic and Cenozoic increased, Spieker (1946,

1956) and Gilluly (1963) warned that the meaning of "Laramide" had become ambiguous. Indeed, Blackstone (1963) suggested that Laramide deformation is still continuing. Eardley (1962) also recognized that Laramide terminology was becoming confused. He suggested limits and subdivided the "Laramide" into early (Late Cretaceous), middle (Paleocene) and late (Eocene) stages (Eardley, 1962, p. 295). More recently, Damon and Mauger (1966) have advocated a return to usage of the "classical Laramide." On the basis of radiometric dates from Arizona and northern Sonora, they suggested that the late Eocene was a time of tectonic quiescence throughout western North America, and that the Laramide ended during middle Eocene time.

The overwhelming evidence in the middle Rocky Mountains is that deformation continued throughout the Eocene (Eardley, 1962; Blackstone, 1973). Uplift of the Uinta Mountains during Duchesnean time can reasonably be considered as Laramide in style, and the Duchesne River Formation thus constitutes a record of very latest Laramide uplift near the end of the Eocene.

Tectonic transition.—Coney (1972) reviewed extensive evidence that the style of tectonic events in North America changed drastically about 40 my ago. This change correlates well with changes in the North Atlantic sea-floor that probably reflect changes in rate and direction of spreading at the mid-Atlantic ridge (Coney, 1972). Important changes in plate interactions during the Cenozoic also occurred in the Pacific Ocean near western North America, but these probably are related to Oligocene and later changes in the Basin and Range province (Atwater, 1970; Scholz and others, 1971; Christiansen and others, 1972). Although rigidity of a body as immense as the North American plate is inconceivable, the synchroneity of tectonic transitions in the middle Rocky Mountains and in the North Atlantic is striking.

The Duchesne River Formation contains evidence of latest Laramide deformation in the Uinta Mountains area. It also contains altered volcaniclastic material that may represent intermediate volcanism in the Basin and Range province. Record of the tectonic transition about 40 my ago may thus be preserved, but with no evidence for an appreciable period of intervening quiescence.

REFERENCES CITED

Allen, J. R. L., 1970, Studies in fluviatile sedimentation; a comparison of fining-upward cyclothems, with special reference to coarse-member composition and interpretation: Jour. Sed. Petrology, v. 40, p. 298–323.

Andersen, D. W., and Picard, M. D., 1972, Stratigraphy of the Duchesne River Formation (Eocene-Oligocene?), northern Uinta Basin, northeastern Utah: Utah Geol. and Mineralog. Survey Bull. 97, 29 p.

Anderson, T. A., 1972, Paleogene nonmarine Gualanday Group, Neiva Basin, Colombia, and regional development of the Colombian Andes: Geol. Soc. America Bull., v. 83, p. 2423–2438.

Armstrong, R. L., 1968, Sevier orogenic belt in Nevada and Utah: *ibid.*, v. 79, p. 429–458.

———, 1970, Geochronology of Tertiary igneous rocks, eastern Basin and Range province, western Utah, eastern Nevada, and vicinity, U.S.A.: Geochim. et Cosmochim. Acta, v. 34, p. 203–232.

———, Ekren, E. B., McKee, E. H., and Noble, D. C., 1969, Space-time relations of Cenozoic silicic volcanism in the Great Basin of the western United States: Am. Jour. Sci., v. 267, p. 478–490.

Atwater, Tanya, 1970, Implications of plate tectonics for the Cenozoic tectonic evolution of western North America: Geol. Soc. America Bull., v. 81, p. 3513–3536.

Axelrod, D. I., 1966, The Eocene Copper Basin flora of northeastern Nevada: Univ. California Pub. Geol. Sci., v. 59, 83 p.

———, and Bailey, H. P., 1969, Paleotemperature analysis of Tertiary floras: Palaeogeography, Palaeoclimatology, Palaeoecology, v. 6, p. 163–195.

Belt, E. S., 1968, Carboniferous continental sedimentation, Atlantic provinces, Canada, *in* Klein, G. de V. (ed.), Late Paleozoic and Mesozoic continental sedimentation, northeastern North America: Geol. Soc. America Special Paper 106, p. 127–176.

Bissell, H. J., 1952, Stratigraphy and structure of northeast Strawberry Valley Quadrangle, Utah: Am. Assoc. Petroleum Geologists Bull., v. 36, p. 575–634.

Black, C. C., and Dawson, M. R.,, 1966, A review of late Eocene mammalian faunas from North America: Am. Jour. Sci., v. 264, p. 321–349.

Blackstone, D. L., Jr., 1963, Development of geologic structure in central Rocky Mountains, *in* Childs, O. E., and Beebe, B. W. (eds.), Backbone of the Americas: Am. Assoc. Petroleum Geologists Mem. 2, p. 160–179.

———, 1973, Late Cretaceous and Cenozoic history of the Laramie Basin region, southeast Wyoming: Geol. Soc. America Abstracts with Programs, v. 5, p. 466.

Blackwelder, Eliot, 1914, A summary of the orogenic epochs in the geologic history of North America: Jour. Geology, v. 22, p. 633–654.

Blatt, Harvey, 1967a, Original characteristics of clastic quartz grains: Jour. Sed. Petrology, v. 37, p. 401–424.

———, 1967b, Provenance determinations and recycling of sediments: *ibid.*, v. 37, p. 1031–1044.
———, and Christie, J. M., 1963, Undulatory extinction in quartz of igneous and metamorphic rocks and its significance in provenance studies of sedimentary rocks: *ibid.*, v. 33, p. 559–579.
Bradley, W. H., 1929, The varves and climate of the Green River epoch: U.S. Geol. Survey Prof. Paper 158-E, p. 87–110.
———, 1931, Origin and microfossils of the oil shale of the Green River Formation of Colorado and Utah: *ibid.*, 168, 58 p.
———, 1948, Limnology and the Eocene lakes of the Rocky Mountain region: Geol. Soc. America Bull., v. 59, p. 635–648.
Bromfield, C. S., and Crittenden, M. D., Jr., 1971, Geologic map of the Park City East Quadrangle, Summit and Wasatch Counties, Utah: U.S. Geol. Survey Map GQ-852.
Bull, W. B., 1964, Alluvial fans and near-surface subsidence in western Fresno County, California: *ibid.*, Prof. Paper 437-A, 71 p.
Burke, J. J., 1934, New Duchesne River rodents and a preliminary survey of the Adjidaumidae: Carnegie Mus. Annals, v. 23, p. 391–398.
Cashion, W. B., 1967, Geology and fuel resources of the Green River Formation, southeastern Uinta Basin, Utah and Colorado: U.S. Geol. Survey Prof. Paper 548, 48 p.
Chamberlin, T. C., 1909, Diastrophism as the ultimate basis of correlation: Jour. Geology, v. 17, p. 685–693.
Chaney, R. W., 1940, Tertiary forests and continental history: Geol. Soc. America Bull., v. 51, p. 469–488.
Christiansen, R. L., and Lipman, P. W., 1972, Cenozoic volcanism and plate-tectonic evolution of the western United States. II. Late Cenozoic: Royal Soc. London Philos. Trans., ser. A, v. 271, p. 249–284.
Clark, John, 1932, A new turtle from the Duchesne Oligocene of the Uinta Basin, northeastern Utah: Carnegie Mus. Annals, v. 21, p. 131–160.
———, 1962, Field interpretations of red beds: Geol. Soc. America Bull., v. 73, p. 423–428.
Coney, P. J., 1972, Cordilleran tectonics and North America plate motion: Am. Jour. Sci., v. 272, p. 603–628.
Crowder, D. F., McKee, E. H., Ross, D. C., and Krauskopf, K. B., 1973, Granitic rocks of the White Mountains area, California-Nevada: age and regional significance: Geol. Soc. America Bull., v. 84, p. 285–296.
Damon, P. E., 1970, Correlation and chronology of ore deposits and volcanic rocks: U.S. Atomic Energy Comm. (Research Div.), Ann. Progress Report C00-689-130, 192 p.
———, and Mauger, R. L., 1966, Epeirogeny-orogeny viewed from the Basin and Range province: Am. Inst. Min., Metall. and Petroleum Engineers Trans., v. 235, p. 99–112.
Dana, J. D., 1895, Manual of geology (4th ed.): Americana Book Co., 1088 p.
Dane, C. H., 1954, Stratigraphic and facies relationships of upper part of Green River Formation and lower part of Uinta Formation in Duchesne, Uinta and Wasatch Counties, Utah: Am. Assoc. Petroleum Geologists Bull., v. 38, p. 405–425.
Dunham, R. J., 1969, Vadose pisolites in Capitan reef (Permian), New Mexico and Texas, *in* Friedman, G. M. (ed.), Depositional environments in carbonate rocks: Soc. Econ. Paleontologists and Mineralogists Special Pub. 14, p. 182–191.
Eardley, A. J., 1952, Wasatch hinterland: Utah Geol. Soc. 8th Ann. Field Conf. Guidebook, p. 52–60.
———, 1962, Structural geology of North America (2nd ed.): Harper and Row, N.Y., 743 p.
———, 1963, Relation of uplifts to thrusts in Rocky Mountains, *in* Childs, O. E., and Beebe, B. W. (eds.), Backbone of the Americas: Am. Assoc. Petroleum Geologists Mem. 2, p. 209–219.
———, 1969, Early Tertiary volcanism near west end of Uinta Mountains: Intermountain Assoc. Geologists 16th Ann. Field Conf. Guidebook, p. 219–220.
Evernden, J. F., Savage, D. E., Curtis, G. H., and James, G. T., 1964, Potassium-argon dates and the Cenozoic mammalian chronology of North America: Am. Jour. Sci., v. 262, p. 145–198.
Fisher, R. A., 1953, Dispersion on a sphere: Roy. Soc. London Proc., ser. A, v. 217, p. 295–305.
Fisher, R. V., 1971, Features of coarse-grained, high-concentration fluids and their deposits: Jour. Sed. Petrology, v. 41, p. 916–927.
Folk, R. L., 1968, Petrology of sedimentary rocks: Hemphill's, Austin, 170 p.
Gazin, C. L., 1959, Paleontological exploration and dating of the early Tertiary deposits in basins adjacent to the Uinta Mountains: Intermountain Assoc. Petroleum Geologists 10th Ann. Field Conf. Guidebook, p. 131–138.
Gilluly, James, 1949, Distribution of mountain building in geologic time: Geol. Soc. America Bull., v. 60, p. 561–590.
———, 1963. The tectonic evolution of the western United States: Geol. Soc. London Quart. Jour. v. 119, p. 133–174.
———, 1973. Steady plate motion and episodic orogeny and magmatism: Geol. Soc. America Bull., v. 84, p. 499–514.
Goddard, E. N. (chm.), 1948, Rock color chart: National Research Council Washington, 6 p.
Goudie, Andrew, 1972, The chemistry of world calcrete deposits: Jour. Geology, v. 80, p. 449–463.
Harms, J. C., and Fahnestock, R. K., 1965, Stratification, bed forms, and flow phenomena (with an example from the Rio Grande), *in* Middleton, G. V. (ed.), Primary sedimentary structures and their hydrodynamic interpretation: Soc. Econ. Paleontologists and Mineralogists Special Pub. 12, p. 84–115.
High, L. R., Jr., and Picard, M. D., 1968, Dendritic surge marks (*"Dendrophycus"*) along modern stream banks: Wyoming Univ. Contr. Geology, v. 7, p. 1–6.
Hopkins, W. S., Jr., Rutter, N. W., and Rouse, G. E., 1972, Geology, paleoecology, and palynology of some Oligocene rocks in the Rocky Mountain trench of British Columbia: Can. Jour. Earth Sci., v. 9, p. 460–470.
Huddle, J. W., Mapel, W. J., and McCann, F. T., 1951, Geology of the Moon Lake area, Duchesne County, Utah: U.S. Geol. Survey Oil and Gas Inv. Map OM-115.

HUDDLE, J. W., AND MCCANN, F. T., 1947, Pre-Tertiary geology of the Duchesne River Area, Duchesne and Wasatch Counties, Utah: *ibid.,* Prelim. Map 75.
JACK, R. N., AND CARMICHAEL, I. S. E., 1969, The chemical "fingerprinting of acid volcanic rocks: Calif. Div. Mines and Geology Special Rept. 100, p. 17–32.
JAMES, N. P., 1972, Holocene and Pleistocene calcareous crust (caliche) profiles: criteria for subaerial exposure: Jour. Sediment. Petrology, v. 42, p. 817–836.
KARCZ, IAAKOV, 1972, Sedimentary structures formed by flash floods in southern Israel: Sed. Geology, v. 7, p. 161–182.
KAY, J. L., 1934, Tertiary formations of the Uinta Basin, Utah: Carnegie Mus. Annals, v. 23, p. 357–371.
KINNEY, D. M., 1955, Geology of the Uinta River-Brush Creek area, Duchesne and Uintah Counties, Utah: U.S. Geol. Survey Bull. 1007, 180 p.
KISTLER, R. W., EVERNDEN, J. F., AND SHAW, H. R., 1971, Sierra Nevada plutonic cycle: part origin of composite granitic batholiths: Geol. Soc. America Bull., v. 82, p. 853–868.
KRYNINE, P. D., 1935, Arkose deposits in the humid tropics: Am. Jour. Sci., v. 29, p. 353–363.
———, 1946, Microscopic morphology of quartz types: Pan-American Cong. Min. and Geol. Engineers, Annals of 2nd Comm., p. 36–49.
———, 1949, The origin of red beds: New York Acad. Sci. Trans., ser. 2, v. 2, p. 60–68.
LANPHERE, M. A., AND REED, B. L., 1973, Timing of Mesozoic and Cenozoic plutonic events in circum-Pacific North America: Geol. Soc. America Abstracts with Programs, v. 5, p. 70.
LIPMAN, P. W., PROSTKA, H. J., AND CHRISTIANSEN, R. L., 1972, Cenozoic volcanism and plate-tectonic evolution of the western United States. I. Early and middle Cenozoic: Royal Soc. London Philos. Trans., ser. A, v. 271, p. 217–248.
MCCORMICK, C. D., AND PICARD, M. D., 1969, Petrology of Gartra Formation (Triassic), Uinta Mountain area, Utah and Colorado: Jour. Sed. Petrology, v. 39. p. 1484–1508.
MCKEE, E. D., AND WEIR, G. W., 1953, Terminology for stratification and cross-stratification in sedimentary rocks: Geol. Soc. America Bull., v. 64, p. 381–390.
MCKEE, E. H., 1971, Tertiary igneous chronology of the Great Basin of western United States—implications for tectonic models: *ibid.,* v. 82, p. 3497–3502.
MONGER, J. W. H., SOUTHER, J. G., AND GABRIELSE, HUBERT, 1972, Evolution of the Canadian Cordillera: a plate-tectonic model: Am. Jour. Sci., v. 272, p. 577–602.
MULLENS, T. E., 1971, Reconnaissance study of the Wasatch, Evanston, and Echo Canyon Formations in part of northern Utah: U.S. Geol. Survey Bull. 1311-D, p. 1–31.
NELSON, M .E., 1971, Stratigraphy and paleontology of Norwood Tuff and Fowkes Formation, northeastern Utah and southwestern Wyoming (Ph.D. thesis): Univ. Utah, Salt Lake City, 169 p.
ORE, H. T., 1964, Some criteria for recognition of braided stream deposits: Wyoming University Contr. Geology, v. 3, p. 1–14.
———, 1965, Characteristic deposits of rapidly aggrading streams: Wyoming Geol. Assoc. 19th Ann. Field Conf. Guidebook, p. 195–201.
OSBORN, H. F., 1895, Fossil mammals of the Uinta Basin: Am. Mus. Nat. History Bull., v. 7, p. 72–105.
PELLETIER, B. R., 1958, Pocono paleocurrents in Pennsylvania and Maryland: Geol. Soc. America Bull., v. 69, p. 1033–1064.
PENNY, J. S., 1969, Late Cretaceous and early Tertiary palynology, *in* Tschudy, R. H., and Scott, R. A. (eds.), Aspects of palynology: Wiley, N.Y., p. 331–376.
PETERSON, O. A., 1932, New species from the Oligocene of the Uinta: Carnegie Mus. Annals, v. 21, p. 61–78.
PICARD, M. D., 1955, Subsurface stratigraphy and lithology of Green River Formation in Uinta Basin, Utah: Am. Assoc. Petroleum Geologists Bull., v. 39, p. 75–102.
———, 1965, Iron oxides and fine-grained rocks of Red Peak and Crow Mountain Sandstone Members, Chugwater (Triassic) Formation, Wyoming: Jour. Sed. Petrology, v. 35, p. 465–479.
———, 1971, Classification of fine-grained sedimentary rocks: Jour. Sed. Petrology, v. 41, p. 179–195.
———, AND HIGH, L. R., JR., 1969, Some sedimentary structures resulting from flash floods: Utah Geol. and Mineralog. Survey Bull. 82, p. 175–190.
———, 1972, Paleoenvironmental reconstructions in an area of rapid facies change, Parachute Creek Member of Green River Formation (Eocene), Uinta Basin, Utah: Geol. Soc. America Bull., v. 83, p. 2689–2708.
———, 1973, Sedimentary structures of ephemeral streams: Elsevier, Amsterdam, 223 p.
POTTER, P. E., AND PRYOR, W. A., 1961, Dispersal centers of Paleozoic and later clastics of the upper Mississippi Valley and adjacent areas: Geol. Soc. America Bull., v. 72, p. 1195–1250.
ROEDER, D. H., 1967, Rocky Mountains: der geologische Aufbau des kanadischen Felsengebirges: Gebrueder Borntraeger, Berlin, 318 p.
RUSSELL, L. S., 1954, The Eocene-Oligocene transition as a time of major orogeny in western North America: Royal Soc. Canada Trans., v. 48, p. 65–69.
SATO, YOSHIAKI, AND DENSON, N. M., 1967, Volcanism and tectonism as reflected by the distribution of nonopaque heavy minerals in some Tertiary rocks of Wyoming and adjacent states: U.S. Geol. Survey Prof. Paper 575-C, p. 42–54.
SCHOLZ, C. H., BARAZANGI, MUAWIA, AND SBAR, M. L., 1971, Late Cenozoic evolution of the Great Basin, western United States, as an ensialic interarc basin: Geol. Soc. America Bull., v. 82, p. 2979–2990.
SCHUMM, S. A., AND LICHTY, R. W., 1963, Channel widening and flood-plain construction along Cimarron River in southwestern Kansas: U.S. Geol. Survey Prof. Paper 352-D, p. 71–88.
SIEVER, RAYMOND, 1949, Trivoli Sandstone of Williamson County, Illinois: Jour. Geology, v. 57, p. 614–618.
SIMPSON, G. G., 1933, Glossary and correlation charts of North American Tertiary mammal-bearing formations: Am. Mus. Nat. History Bull., v. 67, p. 79–121.
———, 1946, The Duchesnean fauna and the Eocene-Oligocene boundary: Am. Jour. Sci., v. 244, p. 52–57.

SMITH, N. D., 1970, The braided stream depositional environment: comparison of the Platte River with some Silurian clastic rocks, north-central Appalachians: Geol. Soc. America Bull., v. 81, p. 2993–3014.
———, 1971, Transverse bars and braiding in the lower Platte River, Nebraska: *ibid.*, v. 82, p. 3407–3420.
SORBY, H. C., 1877, The application of the microscope to geology: Monthly Micros. Jour., v. 17, p. 113–136.
SPIEKER, E. M., 1946, Late Mesozoic and early Cenozoic history of central Utah: U.S. Geol. Survey Prof. Paper 205-D, p. 117–161.
———, 1949, Sedimentary facies and associated diastrophism in the upper Cretaceous of central and eastern Utah, *in* Longwell, C. R. (ed.), Sedimentary facies in geologic history: Geol. Soc. America Mem. 39, p. 55–81.
———, 1956, Mountain-building chronology and nature of geologic time scale: Am. Assoc. Petroleum Geologists Bull., v. 40, p. 1769–1815.
STAGNER, W. L., 1941, The paleogeography of the eastern part of the Uinta Basin during Uinta B (Eocene) time: Carnegie Mus. Annals, v. 28, p. 273–308.
STILLE, HANS, 1936, The present tectonic state of the earth: Am. Assoc. Petroleum Geologists Bull., v. 20, p. 849–880.
STOKES, W. L., AND MADSEN, J. H., JR., 1961, Geologic map of Utah, northeast quarter: Utah Geol. and Mineralog. Survey.
TANNER, W. F., 1959, The importance of modes in cross-bedding data: Jour. Sed. Petrology, v. 29, p. 221–226.
TODD, T. W., AND FOLK, R. L., 1957, Basal Claiborne of Texas, record of Appalachian tectonism during Eocene: Am. Assoc. Petroleum Geologists Bull., v. 41, p. 2545–2566.
TWETO, OGDEN, 1973, Summary of Laramide orogeny in the southern Rocky Mountains: Geol. Soc. America Abstracts with Programs, v. 5, p. 521.
VAN DE GRAAFF, F. R., 1972, Fluvial-deltaic facies of the Castlegate Sandstone (Cretaceous), east-central Utah: Jour. Sed. Petrology, v. 42, p. 558–571.
VAN HOUTEN, F. B., 1948, Origin of red-banded early Cenozoic deposits in Rocky Mountain region: Am. Assoc. Petroleum Geologists Bull., v. 32, p. 2083–2126.
———, 1968, Iron oxides in red beds: Geol. Soc. America Bull., v. 79, p. 399–416.
———, 1969, Molasse facies: records of worldwide crustal stresses: Science, v. 166, p. 1506–1508.
WALKER, T. R., 1967, Formation of red beds in modern and ancient deserts: Geol. Soc. America Bull., v. 78, p. 353–368.
WARNER, M. M., 1963, Sedimentation of the Duchesne River Formation, Uinta Basin, Utah (Ph.D. thesis): State Univ. Iowa, Iowa City, 339 p.
WENTWORTH, C. K., 1922, A scale of grade and class terms for clastic sediments: Jour. Geology, v. 30, p. 377–392.
WILLIAMS, G. E., 1971, Flood deposits of the sand-bed ephemeral streams of central Australia: Sedimentology, v. 17, p. 1–40.
WILLIAMS, P. F., AND RUST, B. R., 1969, The sedimentology of a braided river: Jour. Sed. Petrology, v. 39, p. 649–679.
WILMARTH, M. G., 1938, Lexicon of geologic names of the United States (including Alaska): U.S. Geol. Survey Bull. 896, 2396 p.
WOLMAN, M. G., AND LEOPOLD, L. B., 1957, River flood plains: some observations on their formation: *ibid.*, Prof. Paper 282-C, p. 87–109.
WOOD, H. E. (chm.), 1941, Nomenclature and correlation of the North American continental Tertiary: Geol. Soc. America Bull., v. 52, p. 1–48.

ORIGIN OF LATE CENOZOIC BASINS IN SOUTHERN CALIFORNIA[1]

JOHN C. CROWELL
University of California, Santa Barbara, California

ABSTRACT

Several sedimentary basins in southern California, within and south of the Transverse Ranges, display a history suggestive of a pull-apart or a tipped-wedge origin. Beginning in the Miocene, these basins apparently originated along the soft and splintered margins of the Pacific and Americas plates. Basin walls were formed by both transform faults and by crustal stretching and dip-slip faulting. Basin floors developed on stretched and attenuated marginal rocks, and some floors grew as a complex of volcanic rocks and sediments. As basins enlarged, high-standing blocks are pictured as stretching and separating laterally from terranes that were originally adjacent. Older rocks exposed around basin margins therefore cannot always be extrapolated to depth beneath the basins.

Support for such speculative models comes from accumulating understanding of the modern Salton trough. This narrow graben is now being pulled apart obliquely, with faults of the San Andreas system serving as transforms. With widening, the walls sag and stretch, and margins are inundated by sedimentation that goes on hand in hand with deformation and with volcanism within the basin. The Los Angeles basin apparently started to form as an irregular pull-apart hole in the early midMiocene, and basin-floor volcanism accompanied subsequent voluminous sedimentation. Great thicknesses of Miocene beds and volcanic rocks in the western Santa Monica Mountains probably constitute the displaced northern part of the Los Angeles basin, and were laid down adjacent to high ground from which sediments and large detachment slabs were carried into the growing depression. Basins and intervening banks and ridges in the California Borderland may have originated in a broad right-slip regime where strike-slip faults converge and diverge in plan view to slice the terrane into wedge-shaped segments. Displacement along converging and diverging strike-slip faults bounding such wedges results in shortening and elevation, or in stretching and subsidence, respectively.

INTRODUCTION

As knowledge of the geologic history of late Cenozoic sedimentary basins in southern California has accumulated during the past half century, several genetic models have been proposed to explain their origin. At first the great thicknesses of sediments found in deep basins were credited to vertical tectonics only, in which subcrustal processes brought about subsidence of basins concomitantly with the elevation of adjacent highlands. Erosional debris from the highlands was pictured as washing across the basin margins and directly into the contiguous basins. Only slowly did the concept gather acceptance that major strike slip was significant, and was superimposed upon this pattern of vertical tectonics. For example, Eaton (1926) and Ferguson and Willis (1924) noted that strike slip was primarily responsible for the folds along the Newport-Inglewood zone in the Los Angeles basin, and Vickery (1925) interpreted the pattern of faults and folds east of the San Francisco Bay area in terms of strike slip. In the early fifties, rock sequences offset by many tens of kilometers on the San Andreas fault were recognized by Hill and Dibblee (1953), and strike slip of conglomerates from their source areas was shown to be about 30 kilometers on the San Gabriel fault by Crowell (1952). During the two decades since then many workers have demonstrated great strike-slip components on several California faults, including those associated with major basins. As the data have come in, however, it has grown increasingly clear that other faults have essentially no component of strike slip, and that vertical tectonic movements involving steep flexures at basin margins, normal-slip faults, thrust-slip faults and detachment faults are also common. The record shows as well that deformation has been nearly continuous in southern California as a whole since early in the mid-Tertiary, and that this deformation has not always followed the same pattern.

California at present is being deformed as part of a broad transform zone, the sliced and segmented boundary between the Pacific plate and the Americas plate (Atwater, 1970). The origin of several modern basins, such as those within the Salton Trough and the Gulf of California, is related to their position at or near this plate boundary. Similar origins can be recognized for some more ancient basins. The in-

[1] Studies of the tectonics of southern California have recently been supported by the University of California, Santa Barbara, and the U.S. National Science Foundation, Grant GA 30901. I am also grateful to many students and colleagues for numerous discussions and comments and especially to Arne Junger for suggesting a diagram similar to that of Figure 9.

terpretation of the ocean-floor record including magnetic anomalies west of California reveals a history of major plate interaction across the region back into pre-Tertiary times, but detailed explanations of this interaction before mid-Tertiary are still inconclusive. According to Atwater (1970, fig. 17), this interaction since the mid-Tertiary has included long episodes of strike slip, and right slip has predominated in the vicinity of southern California for about 25 million years. The Americas plate has moved about 1500 km relative to the Pacific plate during this interval.

Only about 300 km of post-Oligocene right slip on the San Andreas fault is now recognized in southern California, however, leaving a difference of 1200 km or so in order to match interpretations of the land record with those from magnetic anomalies on the sea floor. This difference can most easily be accounted for by considering that other faults on land, such as those in the Great Basin and splays of the San Andreas in southern California and the adjacent borderland, took up the difference. In particular, a major right-slip fault may have coincided with the western edge of the continent where it joins the deep Pacific floor at the base of the Patton Escarpment (fig. 1). Despite the fact that right slip of the order needed to match the sea-floor interpretation is still not recognized on faults in southern California, in the borderland, or in the Mojave and Colorado Deserts, the idea that these regions are part of a broad transform zone attracts investigation. In this paper we will therefore accept as a premise the concept that southern California and its borderland have been very mobile laterally during the late Cenozoic as part of a broad and complicated transform zone, but without committing ourselves to the magnitude of total right slip across the soft and broad boundary between plates. We will search for models of basin origin and consider ways to recognize or test them.

BASIN GEOMETRY IN A TRANSFORM REGIME

Several types of basins can be envisaged theoretically along a transform boundary between major tectonic plates, and especially if the boundary is a complex zone of branching faults. Some terranes may be uplifted to make source areas, and others depressed to form basins (Crowell, 1974). If the strike-slip zone is distant from land and cuts the ocean floor only, high-standing blocks along the oceanic transform may not rise into the zone of erosion, so that the associated depressions receive little sediment from them. Near continents, however, and especially along continental transforms, large volumes of sediment may be washed directly to nearby basins. Southern California during the late Cenozoic seems to fit the latter circumstances so that the following geometrical discussion starts with the assumption that the transform zone cuts continental crust. Moreover, in using the term 'transform,' emphasis is placed upon the plate-tectonic concept of major crustal plates moving laterally along a strike-slip zone and upon the relation of the strike-slip zone to spreading centers and subduction zones in order to account for lateral displacements of hundreds of kilometers (Wilson, 1965; Vine and Wilson, 1965). In terms of the geometry of rock units, however, such transform faults are major strike-slip fault zones with nearly vertical fault surfaces and long extent. In this sense, they are synonymous with "wrench faults" or "transcurrent faults." In the examples figured below, right slip rather than left slip is illustrated inasmuch as the San Andreas is a right-slip system.

STRIKE SLIP ALONG A STRAIGHT FAULT

If continental terrane with subdued or near-flat topography is cut by a long and vertical strike-slip fault, no differential elevation or subsidence will result from the deformation (fig. 2). The blocks merely slide by each other. With such a simple system there is little likelihood of fault-branching, and fault zones are straight, narrow, and relatively unbraided. Such a situation seems to prevail today along the straight stretch of the San Andreas between the central Temblor Range and the Gabilan Range (fig. 1, TR and GR). This stretch includes Parkfield and the part of the fault zone exhibiting creep and frequent small earthquakes (Brown and others, 1967).

STRIKE SLIP ALONG A FAULT WITH A GENTLE DOUBLE BEND

Displacement of adjacent blocks along a single dominating strike-slip fault with a gentle double curve displays one of two different geometries. On the one hand, the bend may be in the direction to free or release the blocks as they glide by each other, or on the other, to lock or restrain them. With right slip, for example, if the fault trace curves to the right (clockwise) in looking along the fault toward a displaced feature, the bend will be a *releasing double bend,* and if to the left (counterclockwise), a *restraining double bend* (fig. 3). With left slip, in contrast, a releasing bend curves to the left, and a restraining bend to

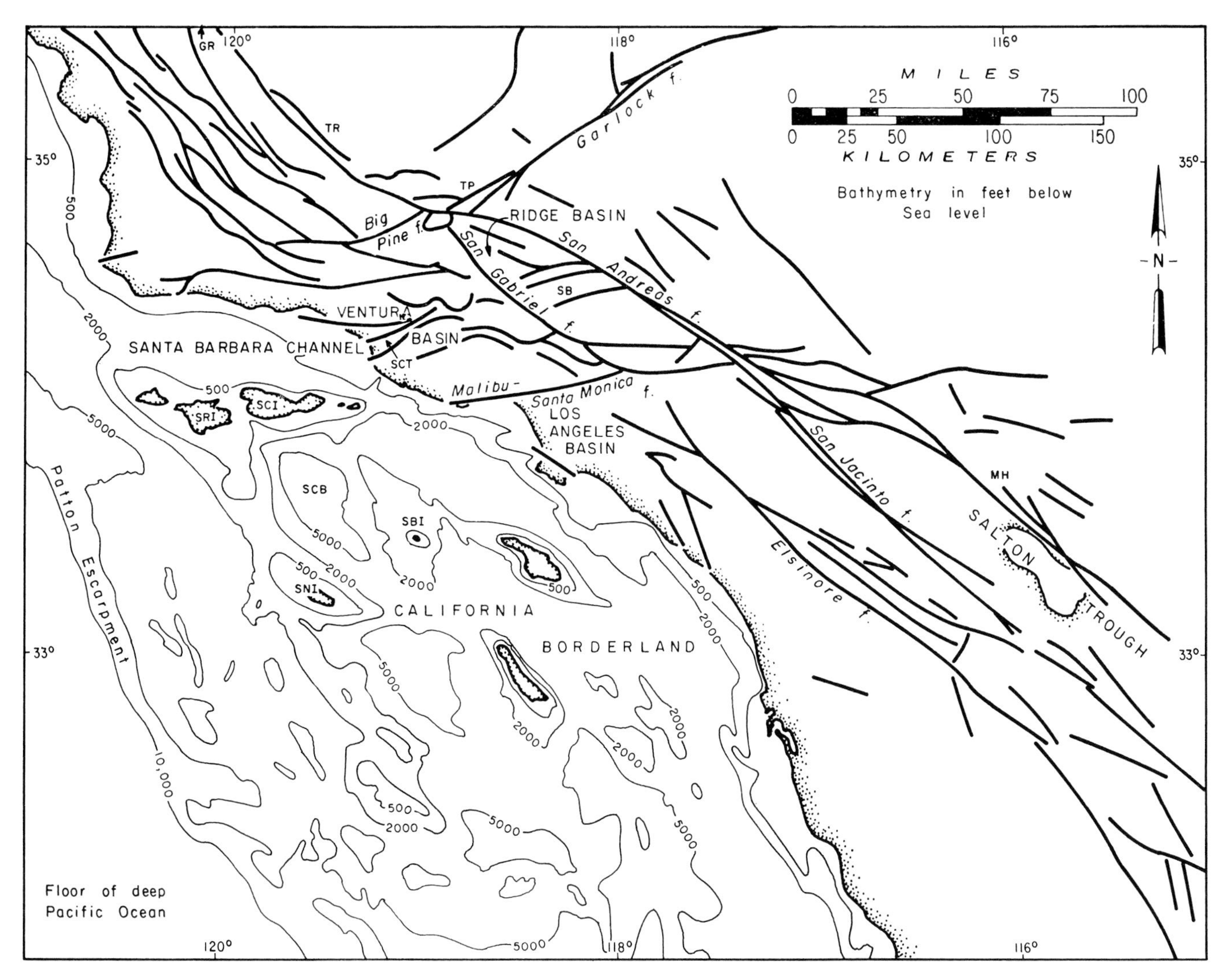

FIG. 1.—Map of southern California and California Borderland showing major onshore faults. Abbreviations: TP, Tejon Pass; SB, Soledad basin; SCT, Santa Clara trough; SRI, Santa Rosa Island; SCI, Santa Cruz Island; SCB, Santa Cruz basin; SBI, Santa Barbara Island; SNI, San Nicolas Island; GR, Gabilan Range; TR, Temblor Range; MH, Mecca Hills.

the right [Note mnemonically that if the two words are the same, a releasing bend results; if different, a restraining bend]. In this discussion the strike-slip fault is visualized as long and extensive, and the bends are relatively local departures in trend. If so, the bends are double in that one curvature takes the fault away from the regional trend and another brings it back into alignment. The direction of shift or strike slip of one block with respect to the other is defined by the strike of the fault on an extensive regional basis. If the fault is considered as a transform fault, this is the direction of relative motion between the major lithospheric plates.

If we consider the blocks as less rigid; that is, as relatively soft and deformable, movement along a fault with a double bend will cause shortening or crowding of the crustal rocks within concavities, and stretching at convexities (fig. 4), in the edges of adjacent blocks. Inasmuch as the crowding and stretching is relieved most easily at the terrane surface, shortening results in elevation of the ground surface, and stretching in subsidence. The maximum effect of these processes occurs near the strike-slip fault and at the point of maximum curvature. As the displacement continues, the centers of elevation or subsidence may move through time. Or on the other hand, one block can remain fixed in shape so that the same terrane continues to be elevated or depressed for long periods as the other block slides by and bends around it. The possibilities range through a continuum from one block remaining fixed in shape as the other participates in all of the bending, to both sharing equally in the bending. And on a single fault this style may change through time.

Ridge Basin, sited adjacent to the Pliocene major strand of the San Andreas fault in the central Transverse Ranges, apparently formed in such a setting (Crowell, 1954; 1962; in press). About 12,000 m (40,000 ft) of both

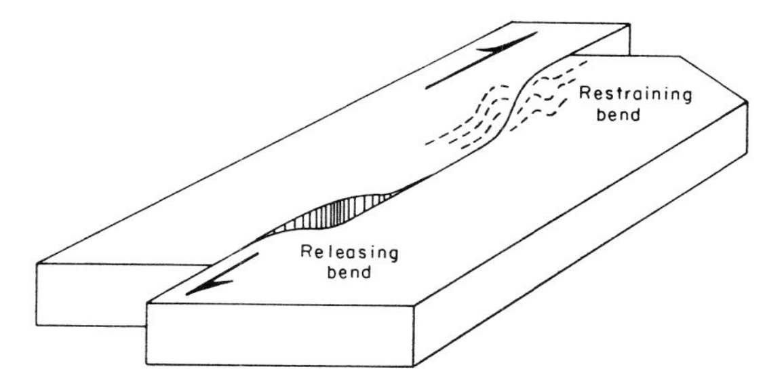

FIG. 3.—Right slip on fault with marked double bends results in pull-aparts at releasing bends and deformation and uplift at restraining bends.

marine and nonmarine sediments accumulated within a narrow basin at the stretched and depressed margin of the Americas Plate as it moved alongside a restricted source area (fig. 4). The lateral movement of the depositional site allowed the accumulation of the vast stratigraphic thickness of sediments by a gradual northwestward overlapping of older strata by younger, but without breaks or unconformities along the axis of the trough as the depocenter migrated. Older strata were carried away laterally as younger ones were deposited opposite the restricted source area. In this case, the uplifted source remained fixed in shape.

STRIKE SLIP ALONG A FAULT WITH A SHARP BEND

A long and straight strike-slip fault with a sharp double bend that sidesteps the fault trend for a relatively short distance can exhibit again two basically different situations. A sharp restraining double bend results in overlap and elevation at the bends (fig. 5), and a releasing double bend results in a pull-apart and subsidence (fig. 6). At restraining bends, geologic structures display oblique shortening, and at releasing bends, stretching or extension.

Restraining double bends bring about overlap

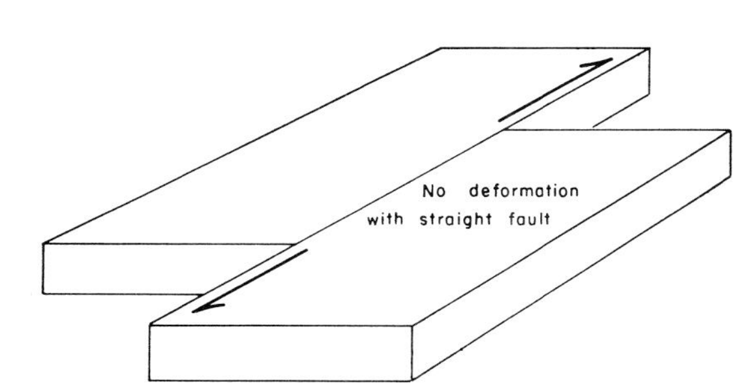

FIG. 2.—Strike slip on straight fault results in no deformation of crustal blocks.

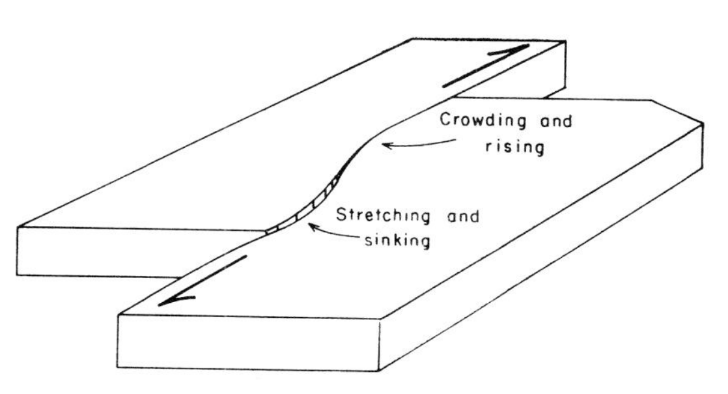

FIG. 4.—Right slip on gently curved fault results in crowding and uplift within convexities of deformable lithospheric plates and stretching and sagging within concavities.

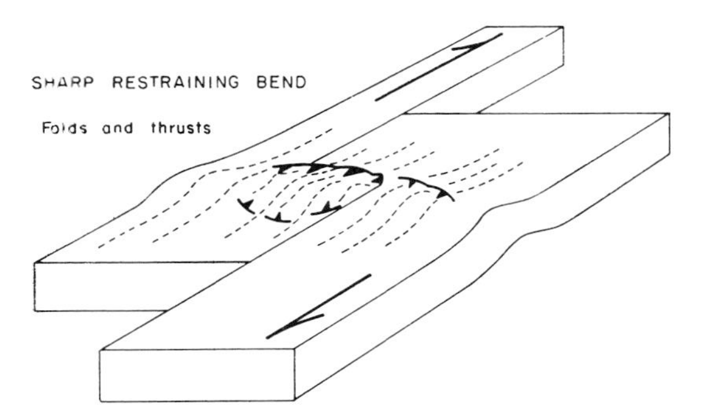

FIG. 5.—Severe deformation at sharp restraining bend results in folds and thrust faults.

of one block over the other, so that, under gravity, the edge of one block is depressed to form a shallow site for sedimentation adjacent to an elevated source area (Crowell, 1974, fig. 6). An example may be the depositional site of the wedge of continental sediments now accumulating on the San Bernardino Plain southwest and adjacent to the San Andreas, which at this place marks the boundary of the San Bernardino Mountains.

In contrast, releasing double bends form deep and narrow sedimentary basins at the pull-apart. These range in scale from small sag ponds within a restricted strike-slip zone floored by local country rock to true rhomobchasms (Carey, 1958), such as those in the Gulf of California, floored by new lava above a spreading center or diapiric volcanic complex. Large and complex pull-aparts are treated more fully below.

BRANCHING AND BRAIDED STRIKE-SLIP ZONES

Strike-slip systems, such as those in California, consists of long and straight master faults with lesser branching faults or splays leading off from them. In addition, many of the major faults within the system are fault zones several kilometers wide containing fault slices and folds (Saul, 1967). Major splays lead away from places where there is a slight change in strike of the master fault. In fact, where the Sunol-Calaveras system extends northward and away from the San Andreas near Hollister, the San Andreas northwest of the juncture is not now as active seismically and a crustal wedge may be forming (Burford and Savage, 1972). The inference is that the local northwestern part of the San Andreas is in the process of being supplanted as the major transform break by the Sunol-Calaveras in response to adjustments between the Pacific and Americas plates.

The splay may either continue on the new trend, or rejoin the original master fault to make a wedge or slice (fig. 7). Wedges range in size up to more than a hundred kilometers in length, such as the one between the San Gabriel and San Andreas faults within the Transverse Ranges. If huge wedges of this type are tipped longitudinally as displacement on the transform system continues, one end may subside to form a depositional site, and the other may elevate to provide a source area. Some further generalizations concerning broad zones of slices and wedges are discussed below.

Braided fault zones consisting of anastomosing faults and obliquely trending folds, such as the Sunol-Calaveras fault zone (Saul, 1967), may display local complexities that have only obscure kinematic relations to strike-slip origin. Individual faults exhibit dip slip and dip separations. Some originated as wedges within the zone were squeezed upward and others, during sagging of wedges downward. Clay-model experiments, such as those illustrated by Wilcox and others (1973), show these complicated patterns very well. Rocks on a regional scale

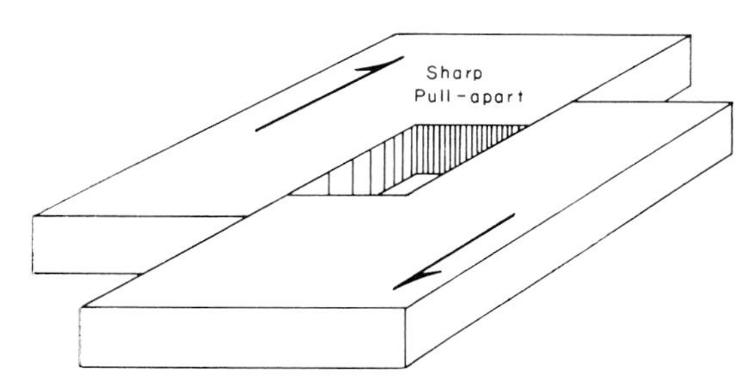

FIG. 6.—A sharp pull-apart on a right-slip fault.

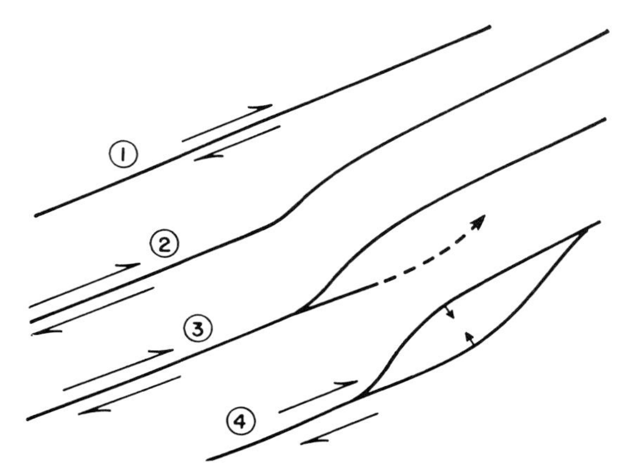

FIG. 7.—Diagrammatic map showing progressive development of fault splays and wedges on a right-slip fault. Straight fault (1) gradually develops a bend through time (2 and 3) and eventually forms a fault wedge (4).

are also weak, similar to clay, so that when a crustal block is pressed upward beside a nearly vertical fault, the hanging well sags outward and downward; the field geologist will correctly map it as a thrust fault. Such an origin for local complexities has been described, especially in New Zealand where horsts and grabens are associated with strike-slip zones (Kingma 1958; Lensen, 1958). Similar braided zones are recognizable in Maritime Canada (Belt, 1968), northwestern Scotland (Kennedy, 1946; Dearnley, 1962), along the Dead Sea fault zone of the Levant (Quennell, 1958; Freund, 1965), and along the north coast of South America in Colombia, Venezuela, and Trinidad (Malfait and Dinkelman, 1972; Wilcox and others, 1973; Crowell, 1974).

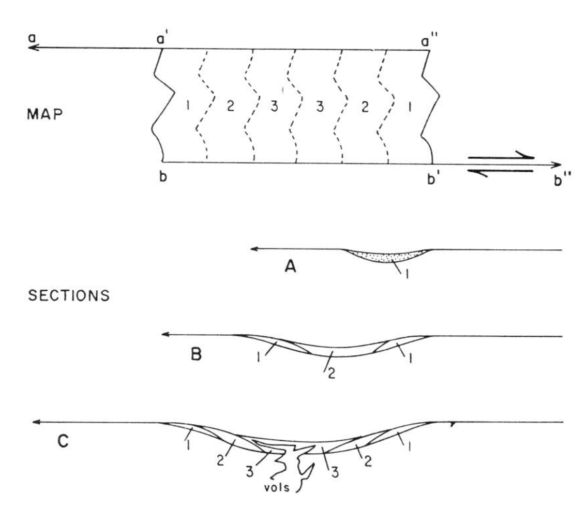

FIG. 8.—Sketch map and sections of pull-apart basin; see text for discussion.

PULL-APART BASINS

At the present time the Gulf of California and Salton Trough are widening and lengthening as continental terrane to the west moves obliquely away from the mainland part of North America (Wegener, 1924; Carey, 1958, Fig. 42; Hamilton, 1961). This process is envisaged as the result of sea-floor spreading along the segmented and offset parts of the East Pacific Rise as it enters the Gulf of California at its southern end (Larson and others, 1968; Moore and Buffington, 1968; Larson, 1972; Larson and others, 1972). Geologic studies and geophysical surveys at the head of the Gulf and within the Salton Trough suggest that the Salton Trough lies above a series of spreading centers or diapiric masses with volcanic rocks at depth, and continental rocks, if any, are attenuated and fragmented near the center of the structure (Elders and others, 1972; Sumner, 1972; Henyey and Bischoff, 1973; Karig and Jensky, 1973). It is therefore probably a true rhombochasm, or chain of them, that has opened while abundant sediment has flooded into the widening hole (Crowell, 1974). Details of the structure along the northeastern border of the Salton Trough, for example, fit reasonably well the idea that they are the consequence of right slip at a steep basin margin with high-standing continental rocks on the northeast and deep quasi-oceanic crust on the southwest. In the Mecca Hills (fig. 1, MH) the deformed sedimentary section of Neogene age thickens rapidly toward the trough and the San Andreas zone there consists of braided faults of several ages, with complex folds and thrusts arranged between them (Dibblee, 1954; Hays, 1957; Crowell, 1962).

According to a pull-apart model, prisms of the oldest sediments lying upon the original basin floor should occur around the margins only, but drilling and geophysical studies are not yet complete enough for confirmation. In the paragraphs below I will consider first some simple theoretical models and then geological processes and circumstances that complicate and modify them. The latter include especially the relative rates of sedimentation in comparison with rates of subsidence and deformation, changes in orientations and rates of deformation through time, and the strength or weakness of the crust and the magnitudes and arrangements of heterogeneities within it.

The simplest pull-apart model consists of side-stepped parallel transforms (a-a′-a″ and b-b′-b″ in fig. 8) sited above a volcanic center. The transform margins (a-a″ and b-b″) of the basin appear straight and parallel in map view, but the pull-apart margins (a′-b and a″-b′) can have any shape and are here drawn crookedly to emphasize their critical geometric distinction from the transforms. When the pull-apart is born, a′-b and a″-b′ fit together, but as the hole opens, the pull-apart walls sag independently. The walls extend and stretch, so that in time their structures may be very different and fail to match, although details in preexisting rocks will still correlate. The walls along the transform margins of the pull-apart tend to sag also but continued strike-slip on them slices off segments out of line, and a complex braided zone results.

The first sediments laid down within the growing basin occupy a position as in figure 8, stage A, and the later layers are shown in stages B and C. As the hole widens, the margins are first stretched but in time lava comes up from below (stage C) so that down-dip near the center of the basin lavas and shallow intrusions

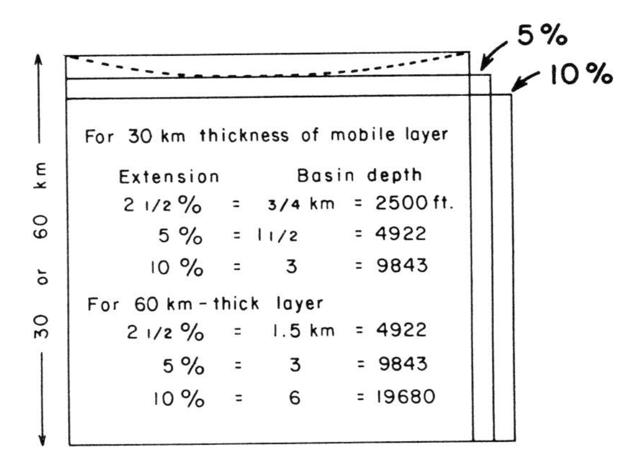

FIG. 9.—Diagram showing crustal cross section, 30 or 60 km thick, with 5 or 10% extension; see text for discussion.

may document extension. The emplacement of such lavas over active spreading centers or over diapirs from the mantle is suggested by gravity and heat-flow data and by the location of geothermal areas and young volcanics in the Salton Trough (Elders and others, 1972). It is also suggested by the distribution of volcanic rocks within the deeper and older parts of the Los Angeles basin, now displaced laterally, uplifted and eroded to view in the western Santa Monica Mountains.

Inasmuch as the crust is weak and easily folded, faulted, stretched, and compressed on a regional scale, some basins probably form in a broad transform system without volcanic flows and intrusions at depth. Most of the basins in the southern California region, including the Gulf of California and the continental borderland, range between 50 and 100 km in a northwest-southeast dimension. Inferences from the depths of earthquake foci along the San Andreas system (Eaton, 1967) and from plate-tectonic models (e.g., Isacks and others, 1968) suggest that although creep takes place below a depth of about 20 km, there is a near-uncoupling of the crust from the upper mantle at or within the low-velocity zone. Under stable cratonic crusts, the thickness of the lithosphere or depth to this zone of uncoupling is of the order of 110 km, under the central Great Basin about 20 km, and in the oceans away from ridges about 75 km (Walcott, 1970).

It may be of interest to estimate the amount of crustal extension needed to form a basin such as those in the California Borderland. For a rough and oversimplified two-dimensional model, if we assume that the lithosphere in southern California has a minimum thickness of 30 km beneath fragmented sialic crustal blocks and a maximum thickness of 60 km and that the average basin has an average depth across the block of 3 km and a northwest-southeast dimension of 60 km, then the extension needed to form the basin is 5 percent for the 60 km lithospheric thickness and 10 percent for the 30 km thickness (fig. 9).

Unfortunately, as yet we have no clear picture whether stretching or "necking down" of the lithosphere as much as 5 or 10 percent is reasonable. We can conclude, however, that stretching of crustal blocks of a few percent may begin the pull-apart process and start the formation of the basin before rupture of the basin floor and the entry of volcanics from below. In fact, in a broad weak transform zone, such as may prevail across the full width of southern California and its borderland, the volcanics may arise irregularly and diapirically into the growing gap after the breaking point is reached. At the same time the basin is being filled from the top by sedimentation.

In figure 8, the sidestepped transforms are shown as ending at a″ and b, but as the hole enlarges, complications at these corners are to be expected, such as continued minor growth along extensions of the transforms. Moreover, the angle between the transform direction and the pull-apart scarps (a′-b and a″-b′) might be very much less than shown in figure 8 so that in the field it would be difficult to locate the basin corners precisely. As the basin deepens, the pull-apart margins may stretch and subside through time, so that successively younger stratigraphic units lap farther and farther basinward leaving behind a record of minor unconformities. If the center of the basin has stretched to the point of rupture, and then has been intruded by volcanics to make a new floor, strata deposited earlier within the basin may be confined to the attenuated margins only and a deep well drilled in the basin center would not penetrate them. Instead, it would pass through younger sediments and into lava flows and associated volcaniclastic rocks. Below these it would drill through fragments only of the earlier basin fillings, and of the basin floor, and finally into diapiric masses of hypabyssal volcanics and volcanic feeder complexes. Such a model at depth suggests that it is unwise to extrapolate strata at the pull-apart margins very far down dip into some basins. The floors of true rhombochasms, for example, consist of new volcanic rocks, and lack the older rocks exposed around their margins, except perhaps as isolated blocks or "floaters." A summary sketch map is shown in figure 10, on which are portrayed possible features associated with pull-apart basins, but no single basin would be expected to possess them all.

FAULT-WEDGE BASINS

In regions such as southern California, and especially within its borderland, rhomboid and lens-shaped basins are associated with similarly shaped high-standing banks and ridges. Such a fragmented portion of the continental plate can be visualized theoretically as forming within a strike-slip regime if the major strike-slip faults converge and diverge in map view. For example, in a right-slip system where two major right-slip faults converge, assuming concurrent or intermittently alternating movement on each, the wedge between the faults will be compressed and elevated where the faults diverge, the block is extended and terrane subsides (fig. 11). Many faults in a broad and anastomosing system probably do not all move at the same time; those that predominate become straighter and longer, whereas some early faults are bent and rotated out of an orientation conducive to easy slip.

Ideas developed by Lensen (1958) to explain horsts and grabens and changing fault dips along strike-slip fault zones in New Zealand can be modified to apply to broad transform

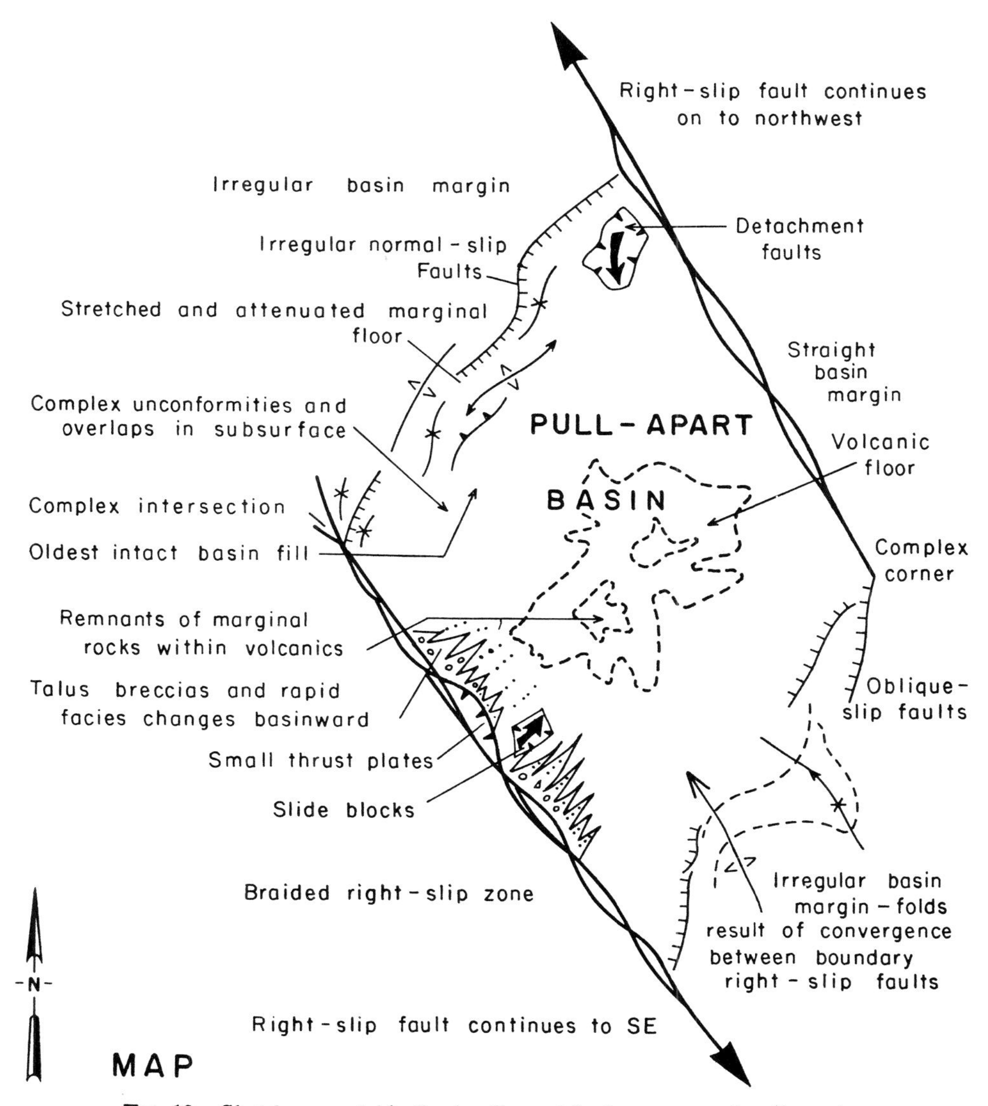

FIG. 10.—Sketch map of idealized pull-apart basin; see text for discussion.

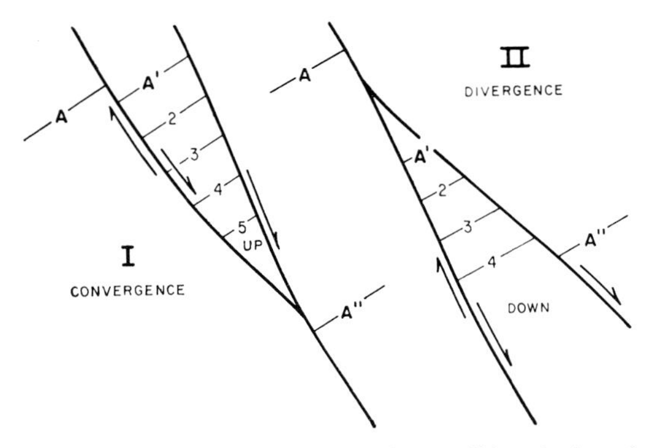

FIG. 11.—Sketch maps showing uplift of tip of fault wedge with convergence of right-slip faults, and subsidence of tip with divergence; see text for discussion.

borders in plate tectonics. If Lensen's concept of the principal horizontal stress is replaced by the concept of the direction of relative motion between two major plates, many of his geometric relations apply. The plate-motion direction will lie at 45° to this direction of principal horizontal stress. Ideally, in plate-tectonic theory, rigid plates glide by each other along a straght transform and the relative motion between the two plates is purely horizontal on a single vertical fault surface. At the present time this direction of relative motion is determined by measurements of the first motions of earthquakes, and by noting the orientation of the long and straight and predominating strike-slip faults that are clearly active.

In a boundary region of weak plate edges between rigid plates, however, braided zones apparently develop. In these systems some faults lie parallel to the direction of relative plate movement whereas others lie at an angle, usually a small angle. In general, those parallel to the plate-movement direction predominate and grow longer; those at an angle may rotate even farther out of alignment. The long predominating faults develop nearly vertical dips; those rotated out of alignment develop dips that depart considerably from the vertical. Strike-slip faults curving or bending away from the plate-movement direction gradually change from pure strike slip upon a vertical fault surface to those with first gentle oblique slip and then steep oblique slip as the fault strike bends more and more. If the curvature carries the fault into a region of extension, the fault becomes a normal oblique-slip fault; if into a region of compression, a reverse oblique-slip fault.

Such geometric relations are especially easy to envisage along braided zones of single major strike-slip faults where all slices occur in surficial rocks. Along such master faults, the faults bounding the slices converge at depth to rejoin the strike of the through-going fault. In regions such as the California Borderland, however, the major faults do not meet at depth, but presumably end in the lower crust in approaching the low-velocity zone. Such a system, including pull-apart basins, highs owing to convergence, and lows owing to divergence between major right-slip faults, is shown in figure 12.

SOUTHERN CALIFORNIA BASINS

Salton Trough, an example of a pull-apart basin, and Ridge Basin, an example of a basin formed by the saggng of weak crust as plates move around a double bend, have been mentioned briefly above. It remains to comment concerning the applicability of models described above to other basins in southern California but some complicating factors need emphasis first. To begin with, the tectonic style across southern California has changed from place to place during the last 25 my so that older basins originated under different tectonic schemes from modern ones. Salton Trough, for example, originated during the past 4 to 6 my and all deformation of Pliocene and Quaternary age can probably be related somehow to this opening. The San Andreas fault in the Salton region, however, is older than this event and was associated with an earlier elongate basin or proto-Gulf (Crowell, 1971; Karig and Jensky, 1973).

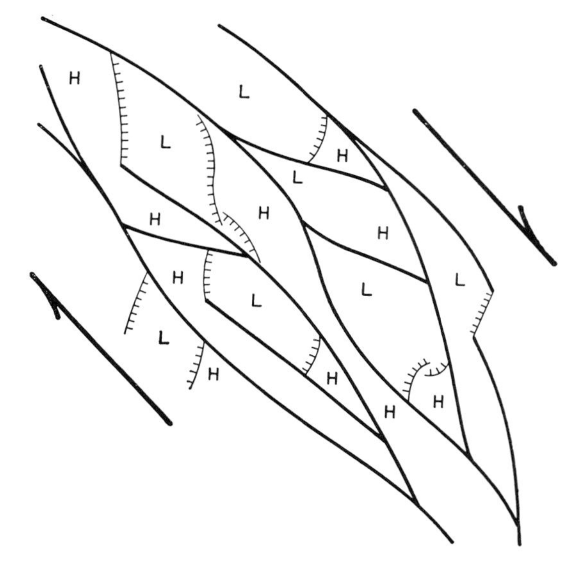

FIG. 12.—Sketch map of region in a transform regime, showing pull-apart basins and tipped fault wedges where right-slip faults converge or diverge; see text for discussion.

During the Quaternary the Transverse Ranges were deformed and uplifted as shown by the age of recent movements on dip-slip faults bounding the range and the age of folding (Crowell, 1971; Dibblee, 1971; Ehlig, in press). In fact, the San Fernando Earthquake of 1971 demonstrates that this deformation is still in progress (Grantz and others, 1971). Moreover, the major bend in the San Andreas fault near Tejon Pass (fig. 1, TP) and the origin or rejuvenation or reorientation of many structures in the region are perhaps all associated with the relative northwestward movement of the Peninsular Ranges toward the Transverse Ranges that accompanied the opening of the Gulf of California. Earlier events in southern California included major left slip in an east-west direction on the combined Malibu Coastal-Santa Monica fault system and Whittier-Elsinore fault system of as much as 90 km in the latest Miocene or early Pliocene (Yerkes and Campbell, 1971; Campbell and Yerkes, 1971), displacement on faults associated with opening of the Los Angeles basin in the early mid-Miocene, and the formation of steep and narrow grabens in the Soledad basin (fig. 1, SB) in the early Miocene (Jahns and Muehlberger, 1954; Crowell, 1968). Deformation in southern California has therefore been locally severe and intermittently continuous during the late Cenozoic as sedimentation and erosion have gone on concomitantly. Older stratigraphic units and structures have been modified by these processes so that we may not easily identify the original basin geometry. Careful palinspastic work, in which the results of younger deformations are first removed, must precede analysis in detail of basin tectonics in the past. This sound procedure, however, is difficult and imposing, and is well beyond the scope of this brief paper.

THE LOS ANGELES BASIN

One of the regions of southern California that displays this type of complicated history is the Los Angeles basin (fig. 13). It contains over 6750 m (22,000 ft) of Miocene and Pliocene sediments in its deepest part and has yielded nearly 6 billion barrels of petroleum to date (Gardett, 1971, table 1). The basin was completely filled with sediments by some time in the Pleistocene and is now being deformed, especiallly along the Newport-Inglewood trend (Harding, 1973; Yeats, 1973; Hill, 1971). Faulting and folding along this trend fit neatly into the concept of deformation along a system of simple right shear in a thick sedimentary section overlying a major fault at depth (Eaton, 1926; Platt and Stuart, in press). This buried fault, however, was primarily active before deposition of the upper Miocene and Pliocene sediments, although it was probably active and instrumental in demarcating the western margin of the basin in early mid-Miocene times. The San Onofre Breccia, for example, was laid down to the east of a fault scarp along this trend during the early part of the middle Miocene; the debris came from a schist terrane, probably exposed to subaerial erosion (Woodford, 1925; Stuart, 1973), that bordered the basin on the west. During the later Miocene, however, the fault was overlapped and marine beds transgressed southwestward.

The southeastern margin of the Los Angeles basin is formed by deformed Cretaceous and lower Tertiary strata overlying Mesozoic sialic basement. Facies trends, particularly in Paleocene beds, show that the paleocontours extended approximately in a north-south direction, with marine waters deepening westward. Nonmarine beds succeeded these in the Oligocene and early Miocene (Sespe and Vaqueros formations) but with approximately the same depositional trends. Similar stratigraphic successions older than middle Miocene are recognizable in terranes surrounding the Los Angeles basin and were disrupted by tectonic processes involved in the origin of the basin (Yerkes and Campbell, 1971; Campbell and Yerkes, 1971). Marked sedimentation within the deepening Los Angeles basin ensued in the middle Miocene, and was quickly followed by both faulting and volcanism in this southeastern region. In the San Joaquin Hills (fig. 13, SJH), for example, irregular north-trending faults displace middle Miocene strata but in turn are intruded by middle Miocene volcanics (Vedder and others, 1957). In addition, some of these faults are also overlapped by upper Miocene beds (Monterey Formation). The continuity and complexity of faulting and deformation along this flank of the Los Angeles basin is emphasized by the fact that some faults of similar trend are clearly younger and truncate the volcanics and Monterey beds. Volcanic rocks are present nearly everywhere in the central part of the Los Angeles basin (Eaton, 1958; Yerkes & others, 1965, fig. 9), and imply a nearly hydrostatic uprising of lava to a compensating level within the basin. Only locally were volcanic rocks extruded at and beyond the basin margins, as in the Glendora region (Shelton, 1955; fig. 13, GV).

During the middle Miocene the Los Angeles basin extended northward beyond the limits of the present Los Angeles Plain. The western Santa Monica Mountains, for example, contain very thick sequences of flyschlike strata of this age that extend downward into the lower Miocene. The beds were probably deposited in fairly

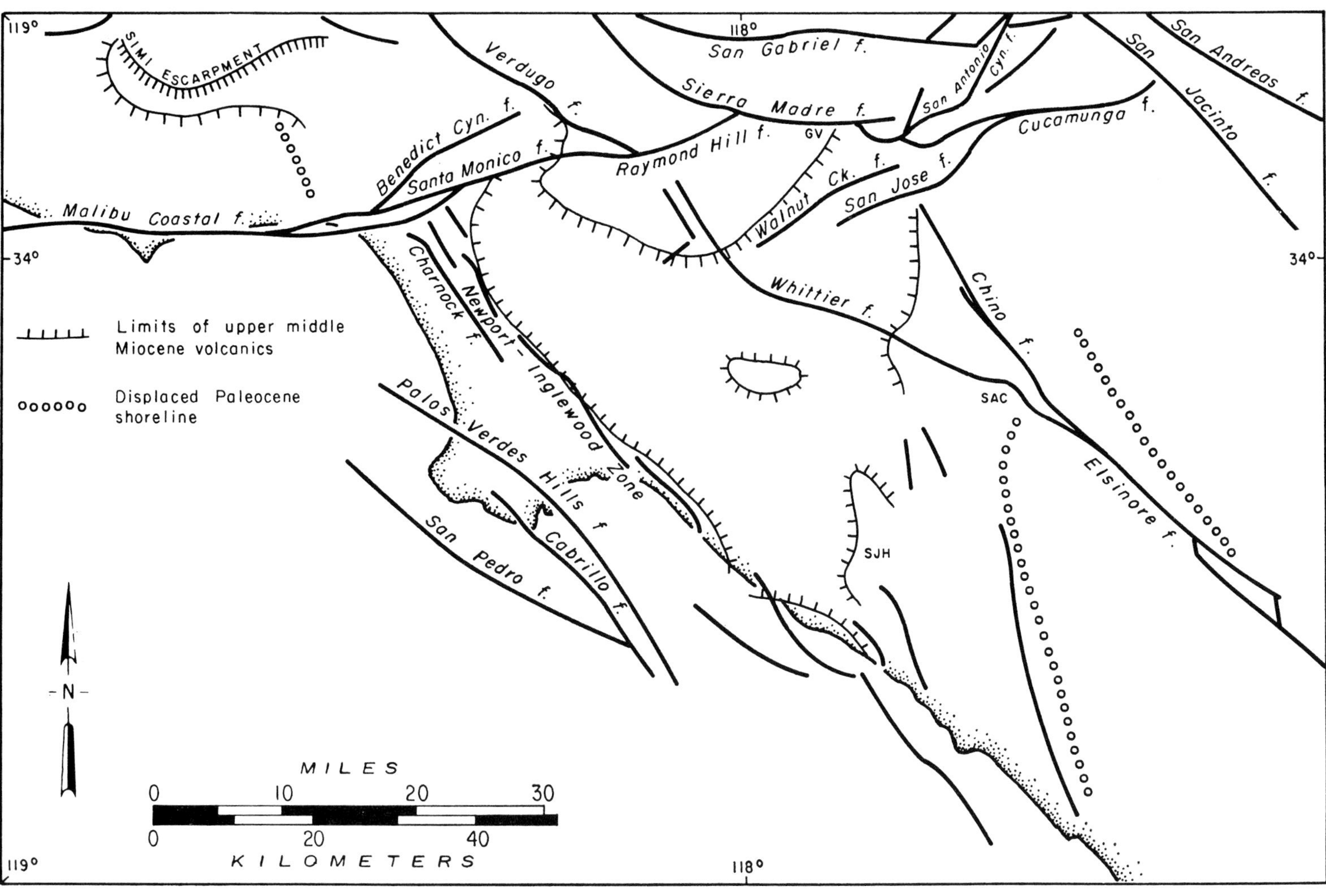

FIG. 13.—Sketch map of the Los Angeles basin region. Abbreviations: GV, Glendora volcanics; SAC, Santa Ana Canyon; SJH, San Joaquin Hills.

deep water with a nearby schist landmass on the southwest (Woodford and Bailey, 1928; Durrell, 1954) and with a coast to the north near the Simi Escarpment (fig. 13). Within this trough the lower and middle Miocene beds reach a total thickness of about 3800 m (12,500 ft) and are intercalated with middle Miocene volcanics aggregateing an aditional 3650 m (12,000 ft) approximately. The middle Miocene section is overlain by about 1400 m (4500 ft) of upper Miocene sandstone and shale. This very thick sequence of clastic sediments, now deformed and uplifted, thins northward very rapidly toward the Simi Escarpment. According to Campbell and others (1966), detachment blocks from the region of this escarpment slid southward into the deep basin during the mid-Miocene. Since these events the whole region has been displaced relatively westward with respect to the main part of the Los Angeles basin by left slip on the Malibu Coastal-Santa Monica fault system (Yerkes and Campbell, 1971; Campbell and Yerkes, 1971). Even more recently, during the Pleistocene, the Santa Monica Mountains were uplifted and deeply eroded. Upper lower and lower middle Miocene sandstone and shale, here considered as laid down in a deep part of the original Los Angeles basin, are now exposed to view along portions of the Malibu Coast. Dike and still complexes with associated flows document extension and perhaps afford an arrested view of the floor of a widening pull-apart basin.

During the Pliocene, the Los Angeles basin deepened but became considerably restricted. Thick sediments accumulated south of the present location of the Santa Monica Mountains and southwest of the Whittier fault zone (fig. 13.) This fault apparently originated near the end of the Miocene because middle Miocene beds are displaced nearly as much as Pliocene (Woodford and others, 1954, p. 75). It apparently formed the northeastern boundary of the Pliocene basin, but may have terminated at the southeast in a "corner" near Santa Ana Canyon (fig. 13, SAC). On the west, Pliocene beds overlapped the Newport-Inglewood fault zone and on the plunging southeastern margin of the basin overlapped as well upon deformed older strata. Subsidence, presumably accompanying extension and stretching and attenuation of crustal rocks, allowed the accumulation of thick sediments in the center of the Los Angeles basin. No volcanics are known to have invaded the basin floor during the Pliocene, either because extension was not sufficient to require it or because a hot spreading center did not then lie beneath it.

By the Pleistocene the Los Angeles basin had filled with sediment and shortly thereafter the beds were deformed, especially along the northeastern margin, demarcated primarily by the Whittier fault zone, and over the buried Newport-Inglewood fault zone (Harding, 1973; Yeats, 1973). The Elsinore fault system originated as a major strike-slip fault late in the Pliocene or in the Pleistocene (Gray, 1961), perhaps associated with the relative northwestward movement of the Peninsular Ranges as the Gulf of California opened. In addition, the San Jacinto fault to the east, with a total right slip of about 24 km (Sharp, 1967), may have resulted from the same movements. The southeastern part of the Elsinore fault, however, apparently has a right slip of only about 5 km, judging from the offset of an ancient cataclastic zone (R. V. Sharp, p. 22, in Lamar, 1972). The Elsinore fault is therefore probably younger than much of the movement on the Whittier fault with which it is now connected, and both faults are younger than the original opening of the Los Angeles basin in the early middle Miocene.

Bouguer gravity anomalies over the Los Angeles basin now show a broad and deep depression corresponding roughly with the thickest section of upper Miocene and Pliocene strata (McCulloh, 1960). The gravity measurements apparently reveal the thick mass of light sediments in the basin and not a high-standing diapir or volcanic complex derived from the upper mantle similar to that inferred to lie beneath the Salton Trough from high values of gravity and other data in that region (Elders and others, 1972, Fig. 3). More complete deep seismic and other geophysical data are needed before we can understand the structure at depth below the Los Angeles region, and before we can reconstruct the deeper structures formed by tectonic events in the late Cenozoic leading up to the present.

The history of the Los Angeles basin as we now understand it includes birth accompanied by rapid deepening during the early middle Miocene followed by thick sedimentation and volcanism, especially in the central part. Detachment faulting took place along the northern margin in the mid-Miocene. Truncation and left displacement of the irregular northern portion of the basin occurred in the late Miocene or after, and were accompanied by death of the Newport-Inglewood zone and nearly simultaneous birth of the Whittier fault zone. Thick sedimentation in the restricted Los Angeles basin followed during the Pliocene. The history ends with Pleistocene filling and northwest-southeast compression manifested by birth of the Elsinore fault and complex right-slip reactivation along the Whittier fault zone. Although it is speculative, perhaps the original basin was born as a pull-apart over a "hot spot"; this may have been associated with the passage of the East Pacific Rise along the coast, or with a local hot "plume."' Later events in the Los Angeles basin, however, are less clearly related to a pull-apart origin.

OTHER BASINS

The California Borderland consists of a series of irregular topographic highs, some surmounted by islands, separated by basins (fig. 1). Those basins near the coast and accessible to debris from rivers in general contain more sediment than those far offshore and relatively isolated from land sources (Emery, 1960, p. 53; Moore, 1969, pl. 14). Extrapolation seaward of the land geology and geologic history into this region suggests a complex history similar to that just reviewed for the Los Angeles basin. Data to document the complicated history are however missing from the published record. In general, the pattern of escarpments, the shapes of banks, and ridges and islands, and the configuration of basins, suggests a pull-apart origin for some of the basins. This pattern is similar to that shown here in figure 12, consisting of diverging strike-slip faults in a broad and soft transform zone. The shape of the Santa Cruz basin (fig. 1, SCB) suggests that it is a pull apart formed when the ridge surmounted by Santa Barbara Island (fig. 1, SBI) was stretched away from the ridge connecting Santa Rosa Island and San Nicolas Island (fig. 1, SRI and SNI). Although other analogies between the style of deformation portrayed on figure 13 and the California Borderland can be recognized, it is not profitable to speculate further here in the absence of specific information. Nonetheless, such a pull-apart origin for some basins, associated with convergence and divergence along major strike-slip faults accompanied by attendant rising and tipping of fault-bounded wedges, should be entertained by geologists and geophysicists investigating the region.

The Ventura basin, and its offshore extension beneath the Santa Barbara Channel (fig. 1), may also have subsided in a regime of stretching. During the Pliocene, for example, the Santa Clara trough (fig. 1, SCT) received more than 4600 m (15,000 ft) of sediments that were then severely compressed in a north-south direction during the Pleistocene (Nagle and Parker, 1971, Fig. 12). In the absence of known Pliocene volcanism at depth, however, there is little reason to suggest that the trough formed as a true sphenochasm or rhombochasm (Carey, 1958).

SUMMARY AND RECOMMENDATIONS

Understanding of the origin of the Salton trough and other basins along the San Andreas transform, now active, leads to the speculation that more ancient basins in southern California may have originated in a similar way. The geologic record is so complicated, however, that in this paper we have focused attention on what we ought to look for in order to find analogies rather than on documentation. In a strike-slip regime, pull-apart basins or rhombochasms will display straight margins where they are parallel to the transform direction, and irregular borders along the pull-apart margins. The transform margins although generally straight will be complicated in detail, and will consist of braided zones, slices, thrust blocks, and detachment faults where high terranes are structurally unsupported against low terranes. Pull-apart margins may display similar features, but even more irregularly. If pull-aparts stretch enough, or are sited above hot spreading centers or diapirs of magma from the upper mantle, volcanic rocks may enter their floors as dikes, sills, irregular bodies, and flows. Under such circumstances the floor of the basin may consist of volcanic rocks and young sediments deposited in the depression, and old or marginal rocks may be absent or present only as isolated blocks or "floaters." Structure and strata exposed around the margins of such pull-aparts cannot be extrapolated to the basin floor with confidence.

A system of anatomosing transforms, converging and diverging in map view, may give rise on a regional scale to wedge-shaped basins separated by uplands. In a soft crust, convergence of faults will result in squeezing of the terrane between the faults and in its deformation and uplift owing to isostatic compensation. Divergence of strike-slip faults will result in stretching and sagging, and the development of down-tipped triangular basins and pull-apart basins. Many of the latter will be rhombic in shape, and some may have floors composed of volcanics intermixed with infilled sediments with few or no remnants of previously existing crustal rocks. Inasmuch as much stretching and sagging and squeezing and uplift goes on hand in hand with sedimentation, complicated unconformities and overlaps are to be expected within the basins, and especially around their margins. Kinematics interpreted from local structures may therefore reveal only remotely their connection to a broad strike-slip regime.

REFERENCES CITED

ATWATER, TANYA, 1970, Implications of plate tectonics for the Cenozoic tectonic evolution of western North America: Geol. Soc. America Bull., v. 81, p. 3513–3536.

BELT, E. S., 1968, Post-Acadian rifts and related facies, eastern Canada, *in* Zen, E-An, White, W. S., Hadley, J. B., and Thompson, J. B., Jr. (eds.), Studies of Appalachian geology, northern and maritime: Wiley Interscience, N.Y., p. 95–113.

BROWN, R. D., JR., VEDDER, J. G., WALLACE, R. E., ROTH, E. F., YERKES, R. F., CASTLE, R. O., WAANANEN, A. O., PAGE, R. W., AND EATON, J. P., 1967, The Parkfield-Cholame, California earthquakes of June-August, 1966: U.S. Geol. Survey Prof. Paper 579, 66 p.

BURFORD, R. O., AND SAVAGE, J. C., 1972, Tectonic evolution of a crustal wedge caught within a transform fault system (abs.): Geol. Soc. America Abstracts with Programs, v. 4, p. 134.

CAMPBELL, R. H., AND YERKES, R. F., 1971, Cenozoic evolution of the Santa Monica Mountains-Los Angeles basin area: II. Relation to plate tectonics of the northeast Pacific Ocean: *ibid.*, v. 3, p. 92.

CAMPBELL, R. H., YERKES, R. F., AND WENTWORTH, C. M., 1966, Detachment faults in the central Santa Monica Mountains, California: U.S. Geol. Survey Prof. Paper 550-C, p. C1–C11.

CAREY, S. W., 1958, The tectonic approach to continental drift, *in* Carey, S. W. (ed.), Continental drift: Univ. Tasmania Geology Dept. Symposium No. 2, p. 177–355.

CROWELL, J. C., 1952, Probable large lateral displacement on the San Gabriel fault, southern California: Am. Assoc. Petroleum Geologists Bull., v. 36, p. 2026–2035.

———, 1954, Geology of the Ridge basin area, Los Angeles and Ventura Counties, California: Calif. Division Mines Bull. 170, map sheet 7.

———, 1962, Displacement along the San Andreas fault, California: Geol. Soc. America Special Paper 71, 61 p.

———, 1968, Movement histories of faults in the Transverse Ranges and speculations on the tectonic history of California, *in* Dickinson, W. R. and Grantz, Arthur (eds.), Proceedings of conference on geologic problems of San Andreas fault system: Stanford Univ. Pub. Geol. Sci., v. 12, p. 323–341.

———, 1971, Tectonic problems of the Transverse Ranges, California: Geol. Soc. America Abstracts with Programs, v. 3, p. 106.

———, 1974, Sedimentation along the San Andreas fault, California, *in* Dott, R. H., and Shaver, R. H. (eds.), Modern and Ancient geosynclinal sedimentation: Soc. Econ. Paleontologists and Mineralogists Special Pub. 19, p. 292–303.

DEARNLEY, RAYMOND, 1962, An outline of the Lewisian complex of the Outer Hebrides in relation to that of the Scottish mainland: Quart. Jour. Geol. Soc. London, v. 118, p. 143–176.

DIBBLEE, T. W., JR., 1954, Geology of the Imperial Valley Region, California: Calif. Division Mines Bull. 170, p. 21–28.

———, 1971, Geologic environment and tectonic development of the San Bernardino Mountains, California: Geol. Soc. America Abstracts with Programs, v. 3, p. 109–110.

DURRELL, CORDELL, 1954, Geology of the Santa Monica Mountains, Los Angeles and Ventura Counties, California: Calif. Division Mines Bull. 170, map sheet 8.

EATON, G. P., 1958, Miocene volcanic activity in the Los Angeles basin, California, *in* Higgins, J. W. (ed.), A guide to the geology and oil fields of the Los Angeles and Ventura region: Pacific Sec. Am. Assoc. Petroleum Geologists, Los Angeles, p. 55–58.

EATON, J. E., 1926, A contribution to the geology of Los Angeles basin, California: Am. Assoc. Petroleum Geologists Bull., v. 10, p. 753–767.

EATON, J. P., 1967, Instrumental seismic studies, Parkfield-Cholame, California earthquake of June-August 1966: U.S. Geol. Survey Prof. Paper 579, p. 57–65.

EHLIG, P. L., in press, Geologic framework of the San Gabriel Mountains: Calif. Division Mines and Geology Report on San Fernando, California earthquake of 1971.

ELDERS, W. A., REX, R. W., MEIDAV, TSVI, ROBINSON, P. T., AND BIEHLER, SHAWN, 1972, Crustal spreading in southern California: Science, v. 178, p. 15–24.

EMERY, K. O., 1960, The sea off southern California: Wiley, N.Y., 366 p.

FERGUSON, R. N., AND WILLIS, C. G., 1924, Dynamics of oil-field structure in southern California: Am. Assoc. Petroleum Geologists Bull., v. 8, p. 576–583.

FREUND, RAPHAEL, 1965, A model of the structural development of Israel and adjacent areas since upper Cretaceous times: Geol. Mag., v. 102, p. 189–205.

GARDETT, P. H., 1971, Petroleum potential of Los Angeles Basin, California, *in* Cram, I. H. (ed.), Future petroleum provinces of the United States—their geology and potential: Am. Assoc. Petroleum Geologists Mem. 15, v. 1, p. 298–308.

GRANTZ, ARTHUR AND OTHERS, 1971, The San Fernando, California earthquake of February 9, 1971: U.S. Geol. Survey Prof. Paper 733, 254 p.

GRAY, C. H., JR., 1961, Geology and mineral resources of the Corona South Quadrangle, California: Calif. Division Mines Bull. 178, 120 p.

HAMILTON, WARREN, 1961, Origin of the Gulf of California: Geol. Soc. America Bull., v. 72, p. 1307–1318.

HARDING, T. P., 1973, Newport-Inglewood trend, California: Am. Assoc. Petroleum Geologists Bull., v. 57, p. 97–116.

HAYS, W. H., 1957, Geology of part of the Cottonwood Springs quadrangle, Riverside County, California (Ph.D. thesis): Yale Univ., New Haven, 324 p.

HENYEY, T. L., AND BISCHOFF, J. L., 1973, Tectonic elements of the northern part of the Gulf of California: Geol. Soc. America Bull., v. 84, p. 315–330.

HILL, M. L., 1971, Newport-Inglewood zone and Mesozoic subduction, California: *ibid.*, v. 82, p. 2957–2962.

———, AND DIBBLEE, T. W., JR., 1953, San Andreas, Garlock and Big Pine faults, California: *ibid.*, v. 64, p. 443–458.

Isacks, Bryan, Oliver, Jack, and Sykes, L. R., 1968, Seismology and the new global tectonics: Jour. Geophys. Research, v. 73, p. 5855–5899.

Jahns, R. H. and Muehlberger, W. R., 1954, Geology of Soledad Basin, Los Angeles County: Calif. Division Mines Bull. 170, Map Sheet 6.

Karig, D. E., and Jensky, Wallace, 1972, The proto-Gulf of California: Earth and Planetary Sci. Letters, v. 17, p. 169–174.

Kennedy, W. Q., 1946, The Great Glen fault: Quart. Jour. Geol. Soc. London, v. 102, p. 41–76.

Kingma, J. T., 1958, Possible origin of piercement structures, local uncomformities, and secondary basins in the Eastern Geosyncline, New Zealand: New Zealand Jour. Geology and Geophysics, v. 1, p. 269–274.

Lamar, D. L., 1972, Microseismicity and recent tectonic activity in Whittier fault area, California: U.S. Geol. Survey National Center for Earthquake Research Final Technical Report, 44 p.

Larson, P. A., Mudie, J. D., and Larson, R. L., 1972, Magnetic anomalies and fracture-zone trends in the Gulf of California: Geol. Soc. America Bull., v. 83, p. 3361–3368.

Larson, R. L., 1972, Bathymetry, magnetic anomalies, and plate tectonic history of the mouth of the Gulf of California: *ibid.,* v. 83, p. 3345–3360.

Larson, R. L., Menard, H. W., and Smith, S. M., 1968, Gulf of California, a result of ocean-floor spreading and transform faulting: Science, v. 161, p. 781–784.

Lensen, G. J., 1958, A method of horst and graben formation: Jour. Geology, v. 66, p. 579–587.

Malfait, B. T., and Dinkelman, M. G., 1972, Circum-Caribbean tectonic and igneous activity and the evolution of the Caribbean Plate: Geol. Soc. America Bull., v. 83, p. 251–272.

McCulloh, T. H., 1960, Gravity variations and the geology of the Los Angeles basin, California: U.S. Geol. Survey Prof. Paper 400-B, p. B320–B325.

Moore, D. G., 1969, Reflection profiling studies of the California continental borderland: Geol. Soc. America Special Paper 107, 138 p.

———, and Buffington, E. C., 1968, Transform faulting and growth of the Gulf of California since the late Pliocene: Science, v. 161, p. 1238–1241.

Nagle, H. E. and Parker, E. S., 1971, Future oil and gas potential of onshore Ventura basin, California: *in* Cram, I. H. (ed.), Future Petroleum Provinces of the United States—their geology and potential: Am. Assoc. Petroleum Geologists, Mem. 15, v. 1, p. 254–297.

Platt, J. P. and Stuart, C. J., in press, Newport-Inglewood fault zone, Los Angeles Basin, California (Discussion) : *ibid.,* Bull.

Quennell, A. M., 1958, The structural and geomorphic evolution of the Dead Sea rift: Quart. Jour. Geol. Soc. London, v. 114, p. 1–24.

Saul, R. B., 1967, The Calaveras fault zone: Calif. Division Mines and Geology Mineral Information Service, v. 20, p. 33–37.

Sharp, R. V., 1967, San Jacinto fault zone in the Peninsular Ranges of southern California: Geol. Soc. America Bull., v. 78, p. 705–730.

Shelton, J. S., 1955, Glendora volcanic rocks, Los Angeles basin, California: *ibid.,* v. 66, p. 45–89.

Stuart, C. J., 1973, Stratigraphy of the San Onofre Breccia, Laguna Beach area, California: *ibid.,* Abstracts with Programs, v. 5, p. 112.

Sumner, J. R., 1972, Tectonic significance of gravity and aeromagnetic investigations at the head of the Gulf of California: *ibid.,* Bull., v. 83, p. 3103–3120.

Vickery, F. P., 1925, The structural dynamics of the Livermore region: Jour. Geology, v. 33, p. 608–628.

Vedder, J. G., Yerkes, R. F., and Schoellhamer, J. E., 1957, Geologic map of the San Joaquin Hills-San Capistrano area, Orange County, California: U.S. Geol. Survey Oil and Gas Inv. Map OM-193.

Vine, F. J., and Wilson, J. T., 1965, Magnetic anomalies over a young oceanic ridge off Vancouver Island: Science, v. 150, p. 485–489.

Walcott, R. I., 1970, Flexural rigidity, thickness, and viscosity of the lithosphere: Jour. Geophys. Research, v. 75, p. 3941–3954.

Wegener, Alfred, 1924, The origin of continents and oceans: Dover, N.Y. [1966 reprint], 246 p.

Wilcox, R. E., Harding, T. P., and Seely, D. R., 1973, Basic wrench tectonics: Am. Assoc. Petroleum Geologists Bull., v. 57, p. 74–96.

Wilson, J. T., 1965, A new class of faults and their bearing on continental drift: Nature, v. 207, p. 343–347.

Woodford, A. O., 1925, The San Onofre Breccia, its nature and origin: Univ. Calif. Pub. Dept. Geol. Sci. Bull., v. 15, p. 159–280.

———, and Bailey, T. L., 1928, Northwestern continuation of the San Onofre Breccia: *ibid.,* v. 17, p. 187–191.

———, Schoellhamer, J. E., Vedder, J. G., and Yerkes, R. F., 1954, Geology of the Los Angeles basin, California: Calif. Division Mines Bull. 170, p. 65–81.

Yeats, R. S., 1973, Newport-Inglewood fault zone, Los Angeles basin, California. Am. Assoc. Petroleum Geologists Bull., v. 57, p. 117–135.

Yerkes, R. F., and Campbell, R. H., 1971, Cenozoic evolution of the Santa Monica Mountains-Los Angeles basin area: I. Constraints on tectonic models: Geol. Soc. America Abstracts with Programs, v. 3, p. 222.

Yerkes, R. F., McCulloh, T. H., Schoellhamer, J. E., and Vedder, J. G., 1965, Geology of the Los Angeles basin, California—an introduction: U.S. Geol. Survey Prof. Paper 420-A, 57 p.